P. HERZOG

Cementing

Dwight K. Smith

Cementing Coordinator

Halliburton Services

Second Printing

Henry L. Doherty Memorial Fund of AIME

Society of Petroleum Engineers of AIME

New York 1976 Dallas

DEDICATION

This book would not have been possible without the efforts of many people. In acknowledgment of these efforts, I should like to make the following dedication:

To the three women in my life — Lois, Becky, and Beverly.

To each individual who has ever pumped or followed a plug with a measuring line on a cementing job.

To the many people in the Halliburton organization without whose help this book would never have been written.

ISBN 0-89520-203-4

Contents

1. Introduction **1**

1.1 Scope of the Monograph 1
1.2 Objectives of the Monograph 1
1.3 The Cementing Procedure 1
1.4 Historical Background 1

2. The Manufacture, Chemistry, and Classification of Oilwell Cements **6**

2.1 Introduction 6
2.2 Manufacture of Cement 6
2.3 Chemistry of Cements 6
2.4 Classifications of Cement 6
2.5 Properties of Cements Covered by API Specifications 9
2.6 Cement Standards Outside the U.S. 9
2.7 Specialty Cements 9

3. Cementing Additives **16**

3.1 Introduction 16
3.2 Cement Accelerators 16
3.3 Light-Weight Additives 19
3.4 Heavy-Weight Additives 23
3.5 Cement Retarders 23
3.6 Additives for Controlling Lost Circulation 24
3.7 Filtration-Control Additives for Cements 24
3.8 Cement Dispersants, or Friction Reducers 25
3.9 Uses of Salt Cements 26
3.10 Special Additives for Cement 27

4. Factors That Influence Cement Slurry Design **33**

4.1 Introduction 33
4.2 Pressure, Temperature, and Pumping Time 33
4.3 Viscosity and Water Content of Cement Slurries 34
4.4 Thickening Time 36
4.5 Strength of Cement To Support Pipe 37
4.6 Mixing Waters 38
4.7 Sensitivity to Drilling Fluids and Drilling-Fluid Additives 38
4.8 Slurry Density 39
4.9 Lost Circulation 39
4.10 Heat of Hydration 39
4.11 Permeability 41
4.12 Filtration Control 41
4.13 Resistance to Down-Hole Brines 42
4.14 Conclusion 42

5. Hole and Casing Considerations **44**

5.1 Introduction 44
5.2 Casing String Design 44
5.3 Wellbore Conditioning and Running Casing 46
5.4 Casing-Landing Procedures 46
5.5 Loss of Casing Down Hole 47
5.6 Summary 47

6. Surface and Subsurface Casing Equipment **49**

6.1 Introduction 49
6.2 Floating and Guiding Equipment 49
6.3 Formation Packer Collars and Shoes 51
6.4 Stage-Cementing Tools 51
6.5 Plug Containers and Cementing Plugs 52
6.6 Casing Centralizers 53
6.7 Casing Scratchers 54
6.8 Special Equipment 55
6.9 Summary 55

7. Primary Cementing **57**

7.1 Introduction 57
7.2 Considerations in Planning a Cementing Job 57
7.3 Considerations During Cementing 59
7.4 Placement Techniques 65
7.5 Displacement — The Critical Period 66
7.6 Cementing Multiple Strings 68
7.7 Cementing Directional Holes 69
7.8 Gas Leakage After Cementing 69
7.9 Cementing Through Soluble Formations 71
7.10 Considerations After Cementing 72
7.11 Summary 73

8. Deep-Well Cementing **74**

8.1 Introduction 74
8.2 Cementing Considerations for Deep Wells 74
8.3 Use of Liners in Deep Wells 76
8.4 Equipment Used in Hanging Liners 77
8.5 Cementing Liners Through Fractured Formations 78
8.6 Cementing Liners Through Abnormal-Pressure Formations 79
8.7 Cementing Liners in Wells With Low Fluid Levels 80
8.8 Factors To Consider in Designing Slurries for Deep Wells 81
8.9 Summary Check Lists for Running and Cementing Liners in Deep Wells 82

9. Squeeze Cementing **85**

9.1 Introduction 85
9.2 Where Squeezing Is Required 85
9.3 Squeeze Terminology 86
9.4 Squeeze Techniques 87
9.5 Squeeze Pressure Requirements 88
9.6 Squeezing Fractured Zones 89
9.7 Erroneous Squeeze-Cementing Theories 89
9.8 Job Planning 90
9.9 Slurry Design 90
9.10 Squeeze Packers 92
9.11 Squeeze Pressure Calculations 93
9.12 WOC Time 93
9.13 Testing Squeeze Jobs 94
9.14 Summary 94
9.15 Helpful Formulas for Squeeze Cementing 95

10. Open-Hole Cement Plugs **97**

10.1 Introduction 97
10.2 Uses of Cement Plugs 97
10.3 Placement Precautions 98
10.4 The Mud System 98
10.5 Cement Volume and Slurry Design 99
10.6 Placement Techniques 99
10.7 Testing Cement Plugs 101
10.8 Barite Plugs 101
10.9 Summary 101

11. Flow Calculations **104**

11.1 Introduction 104
11.2 The Flow Properties of Wellbore Fluids 104
11.3 Instruments Used To Predict Fluid Flow Properties 107
11.4 Displacement Theories — Plug Flow vs Turbulent Flow 108
11.5 Equations Used in Flow Calculations 109
11.6 Summary 110

12. Bonding, Logging, and Perforating **112**

12.1 Introduction 112
12.2 Bonding Considerations 112
12.3 Bonding of Cement to Pipe 112
12.4 Bonding of Cement to Formation 114
12.5 Methods of Locating Cement Behind the Pipe 114
12.6 Perforating — Effects on the Cement Sheath 118
12.7 Perforating Devices and Methods 118
12.8 Perforating in Gas-Producing Zones 120
12.9 Factors Influencing Perforation 121
12.10 Summary 122

13. Regulations **124**

13.1 Introduction 124
13.2 Regulatory Bodies Controlling the Cementing of Wells 124
13.3 Typical Regulations 124
13.4 Permits 127
13.5 Enforcement and Penalties 127
13.6 Summary 127

14. Special Cementing Applications **129**

14.1 Introduction 129
14.2 Large-Hole Cementing 129
14.3 Water-Well Cementing 130
14.4 Waste-Disposal Wells 131
14.5 Steam-Well Cementing 132
14.6 Cementing in Permafrost Environments 133

Appendix A: Examples of Primary Cementing Jobs **138**

A.1 Job No. 1 for XYZ Oil Company 138
A.2 Job No. 2 for XYZ Oil Company 138
A.3 Job No. 3 for XYZ Oil Company 139

Appendix B: Examples of Squeeze Jobs **140**

B.1 Job No. 1 for XYZ Oil Company 140
B.2 Job No. 2 for XYZ Oil Company 140

Appendix C: Calculations for Down-Hole Plugging **142**

C.1 Example Balanced-Cement-Plug Job for XYZ Oil Company 142
C.2 Sacks of Cement for a Given Plugging Operation 142

Appendix D: Flow Calculations for Example Primary Cementing Jobs **143**

D.1 Job No. 1 for XYZ Oil Company 143
D.2 Job No. 2 for XYZ Oil Company 144

Nomenclature and Abbreviations **145**

Metric Conversion Tables **146**

Bibliography **154**

Subject-Author Index **175**

SPE Monograph Series

The Monograph Series of the Society of Petroleum Engineers of AIME was established in 1965 by action of the SPE Board of Directors. The Series is intended to provide members with an authoritative, up-to-date treatment of the fundamental principles and state of the art in selected fields of technology. The work is directed by the Society's Monograph Committee, one of 40 national committees, through a Committee member designated as Monograph Coordinator. Technical evaluation is provided by the Monograph Review Subcommittee. Below is a listing of those who have been most closely involved with the preparation of this book.

Monograph Coordinators

Joseph W. Martinelli, Gulf Energy and Minerals Co. — U.S., Houston

Roscoe C. Clark, Continental Oil Co., Houston

Monograph Review Subcommittee

George H. Bruce, chairman, Esso Exploration, Inc., Houston

Joseph U. Messenger, Mobil Research and Development Corp., Dallas

Farrile S. Young, Jr., Baroid Div. N L Industries, Inc., Houston

SPE Monograph Staff

Thomas A. Sullivan
Technical Services
Manager-Editor

Ann Gibson
Production Manager

Sally A. Wiley
Project Editor

Preface

The Cementing Monograph is the fourth in a series of books on petroleum technology published by the Society of Petroleum Engineers of AIME. It is a composite review of the technical literature on cementing. Basic principles, materials, and techniques of cementing are reviewed and illustrated, and the applicability and limitations of the various procedures are discussed.

The Monograph series is designed to provide the Society with a state-of-the-art treatment of the fundamental principles in a select field of technology. This particular Monograph brings together the published results of many investigations and the thinking of many persons involved in research and field operations dealing with oilwell cementing. The material is presented in a form that will provide a basic background on the subject to engineers who are not directly involved in drilling and cementing. For those engineers who are directly engaged in the cementing process, it contains an up-to-date review of the literature and an extensive bibliography.

In writing a book of this type, an author is inevitably indebted to more people than he is aware of. The published works that he has read and the discussions that have molded his ideas and opinions often are not fully acknowledged. Any such oversights that I may have committed are regretted and unintentional.

I should like to accord special recognition to the technical effort of all the members of the API Standardization Group since its organization. In particular, I should like to recognize the Chairmen, who have directed much of the technical effort that has led to cement standardization: Carl Dawson, Standard Oil Co. of California; Walter Rogers, Gulf Oil Co. — U.S.; George Howard, Amoco Production Co.; Francis Anderson, Halliburton Services; Bill Bearden, Amoco Production Co.; Bob Scott, Standard Oil Co. of California; and Frank Shell, Phillips Petroleum Co.

Members of the Society's Monograph Committee have also played a very significant role in selecting the content of this Monograph. Their thorough review and constructive suggestions were a valuable help in achieving the balance in coverage of the subject.

I am also grateful to David Riley and Dan Adamson of the Dallas staff of the Society of Petroleum Engineers for their confidence and for their "loyalty to the cause" during the many months of preparation of this publication. Gratitude is extended to Sally Wiley for her editing of this manuscript and for straightening out the circumlocutions of a would-be writer.

My special thanks to the management of Halliburton Services for their cooperation in making this work on cementing possible.

Duncan, Oklahoma
May, 1976

Dwight K. Smith

Chapter 1

Introduction

1.1 Scope of the Monograph

The oilwell cementing process is used throughout the world, and it has grown in complexity, with many people, organizations, and technologies contributing to the state of the art. To help the practicing engineer with planning and job evaluation, this monograph has been written as a comprehensive reference with information about the variety of materials and techniques used in well cementing.

Chapters are devoted to cements, additives, testing, job planning, and job execution of primary cementing, liner cementing, squeeze cementing, and plugging operations. The importance of planning in achieving zonal isolation is highlighted. Coverage is also given to mechanical and pumping equipment, mixers, bulk handling systems, and various subsurface tools used to place cement properly. The book is assembled in the logical sequence of field cementing operations to provide the petroleum engineer with a working knowledge of better cementing practices.

1.2 Objectives of the Monograph

This monograph has two purposes:

1. To provide the petroleum engineer responsible for the cementing process with information that will help him to judge the merits of various cementing techniques and to know what results can be expected.
2. To provide a comprehensive review of the state of the art.

1.3 The Cementing Procedure

Oilwell cementing is the process of mixing a slurry of cement and water and pumping it down through steel casing to critical points in the annulus around the casing or in the open hole below the casing string (Fig. 1.1).

The two principal functions of the primary cementing process are to restrict fluid movement between formations and to bond and support the casing.

In addition to isolating oil-, gas-, and water-producing zones, cement also aids in (1) protecting the casing from corrosion, (2) preventing blowouts by quickly forming a seal, (3) protecting the casing from shock loads in drilling deeper, and (4) sealing off zones of lost-circulation, or thief zones.

1.4 Historical Background

Early Jobs

The U. S. petroleum industry traditionally dates its beginning with the drilling of the Drake well in 1859; yet is was not until 1903 that a cement slurry was used to shut off down-hole water just above an oil sand in the Lompoc field in California. Frank F. Hill, with the Union Oil Co., is credited with mixing and dumping, by means of a bailer, a slurry consisting of 50 sacks of neat portland cement.[1,2] After 28 days the cement was drilled from the hole, and the well was completed by drilling through the oil sand; the water zone had been effectively isolated. This became an accepted practice and soon spread to other California fields wherever similar difficulties were encountered.

The early dump bailer and tubing techniques[3] were soon replaced with a two-plug cementing method introduced into the California fields by A. A. Perkins in 1910. It was with Perkins' method that the modern oilwell cementing process was born. The first plugs, or spacers, were of cast iron and contained belting discs that functioned as wipers for mud on the casing. When cement was displaced from the pipe by steam, the plug stopped, causing a pressure increase that shut off the steam pump.

The patent issued to Perkins specified the use of two plugs. The courts later ruled that the patent includes any barrier that prevents the cement from mixing with contaminants, whether the barrier is used ahead of or behind the cement.[4]

The services of the Perkins Co. were not available outside the California area, so elsewhere the cementing process had different beginnings. In Oklahoma, it was

TABLE 1.1 — DEVELOPMENT OF PUMPS FOR OILWELL SERVICING

Type of Pump	Service Era	Pressure (psi)	Volume (bbl/min)	hp	lb_m/hp
Steam duplex	1921-1940	2,250	6	60	32
Steam duplex	1936-1947	3,500	9	100	24
Power-driven duplex	1939-1955	4,000	7	135	23
Vertical double-acting duplex	1939-1954	6,000	8	200	24
Opposed-piston pendulum	Experimental pump	10,000	6	200	40
Plunger triplex	1947-	10,000	10	330	14.5
Plunger triplex	1957-	20,000	24	600	9
Plunger triplex	1965-	12,000	13	400	9.2
Plunger triplex	1975-	10,000	6	250	10

introduced by Erle P. Halliburton in 1920 in the Hewitt field, Carter County.

The practice in Oklahoma was to set casing on top of the sand. In rotary drilled holes the casing was frequently set high to avoid drilling into the producing formation.[5] A blowout on Skelly's No. 1 Dillard occurred while casing was being run into a hole drilled into the oil sand. Efforts to control it failed until Halliburton, using crude mixing and cementing equipment, pumped some 250 sacks of portland cement and water into the casing. After a wait of 10 days, the cement was drilled out, and the well was produced without excessive water or gas production. During the following months 61 wells were cemented by this technique.[6] (See Fig. 1.2.)

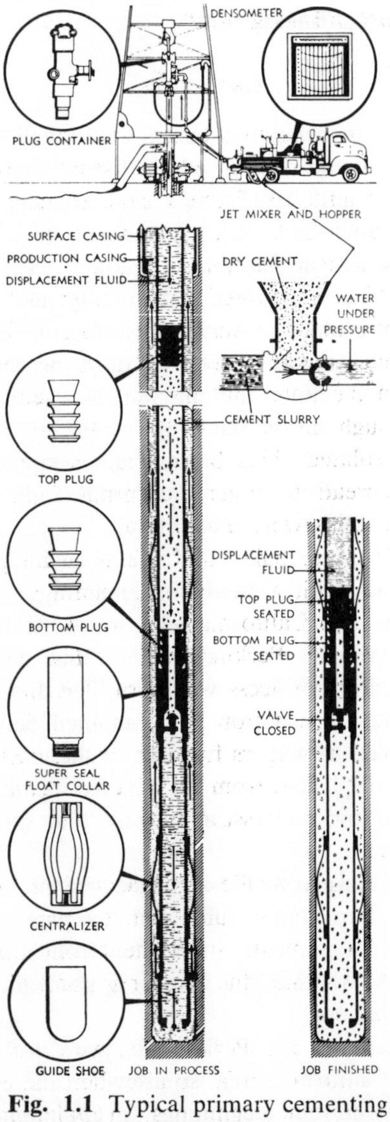

Fig. 1.1 Typical primary cementing job.

Bulk Handling and Additives

Before 1940, wells were cemented with sack cement (Fig. 1.3). Very few additives were used. There were no additives at all in 1930, and only one cement. In 1940, there were two types of cement, and three additives had been developed. Twenty-five years more saw 8 API Classes of cement and 38 additives put on the market. By 1975, although the number of API Classes of cement in common use had decreased to 4, the number of additives had increased to 44.

With the introduction of bulk cement in 1940, the

Fig. 1.2 Cementing in the early 1920's — Hewitt field, Oklahoma.

Fig. 1.3 Early-day cementing using sack cement.

handling of additives became more practical, waste was eliminated, and manpower savings were realized. The first bulk cement station for eliminating sack cement was constructed near Salem, Ill., in 1940. Other early-day stations were constructed in California and Texas. These stations transferred bulk cement from railroad cars to overhead tanks, which dumped cement directly into bulk-transport trucks. Bulk cement handling became well established during the 1940's, but the modern era of bulk handling did not begin until pozzolans were introduced in 1949.

Standardization

In 1937 the American Petroleum Institute (API) established the first committee to study cements. There already existed several cement testing laboratories equipped with strength-measuring apparatus and stirring devices to determine the fluidity or pumpability of cement slurries at down-hole temperatures.[7-10] One of the more innovative devices for evaluating cements was the pressure temperature thickening-time tester developed in 1939 by Farris with the Stanolind Oil & Gas Co.[11]

With the establishment of cement-testing laboratories, many new developments occurred in oilwell cements between 1937 and 1950.[12] During that period, a need arose for standardization of cement testing. To fulfill that need, the Mid-Continent API Committee on Oil Well Cements in 1948 prepared the first draft of *API Code 32*.[13] That code was first published in 1952 and has since been periodically modified by a national API committee on cement standardization, formed in 1953.

Standardization studies are published annually in two booklets. API Specifications are published in *Volume 10A*, and "API Recommended Practice for Testing Oil-Well Cements and Cement Additives" is published in *Volume 10B*. These volumes are now in the 19th and 20th editions, respectively.

Cementing Equipment

Through the years there has been a continuous change in pumping equipment to make it more portable and provide greater horsepower for handling higher pressures. (See Table 1.1 and Fig. 1.4.) To improve primary cementing jobs, a variety of mechanical devices have been used to more effectively place a uniform sheath of cement around the pipe.[14-17] These devices include cementing plugs, measuring lines, centralizers, scratchers, floating equipment, and stage collars.

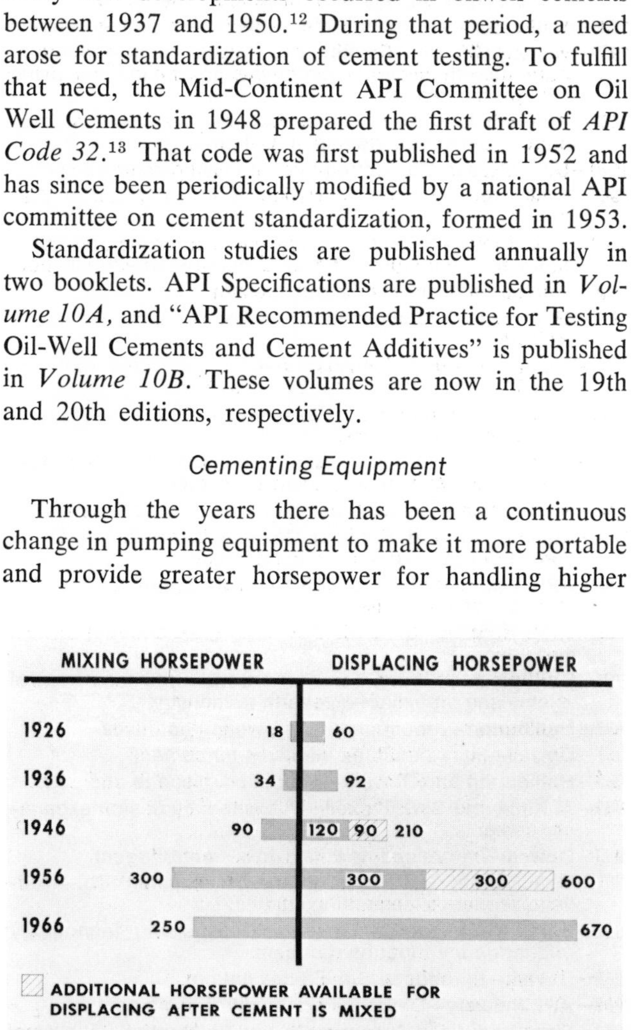

Fig. 1.4 Horsepower of cementing trucks.

Field Practices — Primary Cementing

As wells have become deeper and technology has advanced, cementing practices have changed.[18-21] In the 1910 to 1920 period, wells were considered deep at 2,000 to 3,000 ft. In the later 1920's there were several fields developed below 6,000 ft. Higher temperatures and pressures caused cementing problems. Cements used at 2,000 ft were not practical at greater depths because they tended to set prematurely.

Field placement was a matter of trial and error since laboratory testing equipment was still undeveloped. To retard the cement for use at higher temperatures, tons of ice were sometimes put in the drilling mud to cool the hole. This approach was not completely successful. A more reliable one was to mix and pump the cement as quickly as possible.

The time spent waiting for cement to set was considered unproductive. When cementing failures occurred, short waiting-on-cement (WOC) time or bad cement was given as the cause. Cement accelerators were sold under a variety of trade names, but most of them were calcium chloride solutions. WOC times were reduced as cement composition, testing procedures, and chemical acceleration became better understood. At first, 72 hours was generally considered sufficient for cement to set around the shoe joint, and oil industry regulatory bodies adopted this period almost universally. Then in 1946, Farris published his findings on the influence of time and pressure on the bonding properties of cement.[22] As field experience confirmed the validity of those findings, the regulatory bodies reduced WOC times to 24 to 36 hours.

The success of early cementing jobs was evaluated on the basis of a water shutoff test.[23,24] If no water was found on the test, the cement job was ruled successful. But failures were frequent. Studies of those early jobs revealed that cement should reach a certain strength or hardness if a successful job is to be achieved. Cementing studies of Gulf Coast wells were published by Humble in 1928.[6] Cores taken from a large number of deep wells indicated a high frequency of cement failures as a result of mud contamination. To improve the quality of cement, attention was given to conditioning the mud, to circulating the hole before cementing, and to placing a water spacer between the mud and the cement.

Squeezing and Plugging

Procedures and equipment for shutting off water in wells varied considerably in the early days of cementing. From the beginning, pressure was applied to the cement slurry after it was placed in a well. It was reported that as early as 1905 Frank Hill ran tubing and a packer to the bottom of the casing and pumped cement outside the pipe to obtain a better shutoff. Although the method successfully shut off water, the tubing and packer occasionally became stuck when the water was squeezed from the slurry.

On some cementing jobs, cement was dumped on bottom, then the hole was filled with water to apply squeeze pressure. Also, pump pressure was used in fluid-filled holes to obtain an effective water shutoff.

Where large volumes of cement were used, the column of cement and fluid behind the pipe was heavier than the hydrostatic pressure inside the pipe. Therefore, pressure was necessary to hold the cement in place. Long strings of casing were run with backpressure valves, but frequently a backpressure valve would not hold. Pump pressure was applied until the cement had time to set. This was commonly called "squeezing."

The practice of pumping several hundred sacks of cement into a well under high pressure prompted much discussion. It was reasoned that cement slurry (1) displaced mud trapped behind the pipe that had not been removed by the original cement job, or (2) compressed the exposed formation, or (3) fractured the formation along bedding planes.

Drillable cement retainers[25,26] were used as early as 1912, but it was not until 1939 that a retrievable cement

TABLE 1.2 — SUMMARY — SIGNIFICANT DATES IN THE HISTORY OF OILWELL CEMENTING

1903—**F. F. Hill**—Mixed and dumped 50 sacks of neat cement to shut off bottom-hole water.

1910—**A. A. Perkins, Perkins Cementing Co.**—Cemented the first well using the two-plug method in California.

1912—**R. C. Baker, Baker Oil Tools**—Invented the first cement retainer to pack off between casing and tubing.

1914—**F. W. Oatman**—Reported on the use of calcium chloride to accelerate cement and reduce waiting-on-cement time.

1915—**Bureau of Mines, California**—Created a staff to inspect and witness water shutoff tests.

1918—**A. A. Perkins**—Established an office to service wells in the Los Angeles basin.

1919—**E. P. Halliburton**—Established the cementing business in North Texas.

1920—**E. P. Halliburton**—Cemented the first blowout—for W. G. Skelly near Wilson, Okla.

1920—**Quintana Petroleum Co.**—Rotated casing in 50 wells.

1920—**E. P. Halliburton**—Developed the jet mixer.

1921—**J. T. Bachman, Santa Cruz Cement Co.**—Developed early testing techniques for oilwell cements.

1922—**Halliburton**—Was issued a patent in the two-plug cementing method.

1924—**Halliburton**—Licensed Perkins to use the jet mixer.

1924—**Oklahoma Corporation Commission**—Proposed the rule requiring that WOC time be reduced from 10 days to 7 if accelerator was used.

1925—Cement was first packed in a multiwalled paper bag.

1926—**D. Birch, Barnsdall Oil Co.**—Built a body and valve for special casing and float collar.

1927—**Lone Star Cement Co.**—Manufactured the first Incor high-fineness cement, in Indiana.

1927-28—**Humble Oil and Refining Co.**—Made a comprehensive survey of cementing failures along the Gulf Coast.

1929—**Pacific Portland Cement Co.**—Introduced the first retarded cement

1929—**Halliburton**—Set up the first laboratory for evaluating properties of cements.

1930—**Halliburton, Humble Oil and Refining Co., Standard Oil Co. of California**—Instituted research in oilwell cementing.

1930—**H. R. Irvine**—Patented a device to hold centralizers on pipe.

1930—Bentonite was introduced to the oil industry for use in drilling muds and cement.

1932, 1934—**William Lane and Walter Wells**—Introduced gun perforating in California and on the Gulf Coast.

1934—**Schlumberger**—Patented a method for locating the top of cement with a temperature survey instrument.

1934—**B. C. Craft *et al.***—Reported on extensive testing of oilwell cements.[8]

1935—**E. F. Silcox, Standard Oil Co. of California**—Presented a paper on a testing device for measuring thickening time of cement.[7]

1935—**M. M. Kinley**—Invented the caliper survey instrument.

1935—**T. W. Pew**—Patented a method of high-pressure squeeze cementing.

1935—**Universal Atlas Cement Co.**—Introduced Unaflo retarded cement to industry.

1936—**Lone Star Cement Co.**—Introduced Starcor retarded cement.

1937—**J. E. Weiler, Halliburton**—Built dual container device for testing oilwell cements.

1937—**API**—Established committee to study oilwell cements.

1939—**R. F. Farris, Stanolind Oil & Gas Co.**—Constructed the first pressure temperature thickening-time tester.

1939—**Halliburton**—Developed the retrievable squeeze retainer.

1939—**Humble Oil and Refining Co.**—Mixed small amounts of carnotite with cement to determine tops behind the casing with gamma ray log.

1939—**Kenneth Wright and Bruce Barkis**—Used the first commercial cement scratchers in California.

1940—**U. S. Gypsum Co.**—Introduced the first gypsum cement.

1940—**Halliburton**—Purchased Perkins Cementing Co. in California.

1940—**M. M. Kinley**—Ran first caliper surveys on electric cable to determine the quantities of cement required to fill hole.

1940—**Halliburton**—Introduced bulk cement.

1946—**R. F. Farris, Stanolind Oil & Gas Co.**—Published study on WOC time.[22]

1946—**Texas Railroad Commission**—Changed rules reducing WOC time from 72 hours to 24 to 36 hours.

1946—**A. J. Teplitz and W. E. Hassebroek**—Published study of cementing centralizers.[20]

1948—**G. C. Howard and J. B. Clark, Stanolind Oil & Gas Co.**—Published results of displacement studies.[27]

1948—**Halliburton**—Published company paper on salt cement.

1951—**Humble Oil and Refining Co.**—Used the first modified cement for permanent well completion.

1952—**API**—Approved the first edition of *API Code 32* for testing cement used in wells.[13]

1953—**J. M. Bugbee, Shell Oil Co.**—Published material on lost circulation.[28]

1953—**Phillips Petroleum Co.**—Introduced fluid-loss-control agents and diatomaceous earth to industry.

1957—**Halliburton**—Introduced heavy-weight additives.

1957—**Dowell**—Marketed latex additives for cement.

1958—**Halliburton and Dowell**—Introduced gilsonite and coal.

1958—**A. Klein and G. E. Troxell**—Published studies on expanding cements.[30]

1960—**Dowell**—Introduced new fluid-loss-control agent.

1961—**H. J. Beach, Gulf Research and Development Co.**—Published squeeze-cementing studies.[29]

1962—**Service companies**—Developed dispersing technology and introduced friction reducers.

1968—**Dowell**—Introduced Slo Flo cementing.

1968—**API, Industry**—Developed concept of basic cement.

1972—**Esso Production Research Co. and Halliburton**—Published displacement studies.[31]

retainer was introduced to the industry. The Yowell tool, originally used for washing screens and perforations, was redesigned for use as a retrievable cement retainer. Such retainers, which saved both money and time, became widely used where it was not necessary to hold the cement under pressure until it set.

When a perforated formation produced an unexpected volume of water or excess gas, it was squeezed, drilled out, and reperforated. The frequency of squeezing and reperforating was high, particularly along the Gulf Coast, because most operators would "protection squeeze" or "block squeeze" a sand before perforating for completion.[24]

1.5 Summary

Table 1.2 summarizes important events in the history of oilwell cementing.

References

1. "California's Oil," API (1948) 12.
2. "On Tour," Union Oil Co. of California (Nov.-Dec. 1952).
3. Tough, F. B.: "Method of Shutting off Water in Oil and Gas Wells," *Bull. 136,* USBM; *Petroleum Technology* (1918) **46,** 122.
4. Perkins, A. A., and Double, E.: "Method of Cementing Oil Wells," U. S. Patent No. 1,011,484 (Dec. 12, 1911), filed Oct. 27, 1909.
5. Swigert, T. E., and Schwarzenbek, F. X.: "Petroleum Engineering in the Hewitt Oil Field, Oklahoma," USBM, State of Oklahoma, and Ardmore Chamber of Commerce (Jan. 1921).
6. Millikan, C. V.: "Cementing," *History of Petroleum Engineering,* API Div. of Production, Dallas (1961) Chap. 7.
7. Silcox, D. E., and Rule, R. B.: "Special Factors Must Be Considered in Selection, Specification, and Testing of Cement for Oil Wells," *Oil Weekly* (July 29, 1935) **78,** No. 7, 21; "Cement for Oil Wells," *Petroleum Times* (Aug. 24, 1935) **34,** 195-197.
8. Craft, B. C., Johnson, T. J., and Kirkpatrick, H. L.: "Effects of Temperature, Pressure, and Water-Cement Ratio on the Setting Time and Strength of Cement," *Trans.,* AIME (1935) **114,** 62-68.
9. Weiler, J. E.: "Apparatus for Testing Cement," U. S. Patent No. 2,122,765 (July 5, 1938), filed May 15, 1937.
10. Davis, E. L.: "Specifications for Oil-Well Cement," *Drill. and Prod. Prac.,* API (1938) 372.
11. Farris, R. F.: "Effects of Temperature and Pressure on Rheological Properties of Cement Slurries," *Trans.,* AIME (1941) **142,** 117-130 (reprint, Page 306).
12. Robinson, W. W.: "Cement for Oil Wells: Status of Testing Methods and Summary of Properties," *Drill. and Prod. Prac.,* API (1939) 567-591.
13. "Code for Testing Cement Used in Wells," *API Code 32,* 1st ed., API, Dallas (1948).
14. Halliburton, E. P.: "Method and Means for Cementing Oil Wells," U. S. Patent No. 1,369,891 (March 1, 1921), filed June 26, 1920.
15. Halliburton, E. P.: "Method of Hydrating Cement and the Like," U. S. Patent No. 1,486,883 (March 18, 1924), filed June 20, 1922.
16. Burch, D. D.: "Casing Shoe," U. S. Patent No. 1,603,447 (Oct. 19, 1926), filed Feb. 25, 1926.
17. Baker, R. C.: "Plug for Well Casings," U. S. Patent No. 1,392,619 (Nov. 18, 1913), filed Nov. 20, 1911.
18. Mills, B.: "Rotating While Cementing Proves Economical," *Oil Weekly* (Dec. 4, 1939) **95,** No. 13, 14-15.
19. Reistle, C. E., Jr., and Cannon, G. E.: "Cementing Oil Wells," U. S. Patent No. 2,421,434 (June 3, 1947), filed Nov. 27, 1944. See also K. E. Wright: "Rotary Well Bore Cleaner," U. S. Patent No. 2,402,223 (June 18, 1946), filed June 26, 1944.
20. Teplitz, A. J., and Hassebroek, W. E.: "An Investigation of Oil Well Cementing," *Drill. and Prod. Prac.,* API (1946) 76-101; *Pet. Eng. Annual* (1946) 444.
21. Jones, P. H., and Berdine, D.: "Oil-Well Cementing — Factors Influencing Bond Between Cement and Formation," *Drill. and Prod. Prac.,* API (1940) 45-63.
22. Farris, R. F.: "Method of Determining Minimum Waiting-on-Cement Time," *Trans.,* AIME (1946) **165,** 175-188.
23. Oatman, F. W.: "Water Intrusion and Methods of Prevention in California Oil Fields," *Trans.,* AIME (1915) **48,** 627-650.
24. Doherty, W. T., and Manning, M.: "Gulf Coast Cementing Problems," *Oil and Gas J.* (April 4, 1929) 48, *Oil Weekly* (April 12, 1929) **53,** No. 4, 47-48.
25. Baker, R. C.: "Cement Retainer," U. S. Patent No. 1,035,674 (Aug. 13, 1912), filed Jan. 29, 1912.
26. Huber, F. W.: "Method and Composition for Cementing Oil Wells," U. S. Patent No. 1,452,463 April 17, 1923), filed May 24, 1922.
27. Howard, G. C., and Clark, J. B.: "Factors To Be Considered in Obtaining Proper Cementing of Casing," *Drill. and Prod. Prac.,* API (1948) 257-272; *Oil and Gas J.* (Nov. 11, 1948) 243.
28. Bugbee, J. M.: "Lost Circulation — A Major Problem in Exploration and Development," *Drill. and Prod. Prac.,* API (1953) 14-27.
29. Beach, H. J., O'Brien, T. B., and Goins, W. C., Jr.: "Formation Cement Squeezed by Using Low-Water-Loss Cements," *Oil and Gas J.* (May 29 and June 12, 1961).
30. Klein, A., and Troxell, G. E.: "Studies of Calcium Sulfoaluminate Admixtures for Expansive Cements," *Proc.,* ASTM (1958) **58,** 986-1008.
31. Clark, C. R., and Carter, L. G.: "Mud Displacement With Cement Slurries," *J. Pet. Tech.* (July 1973) 775-783.

Chapter 2

The Manufacture, Chemistry, and Classification Of Oilwell Cements

2.1 Introduction

Materials for cementing or bonding rock, brick, and stone in construction date from some of the earliest civilizations. Remains of those early cements can still be found in Europe, Africa, the Middle East, and the Far East. Testimony to their durability is that in some instances cements are still in an excellent state of preservation in Egypt (gypsum cement), Greece (calcined limestone), and Italy (pozzolanic-lime cements). The earliest hydraulic cements — materials that will harden and set when mixed with water — may be found in early Roman docks and marine facilities in the Mediterranean area. Such materials were composed of silicate residues from volcanic eruptions blended with lime and water. These earliest pozzolanic cements may be found near Pozzuoli, Italy.[1]

Cementing technology advanced very little through the Middle Ages until the time of the crusades. History usually credits the discovery of portland cement to Joseph Aspdin, an English mason, who was issued a patent[2] covering a gray rock-like material called "cement" in 1824. This composition, termed hydraulic because it would hydrate and set or harden when reacted under water, was the first of the portland cements as we know them today. (See Table 2.1.)

It would be difficult to imagine drilling and completing wells without cement; yet many wells were completed in the Eastern U. S. long before the first reported cement job was performed in California.

2.2 Manufacture of Cement

The basic raw materials used to manufacture portland cements are limestone (calcium carbonate) and clay or shale. Iron and alumina are frequently added if they are not present in sufficient quantity in the clay or shale.[4] These materials are blended together, either wet or dry, and then fed into a rotary kiln, which fuses the limestone slurry at temperatures from 2600° to 3000°F into a material called cement clinker. Upon cooling, the clinker is pulverized and blended with a small amount of gypsum, which controls the setting time of the finished cement. (See Fig. 2.1.)

2.3 Chemistry of Cements

A typical oxide analysis of portland cements used in wells is given in Table 2.2.

When these clinkered products hydrate with water, they combine to form four major crystalline phases with the chemical formulas and standard designations shown in Tables 2.3 and 2.4.

2.4 Classifications of Cement

Portland cements are usually manufactured to meet certain chemical and physical standards that depend upon their application. In the U. S. there are several agencies that study and write specifications for the manufacture of portland cement.[5,6] These groups include ACI (American Concrete Institute), AASHO (American Association of State Highway Officials), ASTM (American Society for Testing Materials), API (American Petroleum Institute), and various departments of the Federal government. Of these groups, the best known to the oil industry are the ASTM, which deals with cements for construction and building use, and the API, which writes specifications for cements used only in wells. Cement specifications written by either society are prepared by representatives of both users and manufacturers working together for the com-

TABLE 2.1 — DEVELOPMENT OF EARLY CEMENTS

Egypt	Plaster of Paris ($CaSO_4$ + Heat)
Greece	Lime ($CaCO_3$ + Heat)
Roman Empire	Pozzolan-lime reactions
England	Natural cement (1756 John Smeaton)
	Portland cement (1824 Joseph Aspdin)
United States	Portland cement[3,4] (First manufactured 1872)

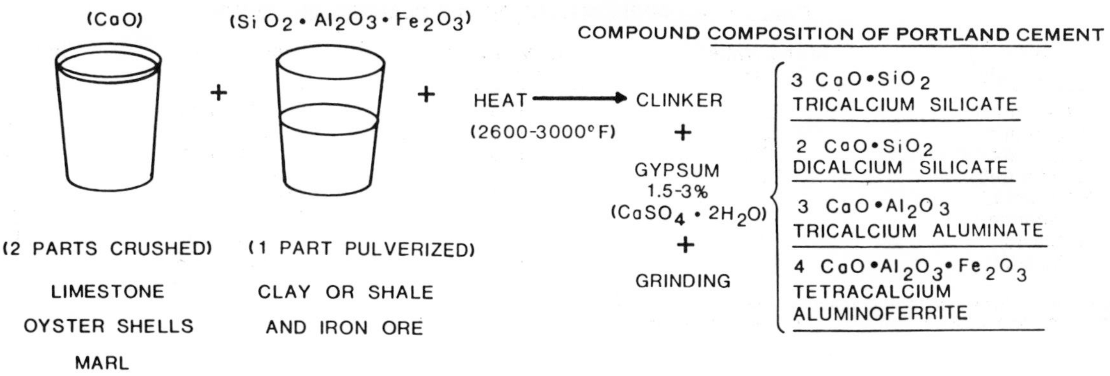

Fig. 2.1 Manufacture of portland cement.[3]

mon interest of their industry.

The ASTM specifications provide for five types of portland cement: Types I, II, II, IV, and V. Cements manufactured for use in wells are subject to wide ranges of temperature and pressure and differ considerably from the ASTM types that are manufactured for use at atmospheric conditions. For these reasons the API provides specifications covering eight classes of oilwell cements, designated Classes A, B, C, D, E, F, G, and H.

API Classes A, B, and C correspond to ASTM Types I, II, and III; ASTM Types IV and V have no corresponding API Classes.

API Classifications

The oil industry purchases cements manufactured predominantly in accordance with API classifications as published in *API Standards 10A,* "Specifications for Oil-Well Cements and Cement Additives."[6] These standards have been published annually by the American Petroleum Institute in Dallas, Tex., since 1953, when the first national standards on cements for use in wells were issued. These specifications are reviewed annually and revised according to the needs of the oil industry. The different classes of API cements for use at downhole temperatures and pressures are defined below.

TABLE 2.2 — TYPICAL OXIDE ANALYSIS OF PORTLAND CEMENTS
(API Class G or H basic cement)

Oxide	Percent
Silicon dioxide (SiO_2)	22.43
Calcium oxide (CaO)	64.77
Iron oxide (Fe_2O_3)	4.10
Aluminum oxide (Al_2O_3)	4.76
Magnesium oxide (MgO)	1.14
Sulfur trioxide (SO_3)	1.67
Potassium oxide (K_2O)	0.08
Lost on ignition	0.54

TABLE 2.3 — CHEMICAL COMPOUNDS FOUND IN SET PORTLAND CEMENT[5]

Compound	Formula	Standard Designation
Tricalcium aluminate	$3CaO \cdot Al_2O_3$	C_3A
Tricalcium silicate	$3CaO \cdot SiO_2$	C_3S
B-dicalcium silicate	$2CaO \cdot SiO_2$	C_2S
Tetracalcium aluminoferrite	$4CaO \cdot Al_2O_3 \cdot Fe_2O_3$	C_4AF

They are as listed in the *API Standards 10A* dated Jan. 1975.

Class A: Intended for use from surface to a depth of 6,000 ft when special properties are not required. Available only in Ordinary type (similar to ASTM C150, Type I).

Class B: Intended for use from surface to a depth of 6,000 ft when conditions require moderate to high sulfate resistance. Available in both Moderate (similar to ASTM C150, Type II) and High Sulfate Resistant types.

Class C: Intended for use from surface to a depth of 6,000 ft when conditions require high early strength. Available in Ordinary type and in Moderate and High Sulfate Resistant types.

Class D: Intended for use at depths from 6,000 to 10,000 ft and at moderately high temperatures and pressures. Available in both Moderate and High Sulfate Resistant types.

Class E: Intended for use at depths from 10,000 to 14,000 ft and at high temperatures and pressures. Available in both Moderate and High Sulfate Resistant types.

Class F: Intended for use at depths from 10,000 to 16,000 ft and at extremely high temperatures and pressures. Available in High Sulfate Resistant type.

TABLE 2.4 — TYPICAL COMPOSITION AND PROPERTIES OF API CLASSES OF PORTLAND CEMENT[6]

API Class	Compounds (percentage)				Wagner Fineness (sq cm/gm)
	C_3S	C_2S	C_3A	C_4AF	
A	53	24	8+	8	1,600 to 1,800
B	47	32	5−	12	1,600 to 1,800
C	58	16	8	8	1,800 to 2,200
D & E	26	54	2	12	1,200 to 1,500
G & H	50	30	5	12	1,600 to 1,800

Property	How Achieved
High early strength	By increasing the C_3S content, grinding finer.
Better retardation	By controlling C_3S and C_3A content and grinding coarser.
Low heat of hydration	By limiting the C_3S and C_3A content.
Resistance to sulfate attack	By limiting the C_3A content.

TABLE 2.5 — APPLICATIONS OF API CLASSES OF CEMENT

API Classification	Mixing Water (gal/sk)*	Slurry Weight (lb/gal)	Well Depth (ft)	Static Temperature (°F)
A (portland)	5.2	15.6	0 to 6,000	80 to 170
B (portland)	5.2	15.6	0 to 6,000	80 to 170
C (high early)	6.3	14.8	0 to 6,000	80 to 170
D (retarded)	4.3	16.4	6,000 to 12,000	170 to 260
E (retarded)	4.3	16.4	6,000 to 14,000	170 to 290
F (retarded)	4.3	16.2	10,000 to 16,000	230 to 320
G (basic)**	5.0	15.8	0 to 8,000	80 to 200
H (basic)**	4.3	16.4	0 to 8,000	80 to 200

*See Table 2.8 for weights and volumes of cement per sack.
**Can be accelerated or retarded for most well conditions.

2.6 — CHEMICAL REQUIREMENTS FOR API CEMENTS[6]

1	2	3	4	5	6	7
	Cement Class					
	A	B	C	D,E,F	G	H
ORDINARY TYPE (O)						
Magnesium oxide (MgO), maximum, per cent	5.00	....	5.00	....	....	....
Sulfur trioxide (SO_3), maximum, per cent[1]	3.50	....	4.50	....	....	....
Loss on ignition, maximum, per cent	3.00	....	3.00	....	....	....
Insoluble residue, maximum, per cent	0.75	....	0.75	....	....	....
Tricalcium aluminate ($3CaO \cdot Al_2O_3$), maximum, per cent[2]	.....	....	15.00	....	....	....
MODERATE SULFATE-RESISTANT TYPE (MSR)						
Magnesium oxide (MgO), maximum, per cent	.....	5.00	5.00	5.00	5.00	5.00
Sulfur trioxide (SO_3), maximum, per cent	.....	3.00	3.50	2.50	2.50	2.50
Loss on ignition, maximum, per cent	.....	3.00	3.00	3.00	3.00	3.00
Insoluble residue, maximum, per cent	.....	0.75	0.75	0.75	0.75	0.75
Tricalcium silicate ($3CaO \cdot SiO_2$), maximum, per cent[2]	.....	....	....	....	58.00	58.00
Tricalcium silicate ($3CaO \cdot SiO_2$), minimum, per cent[2]	.....	....	....	....	48.00	48.00
Tricalcium aluminate ($3CaO \cdot Al_2O_3$), maximum, per cent[2]	.....	8.00	8.00	8.00	8.00	8.00
Total alkali content expressed as sodium oxide (Na_2O) equivalent, maximum, per cent[3]	.....	....	....	....	0.75	0.75
HIGH SULFATE-RESISTANT TYPE (HSR)						
Magnesium oxide (MgO), maximum, per cent	.....	5.00	5.00	5.00	5.00	....
Sulfur trioxide (SO_3), maximum per cent	.....	3.00	3.50	2.50	2.50	....
Loss on ignition, maximum, per cent	.....	3.00	3.00	3.00	3.00	....
Insoluble residue, maximum, per cent	.....	0.75	0.75	0.75	0.75	....
Tricalcium silicate ($3CaO \cdot SiO_2$), maximum, per cent[2]	.....	....	....	....	65.00	....
Tricalcium silicate ($3CaO \cdot SiO_2$), minimum, per cent[2]	.....	....	....	....	48.00	....
Tricalcium aluminate ($3CaO \cdot Al_2O_3$), maximum, per cent[2]	.....	3.00	3.00	3.00	3.00	....
Tetracalcium aluminoferrite ($4CaO \cdot Al_2O_3 \cdot Fe_2O_3$) plus twice the tricalcium aluminate ($3CaO \cdot Al_2O_3$), maximum, per cent[2]	.....	24.00	24.00	24.00	24.00	....
Total alkali content expressed as sodium oxide (Na_2O) equivalent, maximum, per cent[3]	.....	....	....	....	0.75	....

[1]When the tricalcium aluminate content (expressed as C_3A) of the Class A cement is 8% or less, the maximum SO_3 content shall be 2.50%.

[2]The expressing of chemical limitations by means of calculated assumed compounds does not necessarily mean that the oxides are actually or entirely present as such compounds. When the ratio of the percentages of Al_2O_3 to Fe_2O_3 is 0.64 or less, the C_3A content is zero. When the Al_2O_3 to Fe_2O_3 ratio is greater than 0.64, the compounds shall be calculated as follows:

$$C_3A = (2.65 \times \%\ Al_2O_3) - (1.69 \times \%\ Fe_2O_3)$$

$$C_4AF = 3.04 \times \%\ Fe_2O_3$$

$$C_3S = (4.07 \times \%\ CaO) - (7.60 \times \%\ SiO_2) - (6.72 \times \%\ Al_2O_3) - (1.43 \times \%\ Fe_2O_3) - (2.85 \times \%\ SO_3)$$

When the ratio of Al_2O_3 to Fe_2O_3 is less than 0.64, an iron-alumina-calcium solid solution [expressed as ss ($C_4AF + C_2F$)] is formed and the compounds shall be calculated as follows:

$$ss(C_4AF + C_2F) = (2.10 \times \%\ Al_2O_3) + (1.70 \times \%\ Fe_2O_3) \text{ and } C_3S = (4.07 \times \%\ CaO) - (7.60 \times \%\ SiO_2) - (4.48 \times \%\ Al_2O_3) - (2.86 \times \%\ Fe_2O_3) - (2.85 \times \%\ SO_3)$$

[3]The sodium oxide equivalent (expressed as Na_2O equivalent) shall be calculated by the formula:

$$Na_2O \text{ equivalent} = (0.658 \times \%\ K_2O) + \%\ Na_2O$$

Class G: Intended for use as a basic cement from surface to a depth of 8,000 ft as manufactured. With accelerators and retarders it can be used at a wide range of depths and temperatures. It is specified that no addition except calcium sulfate or water, or both, shall be interground or blended with the clinker during the manufacture of Class G cement. It is available in Moderate and High Sulfate Resistant types.

Class H: Intended for use as a basic cement from surface to a depth of 8,000 ft as manufactured. This cement can be used with accelerators and retarders at a wide range of depths and temperatures. It is specified that no additions except calcium sulfate or water, or both, shall be interground or blended with the clinker during the manufacture of Class H cement. Available only in Moderate Sulfate Resistant type.

Table 2.5 lists the API classes of cement and indicates the depths to which they are applicable.

2.5 Properties of Cement Covered by API Specifications

In well completion operations, cements are almost universally used to displace the drilling mud and to fill the annular space between the casing and the open hole. To serve this purpose, cements must be designed for wellbore environments varying from those at the surface to those at depths exceeding 30,000 ft, where temperatures range from below freezing in permafrost areas to more than 700°F in wells drilled for geothermal steam production. Specifications do not cover all the properties of cements over such broad ranges of depth and pressure. They do, however, list physical and chemical properties for different classes of cements that will fit most well conditions. These specifications[6] include chemical analysis and physical analysis. The latter comprises (1) water content, (2) fineness, (3) compressive strength, and (4) thickening time.

Although these properties describe cements for specification purposes, oilwell cements should have other properties and characteristics to provide for their necessary functions down hole.[7,8]

The physical and chemical requirements of API Classes of cements as defined in *API Standards 10A* are shown in Tables 2.6 and 2.7. Typical physical properties of the various API classes of cement are shown in Table 2.8.

API Specifications are not enforced by an official agency; however, use of the API monogram indicates that the manufacturer has agreed to make cement according to the specifications outlined in the *API Standards 10A*. Although the API defines eight different classes of cement, only A, B, C, G, and H are available from the manufacturers and distributed in the U. S.

2.6 Cement Standards Outside the U. S.

In cementing wells in countries other than the U. S., or in their territorial water, it may be necessary to use local products. Table 2.9 lists classifications that have been established in various countries for the most common types of portland cement used for construction.[10]

For some cements, additional classifications have been made — for example, OCI (Ordinary Portland Cement Type I), OCII, OCIII. However, such classifications cause problems in fixing a clear dividing line between types, as OC Type II or III can easily be confused with RHC or HSC cement.

In some countries a specific manufacturer may, for speed and simplicity, use a symbol to identify various types of cement. Table 2.10 lists equivalent identifications for various types of portland cements as used by some countries commonly associated with the oil industry.

Listed below are some manufacturers who hold the API monogram and market cements for the oil industry.

Belgium	Ciments Belges
Canada	Canada Ciment Lafarge
	Inland Cement
Colombia	Cementos Especiales
Ecuador	La Cemento Nacional C.A.
England	Associated Portland Cement Mfg.
	Cement Lafarge
Germany	Dyckerhoff Zementwerke
Italy	Italcementi
Iran	Tehran Cement
	Doroud Cement
Japan	Ube Cement
	Nihon Cement
Lebanon	Cement Libanais
Norway	Norcem A/S
Thailand	Jalaprathan Cement
Trinidad	Trinidad Cement
Venezuela	C. A. Venezolana de Cementos
Mexico	Cementos Apasco

Note: The German-made cement is available world wide.

2.7 Specialty Cements

A number of cementitious materials, used very effectively for cementing wells, do not fall in any specific API or ASTM classification. While these materials may or may not be sold under a recognized specification, their quality and uniformity are generally controlled by the supplier. These materials include (1) pozzolanic-portland cements, (2) pozzolan-lime cements, (3) resin or plastic cements, (4) gypsum cements, (5) diesel oil cements, (6) expanding cements, (7) refractory cements, (8) latex cement, and (9) cement for permafrost environments.

Pozzolanic Cements

Pozzolans include any siliceous materials, either natural or artificial, processed or unprocessed, that in the presence of lime and water develop cementitious qualities. They can be divided into natural and artificial pozzolans. The natural pozzolans are mostly of volcanic

TABLE 2.7 — PHYSICAL REQUIREMENTS FOR API CEMENTS[6]
(Parenthetical values are in metric units.)

1	2	3	4	5	6	7	8	9	10	11	12	13
Cement Class				A	B	C	D	E	F	G	H	J★
Water, per cent by weight of cement				46	46	56	38	38	38	44	38	★
Soundness (autoclave expansion), maximum, per cent				0.80	0.80	0.80	0.80	0.80	0.80	0.80	0.80	—
Fineness *(specific surface), minimum, cm^2 per g				1500	1600	2200	—	—	—	—	—	—
Free water content, maximum, ml,				—	—	—	—	—	—	3.5**	3.5**	—
Compressive Strength Test, Eight Hour Curing Time	Schedule Number, Table 6.1 RP10B	Curing Temp, F C	Curing Pressure psi kg/cm^2	Minimum Compressive Strength, psi (kg/cm^2)								
	—	100 38	Atmos.	250 (18)	200 (14)	300 (21)	—	—	—	300 (21)	300 (21)	—
	1S	95 35	800 56	—	—	—	—	—	—	1500 (106)	1500 (106)	—
	3S	140 60	3000 211	—	—	—	—	—	—	—	—	—
	6S	230 110	3000 211	—	—	—	500 (35)	—	—	—	—	—
	8S	290 143	3000 211	—	—	—	—	500 (35)	—	—	—	500(35)
	9S	320 160	3000 211	—	—	—	—	—	500 (35)	—	—	—
Compressive Strength Test, Twenty-four Hour Curing Time	Schedule Number, Table 6.1 RP10B	Curing Temp, F C	Curing Pressure psi kg/cm^2	Minimum Compressive Strength, psi (kg/cm^2)								
	—	100 38	Atmos.	1800 (127)	1500 (106)	2000 (141)	—	—	—	—	—	—
	4S	170 77	3000 211	—	—	—	1000 (70)	1000 (70)	—	—	—	—
	6S	230 110	3000 211	—	—	—	2000 (141)	—	1000 (70)	—	—	—
	8S	290 143	3000 211	—	—	—	—	2000 (141)	—	—	—	—
	9S	320 160	3000 211	—	—	—	—	—	1000 (70)	—	—	—
	10S	350 177	3000 211	—	—	—	—	—	—	—	—	1000(70)◆
Pressure Temperature Thickening Time Test	Specification Test Schedule Number Table 7.2 RP10B	Simulated Well Depth, ft m	Maximum Consistency 15-30 Minute Stirring Period, Uc†	Minimum Thickening Time, minutes***								
	1	1,000 310	30	90	90	90	—	—	—	—	—	—
	4	6,000 1830	30	90	90	90	90	—	—	—	—	—
	5	8,000 2440	30	—	—	—	—	—	—	90	90	—
	5	8,000 2440	30	—	—	—	—	—	—	120 max.‡	120 max.‡	—
	6	10,000 3050	30	—	—	—	100	100	100	—	—	180
	8	14,000 4270	30	—	—	—	—	154	—	—	—	—
	9	16,000 4880	30	—	—	—	—	—	190	—	—	180

★Class J cement is tentative. Water as recommended by the manufacturer.
*Determined by Wagner turbidimeter apparatus described in ASTM C 115: Fineness of Portland Cement by the Turbidimeter, current edition of *ASTM Book of Standards*, Part 9.
**Based on 250 ml volume, percentage equivalent of 3.5 ml is 1.4%.
◆Compressive strength after 7 days shall be no less than the 24-hour compressive strength on Schedule 10S.
†Units of slurry consistency (U_c), formerly referred to as "poises".
***Thickening-time requirements are based on 75 percentile values of the **total** cementing times observed in the casing survey, plus a 25 per cent safety facto
‡Maximum thickening-time requirement for Schedule 5 is 120 minutes.

origin. The artificial pozzolans are mainly obtained by the heat treatment of natural materials such as clays, shales, and certain siliceous rocks.

Fly ash is a combustion by-product of coal and is widely used in the oil industry as a pozzolan. This is the only pozzolan covered by API and ASTM specifications.

When portland cement hydrates, calcium hydroxide is liberated. This chemical in itself contributes nothing to strength or water-tightness and can be removed by leaching. When fly ash is present in the cement, it combines with the calcium hydroxide, contributing to both strength and water tightness.

Fly ash has a specific gravity of 2.3 to 2.7, depending upon the source, compared with 3.1 to 3.2 for portland cements. This difference in specific gravity results in a pozzolan cement slurry of lighter weight than slurries of similar consistency made with portland cement. (Table 2.11 lists the API specifications for fly ash.)

Pozzolan-Lime Cements

Pozzolan-lime or silica-lime cements are usually blends of fly ash (silica), hydrated lime, and small quantities of calcium chloride.[11] These products hydrate with water to produce forms of calcium silicate. At low temperatures their reactions are slower than similar reactions in portland cements, and therefore they are generally recommended for primary cementing at temperatures above 140°F.

The merits of this type of cement are ease of retardation, light weight, economy, and strength stability at high temperatures.

TABLE 2.9 — SOME CEMENT CLASSIFICATIONS USED OUTSIDE THE U. S.[10]

Abbreviation	Type of Cement	Similar to ASTM	Similar to API
OC	Ordinary Portland Cement	I	A
RHC	Rapid-Hardening (or High-Early-Strength, or High-Initial-Strength) Portland Cement	III	C
HSC	High-Strength Portland Cement	III	C
LHC	Low Heat (or Slow-Hardening, Low Heat of Hydration) Portland Cement	II	B
SRC	Sulfate-Resisting Portland Cement	V	—
AEC	Air-Entraining Portland Cement	—	—

Resin or Plastic Cements

Resin and plastic cements are specialty materials used for selectively plugging open holes, squeezing perforations, and cementing waste-disposal wells. They are usually mixtures of water, liquid resins, and a catalyst blended with an API Class A, B, G, or H

TABLE 2.8 — PHYSICAL PROPERTIES OF VARIOUS TYPES OF CEMENT[9]

Properties of API Classes of Cement

	Class A	Class C	Classes G and H	Classes D and E
Specific gravity (average)	3.14	3.14	3.15	3.16
Surface area (range), sq cm/gm	1,500–1,900	2,000–2,800	1,400–1,700	1,200–1,600
Weight per sack, lb	94	94	94	94
Bulk volume, cu ft/sk	1	1	1	1
Absolute volume, gal/sk	3.6	3.6	3.58	3.57

Properties of Neat Slurries

	Portland	High Early Strength	API Class G	API Class H	Retarded
Water, gal/sk (API)	5.19	6.32	4.97	4.29	4.29
Slurry weight, lb/gal	15.6	14.8	15.8	16.5	16.5
Slurry volume, cu ft/sk	1.18	1.33	1.14	1.05	1.05

Typical Compressive Strength (psi) at 24 Hours

Temperature (°F)	Pressure (psi)	Portland	High Early Strength	API Class G	API Class H	Retarded
60	0	615	780	440	325	*
80	0	1,470	1,870	1,185	1,065	*
95	800	2,085	2,015	2,540	2,110	*
110	1,600	2,925	2,705	2,915	2,525	*
140	3,000	5,050	3,560	4,200	3,160	3,045
170	3,000	5,920	3,710	4,830	4,485	4,150
200	3,000	*	*	5,110	4,575	4,775

Typical Compressive Strength (psi) at 72 hours

Temperature (°F)	Pressure (psi)	Portland	High Early Strength	API Class G	API Class H	Retarded
60	0	2,870	2,535	—	—	*
80	0	4,130	3,935	—	—	*
95	800	4,670	4,105	—	—	*
110	1,600	5,840	4,780	—	—	*
140	3,000	6,550	4,960	—	7,125	4,000
170	3,000	6,210	4,460	5,685	7,310	5,425
200	3,000	*	*	7,360	9,900	5,920

High Pressure Thickening Time (hours:min.)

Depth (ft)	Temperature (°F) Static	Temperature (°F) Circulating	Portland	High Early Strength	API Class G	API Class H	Retarded
2,000	110	91	4:00+	4:00+	3:00+	3:57	*
4,000	140	103	3:36	3:10	2:30	3:20	4:00+
6,000	170	113	2:25	2:06	2:10	1:57	4:00+
8,000	200	125	1:40*	1:37*	1:44	1:40	4:00+

*Not generally recommended at this temperature.

TABLE 2.10 — EQUIVALENT CEMENT CLASSIFICATIONS OUTSIDE THE U. S.[10]

International Designation	Australia	Canada	France	Japan	United Kingdom	West Germany
OC	Type A Ordinary	Normal Portland	CPA-250 CPA-325	Ordinary Portland	Ordinary Portland (B.S. 12:1958)	Z375
RHC	Type B High Early Strength	High Early Strength	CPA-400 CPA-500	Rapid Hardening Portland	Rapid Hardening	—
HSC	—	—	—	—	—	Z450
LHC	Type C Low Heat of Hydration	—	—	Medium Low Heat Portland	Low Heat Portland (B.S. 1370:1958)	—
SRC	—	Sulfate Resisting	—	—	Sulfate Resisting Portland (B.S. 4027:1955)	Z275
AEC	—	—	—	—	—	—
Designation of Standards	AS A2	CSA A5	NF P15-302	JIS R5210	(B.S. 12; 1370; 4027)	DIN 1164
Year Published	1963	1961	1964	1964	1958 and 1966	1969

cement. A unique property of these cements is that when pressure is applied to the slurry the resin phase may be squeezed into a permeable zone and form a seal within the formation. These specialty cements are used in wells in relatively small volumes. They are effective at temperatures ranging from 60° to 200°F.

Gypsum Cement

Gypsum cements are used for remedial cementing work. Normally, they are available in (1) a hemihydrate form of gypsum ($CaSO_4 \cdot \frac{1}{2} H_2O$), and (2) gypsum (hemihydrate) containing a powdered resin additive.

The unique properties of gypsum cement are its capacity to set rapidly, its high early strength, and its positive expansion (approximately 0.3 percent). Gypsum cements are blended with API Class A, G, or H cement in 8 to 10 percent concentration to produce thixotropic properties. This combination is particularly useful in shallow wells to minimize fall-back after placement. (See Fig. 3.8.)

Because of the solubility of gypsum, it is usually considered a temporary plugging material unless it is placed down the hole where there is no moving water. In fighting lost circulation, gypsum cements are sometimes mixed with equal volumes of portland cements to form a permanent insoluble plugging material. These blends should be used cautiously as they have very rapid setting properties and could set prematurely during placement. (See Section 3.6, concerning lost circulation.)

Diesel Oil Cements

To control water in drilling or in producing wells, diesel oil cement slurries are frequently used.[12] These slurries are basically composed of API Class A, B, G, or H cement mixed in diesel oil or kerosene with a surface active agent. Diesel oil cements have unlimited pumping times, and will not set unless placed in a water-bearing zone; there the slurry absorbs water and sets to a hard, dense cement. The function of the surfactant is to reduce the amount of oil needed to wet the cement particles. Some compositions of diesel oil cement contain an anionic surfactant whose effect is to extend the reaction or thickening time to permit additional penetration into the formation. Diesel oil cement is primarily used to shut off water, but it can also be used to repair casing leaks, to combat certain lost-circulation problems, to plug channels behind the pipe, and to control slurry penetration. (See Fig. 2.2.)

Expanding Cements

For certain down-hole conditions it is desirable to

TABLE 2.11 — API SPECIFICATIONS FOR FLY ASH

Physical Properties[6]	
Specific gravity	2.46
Weight equivalent in absolute volume to 1 sk (94 lb) cement, lb	74
Amount retained on 200-mesh sieve, percent	5.27
Amount retained on 325-mesh sieve, percent	11.74
Chemical Analysis	
Silicon dioxide	43.20
Iron and aluminum oxides	42.93
Calcium oxide	5.92
Magnesium oxide	1.03
Sulfur trioxide	1.70
Carbon dioxide	0.03
Lost on ignition	2.98
Undetermined	2.21

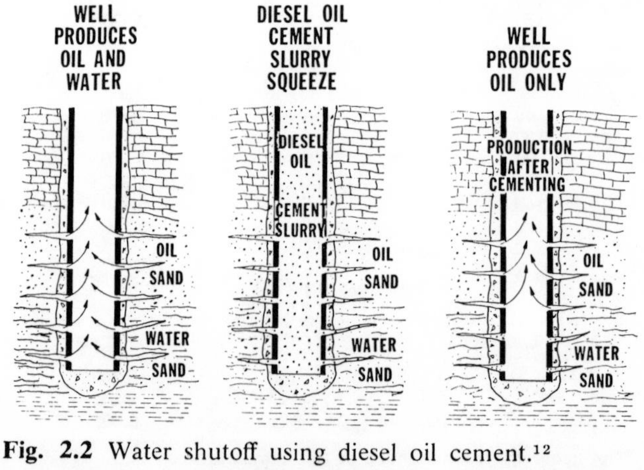

Fig. 2.2 Water shutoff using diesel oil cement.[12]

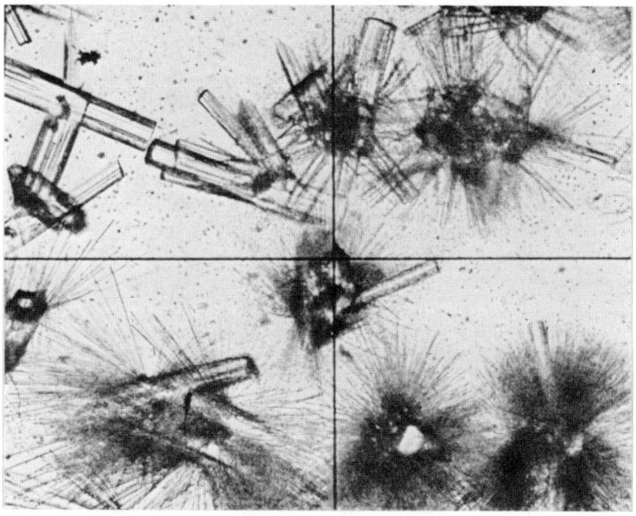

Fig. 2.3 Ettringite crystals in cement.[18]

have a cement that will expand against the filter cake and pipe. For such use the oil industry has evaluated various compositions that expand slightly when set.[13,15] The reactions that cause this expansion are similar to the process described in the cementing literature as Ettringite. Ettringite is the crystal-forming process that takes place between sulfates and the tricalcium aluminate component in portland cement (Fig. 2.3). Commercial expanding cements ($3CaO \cdot Al_2O_3 \cdot 3CaSO_4 \cdot 32H_2O$) are portland types to which have been added an anhydrous calcium sulfoaluminate ($4CaO \cdot 3Al_2O_3 \cdot SO_3$), calcium sulfate ($CaSO_4$), and lime ($CaO$).

Currently there are three types of commercial expanding cement[16]:

Type K,[14] which contains the calcium sulfoaluminate component and is blended with a portland cement by licensed manufacturers. When Type K cement is slurried with water, the reaction created by hydration expansion is approximately 0.05 to 0.20 percent.

Type S, suggested by the Porland Cement Assn., consisting of a high C_3A cement, similar to API Class A, with approximately 10 to 15 percent gypsum. Expansion characteristics are similar to those of Type K.

Type M, which is obtained by adding small quantities of refractory cement to portland cement to produce expansive forces.

There are other formulations of expanding cement:

1. API Class A (portland cement) containing 5 to 10 percent of the hemihydrate forms of gypsum.[17] (The expansion characteristics of API Class A and Class H cements containing gypsum — calcium sulfate — are compared in Table 2.12.)
2. API Class A, G, or H cement containing sodium chloride in concentrations ranging from 5 percent to saturation. The expansion is caused by chlorosilicate reactions. (See Chap. 3 for a discussion of other benefits of salt.)
3. Pozzolan cements. Expansive forces are created when the alkali reacts with Class A, G, or H cement to form sulfoaluminate crystals.

At this time there is no test procedure nor are there specifications in the API or ASTM standards for measuring the expansion forces in cement or concrete. Most laboratories use the expansive bar test, employing a molded 1- × 1- × 10-in. cement specimen. The expansive force is measured shortly after the cement sets for a base reference and then at various time intervals until the maximum expansion is reached. Hydraulic bonding tests have also been used to evaluate the crystal growth of expanding cements.

Calcium Aluminate Cements

Refractory cements are high-alumina cements manufactured by blending bauxite (aluminum ore) and limestone and heating the mixture in reverberatory open hearth furnaces until it is liquefied.[17] Two of the more widely used high-alumina cements are called Lumnite (made by the Universal Atlas Cement Co. in Gary, Ind.), and Ciment Fondu (made in England and

TABLE 2.12 — RESULTS OF LINEAR EXPANSION TESTS

Curing temperature: 100°F.
Curing pressure: atmospheric.
Initial reference time: 5½ hours.

API Class A Cement with Calcium Sulfate (percent by weight of cement)	Salt (percent by weight of cement)	Linear Expansion (percent) After a Cement Curing Time of				
		1 day	3 days	7 days	14 days	28 days
API Class A Cement						
0	0	0.015	0.027	0.034	0.039	0.053
3	0	0.060	0.078	0.087	0.094	0.108
5	0	0.078	0.133	0.142	0.150	0.165
0	18	0.139	0.182	0.196	0.204	0.219
API Class H Cement						
0	0	0.041	0.050	0.059	0.064	0.077
3	0	0.060	0.098	0.108	0.115	0.128
5	0	0.080	0.128	0.145	0.155	0.170
0	18	0.099	0.151	0.167	0.178	0.193

France by the Lafarge Cement Co.). The analyses of these materials differ from those of portland cements since bauxite replaces the clay or shale used in making portland cement. Typical analyses of these refractory cements show that they contain approximately 40 percent lime (CaO), and small amounts of silica and iron. The calcium aluminates in these cements produce high early strength and greater resistance to high temperatures and to attack by corrosive chemicals.

High-alumina cements are used in in-situ combustion wells (firefloods) where temperatures may range from 750° to 2000°F during the burning process.

These products can be accelerated or retarded to fit individual well conditions, but the retardation characteristics will differ from those of portland cements. The addition of portland to a refractory cement will cause a flash set; therefore, when both are handled in the field, they should be stored separately.

Latex Cement

While latex cement is sometimes identified as a special cement, it actually is a blend of API Class A, G, or H with either a liquid or a powdered latex. These latexes are chemically identified as polyvinyl acetate, polyvinyl chloride, or butadiene styrene emulsions. They improve the bonding strength and filtration control of a cement slurry in wells. Liquid latex is added in ratios of approximately 1 gal per sack of cement. Latex in powdered form does not freeze and can be dry blended with cement before it is transported to the wellsite. The properties imparted by liquid latex are shown in Table 2.13.

Permafrost Cement

Special problems occur in cementing conductor and surface casing in a frozen environment. Throughout the Arctic there are ice-bearing formations that extend to depths as great as 3,000 ft (Fig. 2.4). They may be described as frozen earth in some areas and as glacier-like ice blocks in others.[18,19] It is normally desirable to use a quick-setting, low-heat-of-hydration cement that will not melt the permafrost. (See Section 14.6 — Permafrost.)

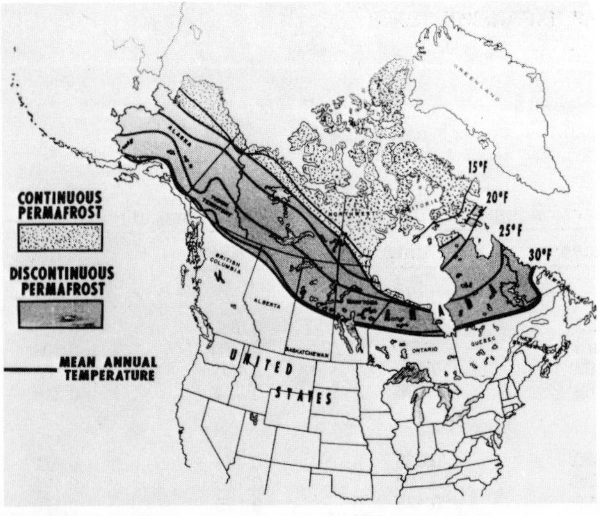

Fig. 2.4 Areas of permafrost in North America.

TABLE 2.13 — EFFECT OF LIQUID LATEX ON PORTLAND CEMENT SLURRY

Latex, gal/sk	0.0	1.0
Water, gal/sk	5.20	5.20
Viscosity, Uc*		
Initial	3	2
After 20 minutes	5	3
Fluid loss, cc/30 min on paper/100 psi	**	17
Slurry weight, lb/gal	15.60	14.43
Slurry volume, cu ft/sk	1.18	1.40

*Uc = Units of consistency; see Section 4.3.
**Dehydrated in 25 seconds.

For such low-temperature conditions, gypsum-cement blends and refractory cement blends[20] have been used very successfully. Gypsum-cement blends can be accelerated or retarded and will set at 15°F before freezing. For surface pipe these slurries are normally designed for 2 to 4 hours' pumpability, yet their strength development is quite rapid and varies little at temperatures between 20° and 80°F.

References

1. "Symposium on Use of Pozzolanic Materials in Mortars and Concretes," *Special Tech. Pub. No. 99,* ASTM, Philadelphia, Pa. (1949).

2. Aspdin, J.: "An Improvement in the Modes of Producing Artificial Stone," British Patent No. 5022 (1824).

3. "Design and Control of Concrete Mixtures," Tech. Bull., Portland Cement Assn., Chicago, Ill. (July 1968).

4. Ludwig, N. D.: "Portland Cements and Their Application in the Oil Industry," *Drill. and Prod. Prac.,* API (1953) 183-209.

5. *ASTM Standards, Part III,* American Society for Testing Materials, Philadelphia, Pa. (1970).

6. "Specifications for Oil-Well Cements and Cement Additives," *API Standards 10A,* 19th ed., API, New York (1974).

7. "Recommended Practice for Testing Oil-Well Cements and Cement Additives," *API RP 10B,* 20th ed., API Div. of Production, Dallas (1974).

8. Hansen, W. C.: "Oil-Well Cements," *Proc.,* Third Intl. Symposium on the Chemistry of Cement, London (1952).

9. *Halliburton Oil Well Cement Manual,* Halliburton Co., Duncan, Okla. (May 1968).

10. "Cement Standards of the World — Portland Cement and Its Derivatives," *CEMBUREAU,* Paris (1967).

11. Smith, D. K.: "A New Material for Deep-Well Cementing," *J. Pet. Tech.* (March 1956) 59-63.

12. Hower, W. F., and Montgomery, P. C.: "New Slurry Effective for Control of Unwanted Water," *Oil and Gas J.* (Oct. 19, 1953).

13. Lafuma, H.: "Expansive Cements," *Proc.,* Third Intl. Symposium on the Chemistry of Cement, London (1952) 581.

14. Klein, A., and Troxell, G. E.: "Studies of Calcium Sulfoaluminate Admixtures for Expansive Cements," *Proc.,* ASTM (1958) **58,** 986-1008.

15. Hansen, W. C.: "Crystal Growth as a Source of Expansion in Portland-Cement Concrete," *Proc.,* ASTM (1963) **63,** 932-945.

16. "Expansive Cement Concretes — Present State of Knowledge," *J. American Concrete Institute* (Aug. 1970) 583.

17. Newman, K.: "The Design of Concrete Mixes with High Alumina Cement," *The Reinforced Concrete Review* (March 1960) **5,** No. 5.

18. White, F. L.: "Setting Cements in Below Freezing Conditions," *Pet. Eng.* (Aug. 1952) B7.

19. Maier, L. F., Carter, M. A., Cunningham, W. C., and Bosley, T. G.: "Cementing Practices in Cold Environments," *J. Pet. Tech.* (Oct. 1971) 1215-1220.

20. Morris, E. F.: "Evaluation of Cement Systems for Permafrost," paper presented at AIME 99th Annual Meeting, Denver, Colo., Feb. 15-19, 1970.

Chapter 3

Cementing Additives

3.1 Introduction

Wells in the oil industry today cover a wider range of depth and temperature conditions than at any other time in history. Cementing compositions are designed regularly for (1) conditions below freezing in the permafrost zones of Alaska and Canada, (2) temperatures up to 500°F in deep oil wells, (3) temperatures of 450° to 500°F in steam wells, and (4) temperatures of 1500° to 2000°F in fireflood wells. Pressures range from atmospheric to 30,000 psi in extremely deep holes. It has been possible to accommodate such a wide range of conditions only through the development of additives to modify the available portland cements for individual well requirements. Today more than 40 additives are used with various API classes of cement to provide optimum slurry characteristics for any downhole condition.

With the advent of a basic cement[1] (API Classes G and H) and bulk blending equipment, the use of additives has become more flexible and simple. Cement slurries can now be tailored for specific well requirements around the world. Practically all cement additives in current use are free-flowing powders that have been dry blended with the cement before it is transported to the well. However, if necessary, most of them can be dispersed in the mixing water at the job site.

Depending on how they are selected, additives can affect the characteristics of cement slurries in a variety of ways. Following are some examples.

1. Density can range from 10.5 to 25 lb/gal.
2. Compressive strength can range from 200 to 20,000 psi.
3. Setting time can be accelerated or retarded to produce a cement that will set within a few seconds or remain fluid for up to 36 hours.
4. Cement filtration can be lowered to as little as 25 cc/30 min. when measured through a 325-mesh screen at a differential pressure of 1,000 psi.
5. The flow properties can be varied over a wide range.
6. Set cement can be made resistant to corrosion by densifying it or by varying its chemical composition.
7. Granular, fibrous, or flake-like bridging agents and gelling agents can be added to control the loss of cement slurries to formations.
8. Resilience can be imparted to set cements by incorporating fine fibers in slurry compositions.
9. Permeability can be controlled in low-temperature wells by densification and at temperatures above 230°F by densification and the use of silica flour.
10. Costs can be reduced, depending upon the well requirements and the properties desired.
11. The set cement can be expanded slightly by the use of gypsum or sodium chloride, or both.
12. The heat of hydration (the heat liberated during the setting process) can be controlled by the use of sand, fly ash, or bentonite in combination with water.

Cement additives are classified as follows:

1. Accelerators,
2. Light-weight additives,
3. Heavy-weight additives,
4. Retarders,
5. Lost-circulation-control agents,
6. Filtration-control agents,
7. Friction reducers, and
8. Specialty materials.

3.2 Cement Accelerators

Cement slurries to be used opposite shallow, low-temperature formations require acceleration to shorten thickening time and to increase early strength, particularly at formation temperatures below 100°F. By using accelerators, basic cements, and good mechanical practices, in as little as 4 hours a strength of 500 psi can be developed. This strength is generally accepted as the minimum for bonding and supporting pipe.[2-4]

The accelerators in common use are listed in Table 3.1.

Calcium Chloride (Table 3.2)

Calcium chloride,[5,6] a very hygroscopic material, is

available in flake and powder forms in the regular 77 percent calcium chloride grade, and in flake form in the anhydrous 96 percent grade. Anhydrous flake form is in more general use because it can absorb some moisture without becoming lumpy, and is easier to store. Normally, 2 to 4 percent calcium chloride, based on the cement, is used, depending on well conditions. In some instances, 4 percent calcium chloride is used with cement mixtures requiring high water ratios, where large volumes of water dilute the concentration of the accelerator.

Sodium Chloride (Table 3.3)

Sodium chloride, common table salt, is an effective accelerator for neat cement at concentrations of 1.5 to 5.0 percent by weight of cement. Two to 3.5 percent gives maximum acceleration, except when slurries of higher water ratio are used.

Low percentages of sodium chloride accelerate, but high concentrations, such as those used to saturate the mixing water, will retard the set of cement (see Section 3.5 on Cement Retarders). Although sodium chloride does not produce the degree of acceleration achieved with calcium chloride, it may be used when some acceleration is desired and calcium chloride is not available.

Gypsum Cement (Table 3.4)

Gypsum cement is composed primarily of a hemihydrate form of calcium sulfate (plaster of Paris). It is used as an accelerator for portland cements at concentrations up to 100 percent, based on cement. Thickening times as short as 5 minutes can be obtained with certain portland-gypsum cement blends.

Sodium Silicate (Table 3.5)

Sodium silicate is used primarily to accelerate cement slurries containing carboxymethyl hydroxyethyl cellulose (CMHEC) retarder.[7]

Cements With Dispersants and Reduced Water

Cement slurries can be accelerated by densifying.

TABLE 3.1 — COMMONLY USED CEMENT ACCELERATORS

Accelerator	Amount Used (percent by weight of cement)	Type of Cement	How Used
Calcium chloride ($CaCl_2$) (flaked, powdered, anhydrous)	2 to 4	Any API Class	Dry or with water
Sodium chloride (NaCl)	3 to 10* 1.5 to 5	Any API Class	Dry or with water
Hemihydrate forms of gypsum (plaster of Paris)	20 to 100	API Class A, B, C, G, or H	Dry only
Sodium silicate (Na_2SiO_2)	1 to 75	API Class A, B, C, G, or H	Dry or with water
Cements with dispersants and reduced water	0.5 to 1.0	API Class A, B, C, G, or H	Dry or with water
Sea water	—	API Class A, B, C, D, E, G, or H	As mixing water

*Percent by weight of water.

TABLE 3.2 — EFFECT OF CALCIUM CHLORIDE UPON THE THICKENING TIME AND COMPRESSIVE STRENGTH OF API CLASS A CEMENT

Water ratio: 5.2 gal/sk.
Slurry weight: 15.6 lb/gal.

Calcium Chloride (percent)	Thickening Time (hours: min.) — API Casing Cementing Tests for Simulated Well Depth (ft) of 1,000	2,000	4,000	6,000	API Squeeze Cementing Tests for Simulated Well Depth (ft) of 1,000	2,000	4,000	6,000
0.0	4:40	4:12	2:30	2:25	3:30	3:29	1:52	0:58
2.0	1:55	1:43	1:26	1:10	1:30	1:20	0:54	0:30
4.0	0:50	0:52	0:50	0:58	0:48	0:53	0:37	0:23

Curing Time (hours)	Calcium Chloride (percent)	Compressive Strength (psi) — At Atmospheric Pressure, and Temperature of 40°F	60°F	80°F	At API Curing Pressure and Temperature of 800 psi 95°F	1,600 psi 110°F
6	0	N.S.	20	75	235	860
12	0	N.S.	70	405	1,065	1,525
24	0	30	940	1,930	2,710	3,680
48	0	505	2,110	3,920	4,820	5,280
6	2	N.S.	460	850	1,170	1,700
12	2	65	785	1,540	2,360	2,850
24	2	415	2,290	3,980	4,450	5,025
6	4	N.S.	755	1,095	1,225	1,720
12	4	15	955	1,675	2,325	2,600
24	4	400	2,420	3,980	4,550	4,540

Note: N.S. = Not Set.

TABLE 3.3 — EFFECT OF SODIUM CHLORIDE UPON THE THICKENING TIME AND COMPRESSIVE STRENGTH OF API CLASS A CEMENT

Water ratio: 5.2 gal/sk.
Slurry weight: 15.6 lb/gal.

Sodium Chloride (percent)	Thickening Time (hours: min.) API Casing Cementing Tests for Simulated Well Depth (ft) of 1,000	2,000	4,000	6,000
0.0	4:40	4:12	2:30	2:25
2.0	3:05	2:27	1:52	1:13
4.0	3:05	2:35	1:35	1:20

Curing Time (hours)	Sodium Chloride (percent)	Compressive Strength (psi) At Atmospheric Pressure, and Temperature of 40°F	80°F	At API Curing Pressure and Temperature of 800 psi 95°F	1,600 psi 110°F
12	0	70	405	1,065	1,525
24	0	940	1,930	2,710	3,680
48	0	2,110	3,920	4,820	5,280
12	2	290	960	1,590	2,600
24	2	1,230	2,260	3,200	3,420
48	2	3,540	3,250	3,900	4,350
12	4	280	1,145	1,530	2,575
24	4	1,390	2,330	3,150	3,400
48	4	3,325	3,500	3,825	4,125

TABLE 3.4 — PROPERTIES OF GYPSUM AND GYPSUM-CEMENT COMPOSITIONS

Gypsum

Hemihydrate Gypsum (lb)	Water (gal)	Slurry Weight (lb/gal)	Slurry Volume (gal)	Setting Time at 60° to 180°F (min.)	Compressive Strength 1 Hour After Setting (psi)
100	4.8	15.0	9.3	50 to 60	2,500

50/50 Gypsum/Cement Mixture

Cement: API Class G.
Water ratio: 5.0 gal/sk.
Slurry weight: 15.3 lb/gal.

Setting Time (min.)	Thickening Time at 80°F (hours:min.)	Compressive Strength (psi) at 72°F After Following Hours: 2	4	8	24
12 to 20	0:23	685	725	730	1,080

TABLE 3.5 — EFFECTS OF SODIUM SILICATE ON THICKENING TIME AND COMPRESSIVE STRENGTH OF API CLASS A CEMENT SLURRIES CONTAINING CARBOXYMETHYL HYDROXYETHYL CELLULOSE (CMHEC)

Water ratio: 6.0 gal/sk.
Slurry weight: 15.0 lb/gal.

CMHEC Retarder (percent)	24-Hour Compressive Strength (psi) With Following Percentage of Sodium Silicate 0	1	3	5	Thickening Time (hours: min.) With Following Percentage of Sodium Silicate 0	1	3	5
At 2,000 ft (bottom-hole static temperature: 110°F)								
0.3	*	*	1,360	1,370	*	*	8:00+	8:00+
0.5	*	*	1,080	1,250	*	*	8:00+	8:00+
0.7	*	*	810	1,080	*	*	8:00+	8:00+
At 8,000 ft (bottom-hole static temperature: 200°F)								
0.3	3,060	3,220	2,630	—	8:00+	6:25	3:19	2:40
0.5	3,340	3,310	2,860	—	8:00+	8:00+	5:45	3:55
0.7	2,530	3,122	2,690	—	8:00+	8:00+	6:00+	4:06

*System not recommended under these conditions without testing individual cement.

TABLE 3.6 — EFFECT OF DENSIFICATION ON THICKENING TIME OF API CLASS G CEMENTS

Water (gal/sk)	Dispersant (percent)	Slurry Weight (lb/gal)	Slurry Volume (cu ft/sk)	Thickening Time* (hours:min.)
5.20	—	15.6	1.18	2:15
3.78	1.0	17.0	0.99	1:40
3.38	1.0	17.5	0.93	1:15

*Using 8,000-ft API casing test.

This is done by adding friction reducer and lowering the amount of mixing water.[8]

The most common densified slurry is API Class A, G, or H cement with 0.75 to 1.0 percent dispersant mixed at 17.5 lb/gal at a water ratio of 3.4 gal/sk. When the slurry is used for a whipstock plug, the addition of 15 to 20 lb of sand per sack of cement mixed at 18 lb/gal with the same water ratio will produce high early strength. When longer pumping times are necessary because of depth or temperature, retarders can be used. In general, the above slurry can achieve relatively good strength within 8 hours at a designated bottom-hole static temperature when designed for a pumping time of 1½ to 2 hours.

The data in Table 3.6 indicate the thickening times attainable by densifying cements.

Sea Water

Sea water is used extensively for mixing cement slurries in marine locations.[9] It contains up to 23,000 ppm of chlorides, which act as an accelerator. Sea water from the open areas of the sea or ocean is quite uniform. However, because it may be diluted by fresh water from rivers, sea water near the shore may not produce the desired acceleration. (Table 3.7 gives data on water from various sources.)

The effect of ocean water upon the thickening time and compressive strengths of slurries of Classes A and H cements compared with those of fresh water is shown in Table 3.8. Where bottom-hole static temperatures exceed 160°F, cement slurries mixed with sea water should be suitably retarded.

3.3 Light-Weight Additives

Neat cement slurries, when prepared from API Class A, B, G, or H cement using the recommended amount of water, will have slurry weight in excess of 15 lb/gal. Many formations will not support long cement columns of this density. Consequently, additives are used to reduce the weight of the slurry.[10,11] The additives also make the slurries cheaper, increase yield, and sometimes lower filter loss. The weight of cement slurries can be reduced by adding water, by adding solids having a low specific gravity, or by adding both.

The materials commonly used in cements as light-weight additives are shown in Table 3.9 in order of their general effectiveness.

Bentonite

Bentonite — sodium montmorillonite — is a colloidal clay mined in Wyoming and South Dakota. It im-

TABLE 3.7 — SEA WATER ANALYSES (WET CHEMICAL)[9]
(Constituents are given in milligrams per liter.)

Constituents	Gulf of Mexico	Cook Inlet Alaska	Trinidad W. I.	Persian Gulf (Kharg Is.)	Gulf of Suez	Sable Island
Chloride	19,000	16,600	19,900	23,000	22,300	18,900
Sulfate	2,500	2,000	2,400	3,100	3,100	2,260
Bicarbonate	127	140	78	171	134	140
Carbonate	12	0	27	24	11	
Sodium and potassium	10,654	9,319	11,170	13,044	12,499	10,690
Magnesium	1,300	1,080	1,300	1,500	1,570	1,199
Calcium	400	360	408	520	464	370
Total dissolved solids	33,993	29,499	35,283	41,359	40,078	33,559
pH	8.2	8.0	8.3	8.2	8.2	7.3
Specific gravity	1.026	1.023	1.027	1.031	1.03	1.022
Temperature, °F	75	71	70	74	75	

TABLE 3.8 — COMPARISON OF EFFECTS OF SEA WATER AND FRESH WATER ON THICKENING TIME AND COMPRESSIVE STRENGTH OF API CLASSES A AND H CEMENT SLURRIES

Water ratio: 5.0 gal/sk.
Curing time: 24 hours.

	Thickening Time (hours: min.) at Well Depth (ft) of		Compressive Strength (psi) at Curing Pressure and Temperature of		
	6,000	8,000	0 psi 50°F	1,600 psi 110°F	3,000 psi 140°F
API Class A Cement					
Fresh water	2:25	1:59	435	3,230	4,025
Sea water	1:33	1:17	520	4,105	4,670
API Class H Cement					
Fresh water	2:59	2:16	380	1,410	2,575
Sea water	1:47	1:20	460	2,500	3,085

TABLE 3.9 — SUMMARY OF LIGHT-WEIGHT CEMENT ADDITIVES

Type of Material	Usual Amount Used
Bentonite	2 to 16 percent*
Blended bentonite cement	
Prehydrated bentonite cement	
Modified bentonite cement	
High-gel salt cement	
Diatomaceous earth	10, 20, 30, or 40 percent*
Natural hydrocarbons	
Gilsonite	1 to 50 lb per sk of cement
Coal	5 to 50 lb per sk of cement
Expanded perlite	5 to 20 lb per sk of cement
Nitrogen	0 to 70 percent (depending on density, temperature, and pressure)
Others	
Artificial pozzolan (fly ash)	74 lb per sk of cement
Pozzolan-bentonite cement	Variable
Sodium silicate	1 to 7.5 lb per sk of cement

*Percent by weight of cement.

TABLE 3.10 — PHYSICAL AND CHEMICAL REQUIREMENTS FOR BENTONITE ACCORDING TO API SPECIFICATIONS[12]

Dry screen analysis	100 percent through U.S. Standard No. 40 Sieve (420 microns)
Wet screen analysis	2.5 percent maximum retained on U.S. Standard No. 200 Sieve (74 microns)
Moisture content (as received)	10 percent, maximum
Viscometer reading*	22, minimum at 600 rpm
Yield point*, lb/100 sq ft	3 × plastic viscosity, maximum
Filtration properties*	15.0 ml, maximum (100 psi paper)
pH*	9.5 maximum

*Based on 22.5 gm bentonite in 350 ml distilled water.

parts viscosity and thixotropic properties to fresh water by swelling to about 10 times its original volume. Bentonite (or gel) was one of the earliest additives used in oilwell cements to decrease slurry weight and to increase slurry volume.[11] The API specifications[12] for bentonite for use in cement are given in Table 3.10. Bentonite can be added to any API class of cement in concentrations from 1 to 16 percent by weight of the cement.[13-15] When dry mixed with the cement (in quantities of 8 to 12 percent) it requires approximately 1.3 gal of water for each 2 percent bentonite. The effect of 1 percent of prehydrated bentonite is the same as that of 3.6 percent dry mixed. With 8- to 12-percent-gel cement, dispersants are often used to reduce viscosity and to obtain flexibility in the amount of water that must be used. The effects of bentonite on the composition and properties of Class H cement slurries are shown in Table 3.11.

Bentonite (gel) is used in formulating the following different kinds of cements:

1. Blended gel cement,
2. Premixed bentonite (prehydrated),[16]
3. Modified cement[13] (patented by Humble Oil and Refining Co.),
4. High-gel salt cement[15] (patented by Gulf Oil Corp.).

High percentages of bentonite in cement reduce the compressive strength and thickening time of both regular and retarded cements. Bentonite and water also lower its resistance to chemical attack from formation waters.

Since API specifications both for the API Classes of cements and for bentonite establish only minimum

TABLE 3.11 — EFFECTS OF BENTONITE ON THE COMPOSITION AND PROPERTIES OF CLASS H CEMENT SLURRIES

Bentonite (percent)	Water Requirement (gal/sk)	Viscosity 0 to 20 minutes (Uc)*	Slurry Weight (lb/gal)	Slurry Volume (cu ft/sk)
0	5.2	4 to 12	15.6	1.18
2	6.5	10 to 20	14.7	1.36
4	7.8	11 to 21	14.1	1.55
6	9.1	13 to 24	13.5	1.73
8	10.4	12 to 19	13.1	1.92

Bentonite (percent)	Thickening Time (hours: min.)**					
	API Casing Cementing Tests for Simulated Well Depth (ft) of			API Squeeze Cementing Tests for Simulated Well Depth (ft) of		
	4,000	6,000	8,000	2,000	4,000	6,000
0	4:04	3:12	2:26	3:58	2:32	1:46
2	3:15	2:27	1:44	3:37	2:17	1:21
4	3:04	2:26	1:43	3:08	1:55	1:20
6	2:52	2:09	1:58	3:19	2:02	1:18
8	2:58	2:17	1:43	3:05	2:08	1:22

Bentonite (percent)	Compressive Strength (psi)							
	After 24 Hours at Temperature (°F) of				After 72 Hours at Temperature (°F) of			
	60	80	95	110	60	80	95	110
0	190	950	1,505	1,950	1,335	2,450	2,805	3,388
2	135	665	1,040	1,300	825	1,600	1,980	2,295
4	90	430	735	830	450	1,015	1,370	1,550
6	50	285	405	545	340	620	890	1,095
8	40	185	255	255	270	395	575	710

*Uc = Units of consistency (see Section 4.3).
**From pressure temperature thickening-time test.

TABLE 3.12 — PROPERTIES AND RETARDER REQUIREMENTS OF MODIFIED CEMENT FOR DIFFERENT DEPTHS

Cement: API Class A.
Percent bentonite: 12.
Slurry weight: 13.0 lb/gal.
Slurry volume: 2.02 cu ft/sk.
Mixing-water ratio: 11.0 gal/sk.

Formation Temperature (°F)	Depth (ft)	Calcium Lignosulfonate (percent)
Below 140	4,000	0.5
140 to 180	4,000 to 7,000	0.6
180 to 220	7,000 to 9,500	0.7
220 to 250	9,500 to 11,500	0.7 to 0.8

Note: Thickening time will be from 2 to 3 hours.
WOC time will be from 12 to 36 hours.

requirements, the properties of different brands or different batches of the same brand of either cement or bentonite can vary. For example, the compressive strength of a bentonite cement prepared from a cement that barely meets the minimum strength specifications (1,800 psi in 24 hours at 100°F) will be lower than that of one prepared from a cement having a strength of 3,500 psi under the same test conditions.

Prehydrated Bentonite

Where bulk equipment is not available for dry blending, it may be necessary to add the bentonite to the water (that is, to prehydrate it).[16] Gel can be prehydrated in about 30 minutes unless it is mixed with a high-shearing-type mixer (in which case it will swell to most of its maximum yield in less than 5 minutes). Allowing the gel to prehydrate for 24 hours before adding cement actually increases the separation of free water from the slurry and should be avoided.

Modified Cements

"Modified cements" are composed of regular portland cement, 8 to 25 percent bentonite, and a dispersant — calcium lignosulfonate.[13] For more detailed composition and properties, see Table 3.12.

Calcium lignosulfonate in a high-gel cement slurry functions as a dispersant and retarder. In addition to lightness, low cost, and increased yields, modified cement slurries have a low filter loss (100 to 120 cc with 12 percent gel; 50 to 67 cc with 25 percent gel)* provided they are batch mixed using a high rate of shear and not mixed through the standard jet mixer. Modified cements are used primarily for permanent well completions and multiple-string completions.

API Classes D and E cements are not recommended for preparing modified cement since they contain a lignin dispersant, which is a chemical retarder.

High-Gel Salt (HGS) Cements

High-gel salt cements[15] consist of portland cement, 12 to 16 percent bentonite, 3.0 to 7.0 percent inorganic salt (sodium chloride, preferably), and 0.1 to 1.5 percent dispersing agent (calcium lignosulfonate). Salt acts as both an accelerator and a dispersant, and the calcium lignosulfonate provides retardation and dispersion. Dissolving the salt in the mixing water makes it more effective. The composition and properties of the commonly used high-gel salt cements are shown in Table 3.13. Because of the dispersing properties of

*As measured according to API, using 100-psi pressure filter paper.

TABLE 3.13 — COMPOSITION AND PROPERTIES OF HIGH-GEL SALT CEMENTS PREPARED WITH API CLASS A CEMENT

Bentonite (percent)	Dispersant** (percent)	Water (gal/sk)	Thickening Time (hours:min.) API Casing Cementing Tests for Simulated Well Depth (ft) of 2,000	4,000	6,000	8,000
3 Percent Salt (12.7 lb/gal)						
16	0.0	13.0	2:48	2:34	2:00	1:12
16	0.1	13.0	3:00+	3:08	2:20	1:23
16	0.2	13.0	3:00+	3:27	2:05	1:16
16	0.4	13.0	3:00+	3:00+	2:54	2:05
16	0.6	13.0	3:00+	3:00+	3:13	2:27
5 Percent Salt						
16	0.0	13.0	2:35	2:30	2:15	1:09
16	0.1	13.0	3:00+	3:11	2:20	1:30
16	0.2	13.0	3:00+	3:00+	2:25	1:22

Salt (percent)	Dispersant** (percent)	Compressive Strength (psi)* At Atmospheric Pressure, and Temperature (°F) of 100	120	140	At 3,000 psi Pressure, and Temperature (°F) of 170	200
3	0.0	620	700	700	725	620
5	0.0	665	705	690	635	640
7	0.0	605	655	650	645	670
3	0.2	515	595	560	770	695
5	0.2	385	520	450	625	710
7	0.2	395	435	445	630	690
3	0.4	335	395	385	785	695
3	0.6	360	440	375	620	870

*In compressive strength tests: bentonite, 15 percent; water, 13 gal/sk; curing time, 24 hours.
**Calcium lignosulfonate.

TABLE 3.14 — EFFECTS OF DIATOMACEOUS EARTH* ON API CLASSES A AND H CEMENTS[23]

Diatomaceous Earth (percent)	Water (gal/sk)	Slurry Weight (lb/gal)	Slurry Volume (cu ft/sk)	Thickening Time (hours:min.) For API Class A Cement at Well Depth (ft) of 4,000	6,000	8,000	For API Class H Cement at Well Depth (ft) of 4,000	6,000	8,000
0	5.2	15.6	1.18	3:36	2:41	1:59	3:50	3:37	4:00+
10	10.2	13.2	1.92	4:00+	3:00+	2:14	4:00+	4:00+	4:00+
20	13.5	12.4	2.42	4:00+	3:00+	2:38	4:00+	4:00+	4:00+
30	18.2	11.7	3.12	4:00+	3:00+	3:00+	4:00+	4:00+	4:00+
40	25.6	11.0	4.19	4:00+	3:00+	3:00+	4:00+	4:00+	4:00+

Compressive Strength of API Class A Cement (psi)

Diatomaceous Earth (percent)	After Curing 24 Hours at Temperature and Pressure of 80°F	95°F 800 psi	110°F 1,600 psi	140°F 3,000 psi	After Curing 72 Hours at Temperature and Pressure of 80°F	95°F 800 psi	110°F 1,600 psi	140°F 3,000 psi
0	1,360	1,560	2,005	2,620	2,890	3,565	4,275	4,325
10	110	360	520	750	440	660	945	1,125
20	70	190	270	710	240	345	645	1,000
40	15	30	50	260	70	150	220	630

*Diacel D.

both salt and retarder, high-gel salt cement slurries are very pumpable even though the recommended water ratio is generally below that usually associated with the above-mentioned quantities of bentonite (12 to 16 percent).

Diatomaceous Earth

Specially graded diatomaceous earth, because it requires a high percentage of water, can be used for making light-weight cements.[17] It will impart about the same properties to cement slurries as will bentonite, but it is much more expensive. Its usefulness lies in the fact that, when used in high percentages, it does not increase the viscosity of the slurry as do expanding clays like bentonite. Table 3.14 lists cement slurry properties obtainable with diatomaceous earth.

Gilsonite

In a cement slurry, gilsonite acts both as a light-weight additive and as a unique lost-circulation agent (see Section 4.9 for further discussion of lost circulation). Gilsonite is a naturally occurring asphaltite that is inert in cement slurries.[18] It is graded in particle size from fine to ¼ in. It has a dry bulk density of 50 lb/cu ft, a water requirement of about 2 gal/cu ft, and a specific gravity of 1.07. Because of this low specific gravity, gilsonite is especially good for reducing density. Also, unlike perlite, it does not absorb water under pressure.[19,20] Gilsonite cement, therefore, has a higher strength at any age than other set cements of the same slurry weight containing other available light-weight or lost-circulation-control additives. Gilsonite does not significantly change the pumping time of most API Classes of cement.

Data in Table 3.15 show the composition and properties of gilsonite cement slurries prepared with Class A, B, or G cement.

Expanded Perlite

Perlite is a volcanic material that is mined, crushed, screened, and expanded by heat to form a cellular product of extremely low bulk weight.[20] It was originally manufactured for creating light-weight concretes. Now it is used in oilwell cements, normally with a small quantity of bentonite (2 to 6 percent) to help prevent segregation of the perlite particles from the cement slurry. Expanded perlite particles contain open and closed pores and matrix. Down the wellbore, the open holes fill with water and some closed pores crush and fill with water. The final density of the cement depends on how many pores remain closed and on how much water is immobilized in the open pores. Because of this water take-up, cement slurries containing perlite are mixed with what might appear to be an excessive amount of water to allow the cement slurry to remain pumpable under down-hole conditions.

Nitrogen

Nitrogen[21] is used ahead of cement to help reduce the bottom-hole hydrostatic pressure during cementing operations. One of the following techniques is used: (1) the nitrogen is introduced into the drilling-mud stream ahead of the cement slurry, or (2) with the hole loaded with mud and circulation established, cir-

TABLE 3.15 — COMPOSITION AND PROPERTIES OF API CLASSES A, B, AND G CEMENT SLURRIES CONTAINING GILSONITE

Gilsonite (lb/sk)	Water (gal/sk)	Slurry Weight (lb/gal)	Slurry Volume (cu ft/sk)
0 Percent Gel			
0	5.2	15.6	1.18
10	5.4	14.7	1.36
20	5.7	14.0	1.55
25	6.0	13.6	1.66
50	7.0	12.5	2.17
4 Percent Gel			
0	7.8	14.1	1.55
10	8.2	13.5	1.75
20	8.6	13.0	1.95
25	8.8	12.8	2.06
50	9.7	12.0	2.53

culation is stopped and a "slug" of nitrogen is introduced before cementing.

3.4 Heavy-Weight Additives

To offset high pressures frequently encountered in deep wells, cement slurries of high density are required. To increase cement slurry density, an additive should (1) have a specific gravity in the range of 4.5 to 5.0, (2) have a low water requirement, (3) not significantly reduce the strength of the cement, (4) have very little effect on pumping time of cement, (5) exhibit a uniform particle-size range from batch to batch, (6) be chemically inert and compatible with other additives, and (7) not interfere with well logging.

The most common materials used for weighting cements are shown in Table 3.16. Of these, hematite has been most widely used because it best fits the physical requirements and achieves the highest effective specific gravity. The physical properties of these agents and the quantities required to obtain a specified weight are given in Table 3.17.

3.5 Cement Retarders

In present-day drilling, bottom-hole static temperatures from 170° to 500°F or more are encountered over a depth range of 6,000 to 25,000 ft. To prevent the cement from setting too quickly, retarders must be added to the neat cement slurries, which can be placed safely to only about 8,000 ft. Increasing temperature hastens thickening more than increasing depth (pressure) does. Retarders must be compatible with the various additives used in cements as well as with the cement itself.

The retarders in commercially available cements (Classes D and E for example) are compounds such as lignins (salts of lignosulfonic acid), gums, starches, weak organic acids, and cellulose derivatives. Sometimes these retarders are not totally compatible with retarders added by the service companies, so the cements should be tested before they are used. It is this problem of compatibility that led to the development of API Classes G and H cements, which are not permitted to contain a chemical retarder as manufactured. These basic cements can be used to 8,000 ft as received, and respond well to retarders for use at depths as great as 30,000 ft.

Additives with high water ratios require additional retarder to achieve a desirable thickening time. This is because (1) materials with large surface areas, which generally have high water requirements, will adsorb part of the retarder, leaving less to retard the cement, and (2) additional water dilutes the retarder and reduces its effectiveness.

The chemicals presently used as retarders are listed in Table 3.18.[22]

Lignin Retarders

Lignin retarders — calcium lignosulfonate and calcium sodium lignosulfonate — are derived from wood. They are generally used over a range of 0.1 to 1.0 percent by weight of a 94-lb sack of cement (Table 3.19).

The lignin retarders have been used very successfully in retarding cement of all API Classes to depths of 12,000 to 14,000 ft or where static bottom-hole temperatures range from 260° to 290°F. (See Table 3.20.) They have also been used to increase the pumpability of API Classes D and E cements in high-temperature wells (300°F and higher), but for this purpose are not so effective as the lignosulfonates modified with organic acids.

TABLE 3.16 — HEAVY-WEIGHT CEMENT ADDITIVES

Material	Amount Used (percent by weight of cement)
Hematite	4 to 104
Ilmenite (iron-titanium oxide)	5 to 100
Barite	10 to 108
Sand	5 to 25
Salt	5 to 16
Cements with dispersants and reduced water	0.05 to 1.75

TABLE 3.17 — DATA ON VARIOUS MATERIALS FOR WEIGHTING API CLASS D OR E CEMENT

	Hematite	Barite	Ottawa Sand	Iron Arsenate
Comparison of Quantities Required				
Slurry Weight (lb/gal)	Pounds per Sack of Cement			
16.2	—	—	—	—
17.0	12	22	28	12
17.5	20	37	51	21
18.0	28	55	79	31
18.5	37	76	—	41
19.0	47	108	—	52
Physical Properties				
Specific gravity	5.02	4.23	2.65	6.98
Water requirements (percent by water)	3	22	0	19
Effective specific gravity with water	4.49	2.67	2.65	3.57
Absolute volume of additive and water (gal/lb)	0.0275	0.0548	0.0456	0.0400

TABLE 3.18 — COMMONLY USED CEMENT RETARDERS

Material	Usual Amount Used
Lignin retarders	0.1 to 1.0 percent*
Calcium lignosulfonate, organic acid	0.1 to 2.5 percent*
Carboxymethyl hydroxyethyl cellulose (CMHEC)	0.1 to 1.5 percent*
Saturated salt	14 to 16 lb per sk of cement
Borax	0.1 to 0.5 percent*

*Percent by weight of cement.

TABLE 3.19 — RETARDING EFFECT OF CALCIUM LIGNOSULFONATE ON API CLASS G OR H CEMENT SLURRIES

Retarder (percent)	Thickening Time (hours:min.) API Casing Cementing Tests for Simulated Well Depth (ft) of 8,000	10,000	12,000	14,000
0.0	1:56	1:26	1:09	1:00
0.2	2:15	2:12	1:38	1:25
0.3	3:38	2:40	2:14	1:58
0.4	4:42	3:36	3:10	2:58

TABLE 3.20 — NORMALLY RECOMMENDED AMOUNTS AND THICKENING TIMES OF CALCIUM LIGNOSULFONATE RETARDER IN API CLASSES G AND H CEMENTS

Depth (ft)	Temperature (°F) Static	Temperature (°F) Circulating	Retarder (percent)	Thickening Time (hours)
Casing Cementing				
4,000 to 6,000	140-170	103-113	0.0	3 to 4
6,000 to 10,000	170-230	113-144	0.0-0.3	3 to 4
10,000 to 14,000	230-290	144-206	0.3-0.6	2 to 4
14,000 to 18,000	290-350	206-300	0.6-1.0	*
Squeeze Cementing				
2,000 to 4,000	110-140	98-116	0.0	3 to 4
4,000 to 6,000	140-170	116-136	0.0-0.3	2 to 4
6,000 to 10,000	170-230	136-186	0.3-0.5	3 to 4
10,000 to 14,000	230-290	186-242	0.5-1.0	2 to 4*
14,000 plus	290 plus	242 plus	1.0 plus	*

*Requires special laboratory testing or the use of modified lignin retarder.

Carboxymethyl Hydroxyethyl Cellulose (CMHEC)

CMHEC, a soluble wood derivative, is a highly effective retarder.[23] It can be used at concentrations up to 0.70 percent without the addition of extra water to control slurry viscosity. Thereafter, from 0.80 to 1.0 gal of water per sack of cement should be added for each percent retarder used. The range of usage is usually from 0.1 to 1.5 percent by weight of the basic cementing composition, yet higher concentrations may be necessary for retardation at temperatures above 300°F. CMHEC is compatible with all API Classes of cement both for retarding and, to some extent, for controlling fluid loss.

Saturated Salt Water

Water saturated with salt and mixed with dry cement provides enough pumpability to place API Class A, G, or H cement to depths of 10,000 to 12,000 ft at temperatures of 230° to 260°F. (See Fig. 3.1.)

For cementing through salt sections, slurries are generally salt saturated, but for most shales and bentonitic sands that are fresh-water sensitive, lower salt concentrations are usually adequate.[24,25]

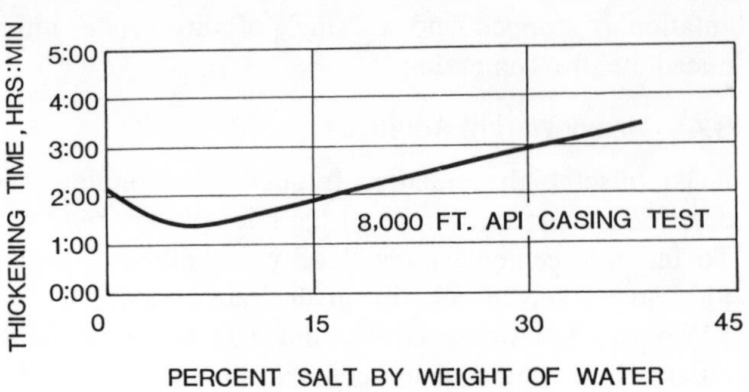

Fig. 3.1 Effect of salt on thickening time and strength of API Class G cement.[24]

3.6 Additives for Controlling Lost Circulation

"Lost circulation" (sometimes called "lost returns") is defined as the loss to induced fractures of either whole drilling fluid or cement slurry used in drilling or completing a well.[26-28] It should not be confused with the volume decrease due to filtration, or the volume required to fill new hole. Usually there are two steps in combating lost circulation.[29-32] The first is to reduce the density of the slurry, and the second is to add a bridging or plugging material. Another technique is to add nitrogen to the mud system. For more data on materials for controlling lost circulation, see Table 3.21.

3.7 Filtration-Control Additives for Cements

The filter loss (see Section 4.12) of cement slurries is lowered with additives to (1) prevent premature dehydration or loss of water against porous zones, particularly in cementing liners, (2) protect sensitive formations, and (3) improve squeeze cementing. A neat slurry of API Class G or H cement has a 30-minute API filter loss in excess of 1,000 cc.

The principal functions of filtration-control additives are (1) to form films or micelles, which control the flow of water from the cement slurry and prevent rapid dehydration, and (2) to improve particle-size distribution, which determines how liquid is held or trapped in the slurry. (See Table 3.22 for a list of filtration-

TABLE 3.21 — MATERIALS COMMONLY ADDED TO CEMENT SLURRIES TO CONTROL LOST CIRCULATION

Type	Material	Nature of Particles	Amount Used	Water Required
Additives for Controlling Lost Circulation				
Granular	Gilsonite	Graded	5 to 50 lb/sk	2 gal/50 lb
	Perlite	Expanded	½ to 1 cu ft/sk	4 gal/cu ft
	Walnut shells	Graded	1 to 5 lb/sk	0.85 gal/50 lb
	Coal	Graded	1 to 10 lb/sk	2 gal/50 lb
Lamellated	Cellophane	Flaked	⅛ to 2 lb/sk	None
Fibrous	Nylon	Short-fibered	⅛ to ¼ lb/sk	None
Formulations of Materials for Controlling Lost Circulation				
Semisolid or flash setting	Gypsum cement	—	—	4.8 gal/100 lb
	Gypsum-portland cement	—	10 to 20% gypsum	5.0 gal/100 lb
	Bentonite cement	—	10 to 25% gel	12 to 16 gal/sk
	Cement + sodium silicate	—	—	(the silicate is mixed with water before adding cement)
Quick gelling	Bentonite-diesel oil	—	—	—

TABLE 3.22 — FILTRATION-CONTROL ADDITIVES

Type and Function of Additive	Recommended Amount	Type of Cement	How Handled
Organic polymers (cellulose). To form micelles.	0.5 to 1.5 percent*	All API Classes	Dry mixed
Organic polymers[36] (dispersants). To improve particle-size distribution **and form micelles in the filter cake.**	0.5 to 1.25 percent*	All API Classes (densified)	Dry mixed or with mixing water
Carboxymethyl hydroxyethyl cellulose. To form micelles.	0.3 to 1.0 percent*	All API Classes	Dry mixed
Latex additives.[35] To form films.	1.0 gal/sk	All API Classes	Dry mixed or with mixing water
Bentonite cement with dispersant. To improve particle-size distribution.	12 to 16 percent gel, 0.7 to 1.0 percent dispersant	API Class A, G, or H	Batch mixed

*Percent by weight of cement.

control additives in current use.[23,33,34])

The two most widely used filtration-control materials are **organic polymers (cellulose) and friction reducers.[33A]**

The high-molecular-weight cellulose compounds will produce low water loss in all types of cementing compositions at concentrations from 0.5 to 1.5 percent by weight of cement. (See Table 3.23.) The water requirement, however, may have to be adjusted to produce the desired viscosity; i.e., an API Class A cement will require 5.6 instead of the usual 5.2 gal of water per sack.

Dispersants, or friction reducers, are commonly added to cement slurries to control filter loss by dispersing and packing the cement particles and thus densifying the slurry. This is especially effective when the water/cement ratio is reduced. The effect that densification of a cement slurry has on its filter loss is shown in Table 3.24.

3.8 Cement Dispersants, or Friction Reducers

Dispersing agents are added to cement slurries to **improve their flow properties.[35A] Dispersed slurries have** lower viscosity and can be pumped in turbulence at lower pressures, thereby minimizing the horsepower required and lessening the chances of lost circulation and premature dehydration.[37-39] Dispersants lower the yield point and gel strength of the slurry. (Table 3.25 lists some commonly used dispersants; Table 3.26 illustrates the effect of dispersants on the critical flow rate — the flow rate required to achieve turbulence — of slurries.)

The dispersants commonly added to cement slurries are polymers, fluid-loss agents in gel cement, and salt (sodium chloride). They are used at low temperatures because they retard the cement only slightly. (See Table 3.27.) Calcium lignosulfonates — organic acid blends — retard substantially and are generally used at higher temperatures.

Polymers (Dispersants, or Thinners)

Polymers, manufactured in powdered form, produce unusual and useful properties in cement systems. They do not significantly accelerate or retard most slurries, but they do markedly reduce apparent viscosity (see Fig. 3.2). They are well suited over a temperature range of 60° to 300°F. Despite their viscosity-reducing property, polymers do not cause excessive free water separation or settling of cement particles from the slurry. They are compatible with nearly all types of cement systems except those containing high concentrations of salt. Although the polymers thin such slurries initially and appear to be effective, they are incompatible with the salt, which can cause them to flocculate, and after 10 to 20 minutes of mixing they cause a rapid increase in viscosity.

Salt (Sodium Chloride)

Common salt, in addition to acting as a weighting agent, an accelerator, and a retarder, can also act as a thinner (dispersant) in many cementing compositions (Fig. 3.3). It is especially effective for reducing the

TABLE 3.23 — EFFECTS OF ORGANIC POLYMERS ON THE FILTER LOSS OF API CLASS H CEMENT

Polymer (percent by weight of cement)	API Fluid Loss at 1,000 psi (cc/30 min.)	Permeability of Filter Cake At 1,000 psi (md)	Time To Form 2-in. Cake (minutes)
0.00	1,200	5.00	0.2
0.50	300	0.54	3.4
0.75	100	0.09	30.0
1.00	50	0.009	100.0

TABLE 3.24 — API FILTER LOSS OF DENSIFIED CEMENT SLURRIES

Cement: API Classes A and G.
API Fluid-Loss Test
Screen: 325 mesh.
Pressure: 1,000 psi.
Temperature: 80°F.

Dispersant (percent)	Fluid Loss (cc/30 min.) at a Water Ratio (gal/sk) of			
	3.78	4.24	4.75	5.2
0.50	490	504	580	690
0.75	310	368	476	530
1.00	174	208	222	286
1.25	118	130	146	224
1.50	72	80	92	—
1.75	50	54	64	—
2.00	36	40	48	—

TABLE 3.25 — COMMONLY USED CEMENT DISPERSANTS

Type of Material	Amount Used (lb per sk of cement)
Polymer: Blend	0.3 to 0.5
Long chain	0.5 to 1.5
Sodium chloride	1 to 16
Calcium lignosulfonate, organic acid (retarder and dispersant)	0.5 to 1.5

apparent viscosity of slurries containing bentonite, diatomaceous earth, or pozzolan.[24,25,42]

3.9 Uses of Salt Cements

Salt is used in cement slurries to bond the set cement more firmly to salt sections (Fig. 3.4) and shales, and to make the set cement expand. The samples in Fig. 3.4 show that the fresh-water slurry has dissolved part of the salt, preventing a bond between the rock and the cement and expanding the hole. Where the salt-saturated slurry has been used, bonding has been achieved and the hole has not been enlarged. This illustrates that in cementing through salt sections, better results can be achieved by cementing with a salt-laden slurry.

Cement slurries containing salt help to protect shale sections from sloughing and heaving during cementing,[43-45] and prevent annular bridging and the lost circulation that can result (Fig. 3.5). A shale that is sensitive to cement filtrate can actually become so soft by being wetted before the cement sets that it will flow, creating channels behind the cement sheath from one perforated zone to another. Cement slurries containing 5 to 20 percent salt have proved effective in the field in minimizing both sloughing and channeling of shale. (An analysis of a typical filtrate from salt-cement slurries is given in Table 3.28.)

When salt water is mixed with cement, foaming sometimes occurs, making it difficult to control slurry

TABLE 3.26 — EFFECT OF DISPERSANTS ON CRITICAL FLOW RATES OF SLURRIES IN TURBULENCE[38,40,41]

Cement: API Class H.
Water ratio: 5.2 gal/sk.
Slurry weight: 15.6 lb/gal.
Slurry properties:

	Neat	With 1 Percent Dispersant
Flow behavior index *(n')**	0.30	0.67
Consistency index *(K')**	0.195	0.004

Pipe Size (in.)	Hole Size (in.)	Critical Velocity (ft/sec) Without Dispersant	Critical Velocity (ft/sec) With Dispersant	Critical Pump Rate (bbl/min.) Without Dispersant	Critical Pump Rate (bbl/min.) With Dispersant
4½	6¾	9.2	2.6	13.6	3.85
4½	7⅞	8.6	2.13	20.9	5.18
5½	7⅞	9.1	2.54	16.9	4.70
5½	8¾	8.6	2.17	23.3	5.85
7	8¾	9.6	2.96	15.5	4.76
8⅝	11	9.1	2.54	24.4	6.90
8⅝	12¼	8.5	2.05	37.0	9.05
9⅝	12¼	9.0	2.41	29.6	8.08

*See Section 11.2.

TABLE 3.27 — EFFECT OF POLYMER DISPERSANTS ON THE THICKENING TIME AND COMPRESSIVE STRENGTH OF API CLASS G CEMENT

Dispersant (percent)	Thickening Time (hours:min.) For API Casing Tests at Well Depth (ft) of 6,000	8,000	Compressive Strength (psi) After Curing 24 Hours at a Temperature (°F) of 80	80*	95	110	140	12 Hours 140°F
API Class G Cement, Neat								
0.0	2:16	1:08	1,480	2,700	1,405	2,375	5,200	2,780
0.5	1:55	1:23	1,425	2,375	1,795	2,350	5,285	2,875
0.75	2:12	1:55	1,565	2,575	1,810	2,775	4,660	2,965
1.0	3:00+	3:00+	1,410	2,440	1,920	2,285	2,595	2,405
1.25	3:00+	3:00+	1,350	2,480	1,895	2,025	2,345	2,260
API Class G Cement, With 18 Percent Salt Water								
0.5	2:05	1:23	—	—	3,265	3,925	4,220	3,995
0.75	2:35	2:10	—	—	2,880	3,595	3,820	3,580
1.0	3:00+	3:00+	—	—	2,555	3,295	3,425	3,285
1.25	3:00+	3:00+	—	—	2,290	2,925	3,125	2,975

*Slurry contains 2 percent calcium chloride.

TABLE 3.28 — ANALYSIS OF CEMENT FILTRATE

Cement: API Class A.
Water ratio: 5.2 gal/sk.

Salt (percent by weight of water)	Specific Gravity	pH	OH^- (mg/l)	Ca^{++} (mg/l)	SO_4^{++} (mg/l)	Cl^- (mg/l)	NaCl Equivalent of Cl_2 in ppm
0	1.008	12.6	1,013	860	4,950	20	0
5	1.048	12.1	853	1,685	7,000	31,250	5,214
10	1.080	12.0	728	2,060	9,600	59,750	10,319
18	1.121	11.8	614	1,675	8,400	105,500	18,610
Sat.	1.206	11.6	274	650	7,400	185,000	30,970

TABLE 3.29 — EFFECTS OF POTASSIUM CHLORIDE ON THE STRENGTH OF CEMENT

Cement: API Class A.
Water ratio: 5.2 gal/sk.

Percent KCl by Weight of Water	Compressive Strength (psi): After Curing 8 Hours at a Temperature of 80°F	After Curing 8 Hours at a Temperature of 100°F	After Curing 24 Hours at a Temperature of 80°F	After Curing 24 Hours at a Temperature of 100°F
0	120	705	1,635	2,820
5	275	1,225	2,600	4,160
10	300	1,225	2,215	4,225
15	235*	885*	2,635*	3,885*
Sat.	50*	200*	1,355*	2,080*

*Excessive slurry velocity.

weight and volume. This can be prevented by adding antifoaming agents to the mixing water or by dry blending salt with the cement. Dry blending also eliminates waste in handling salt at the wellsite.

The use of dry salt in cementing slurry produces similar effects on the properties of cement of all API Classes and on those of pozzolan cement and bentonite cement.

Although the salt generally used with cement is sodium chloride, potassium chloride is also used (see Table 3.29), and in some cases may be more effective at lower concentrations. It has no significantly different effect on cement slurries except at the higher concentrations, where slurry viscosity becomes excessive.

3.10 Special Additives for Cement

The special additives presently used in cement slurries are listed in Table 3.30.

Mud Decontaminants

Paraformaldehyde or a blend of paraformaldehyde and sodium chromate are sometimes used to minimize the cement retarding effects of various drilling-mud chemicals in the event a cement slurry becomes contaminated by intermixing with the drilling fluid.[46] A mud decontaminant consisting of a 60:40 mixture of

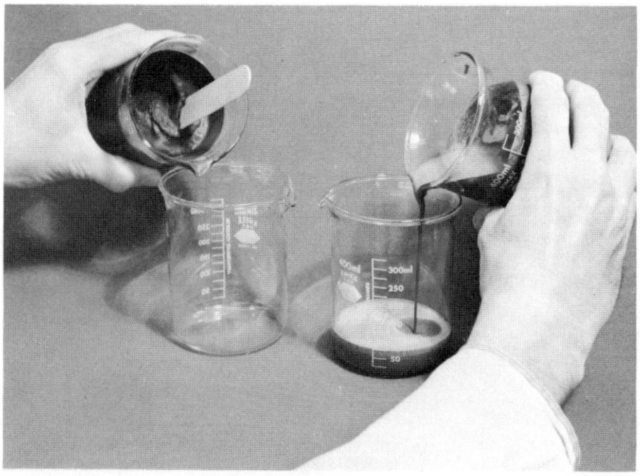

Fig. 3.2 Effect of polymer dispersant on API Class H cement slurry (sample at left — no polymer; right — 1 percent polymer).

TABLE 3.30 — SPECIAL ADDITIVES FOR CEMENTS

Additive	Recommended Quantity
Mud decontaminants	1.0 percent*
Silica flour	30 to 40 percent*
Radioactive tracers	Variable
Dyes	0.1 to 1.0 percent*
Hydrazine	6 gal/1,000 bbl mud
Fibers	0.125 to 0.5 percent*
Gypsum	4 to 10 percent*

*Percent by weight of cement.

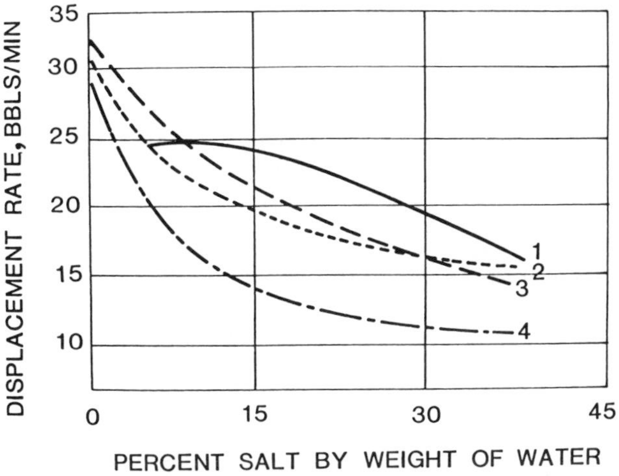

1. API Class A cement.
2. 4-percent-gel cement.
3. 12-percent-gel cement.
4. Pozzolan cement.

Fig. 3.3 Displacement rates of salt-cement slurries in turbulent flow (8¾-in. hole, 5½-in. casing.[24]

Fig. 3.4 Bonding of cement to rock salt. The slurry for the unbonded sample (left) contained fresh water; the bonded sample (right) contained saturated salt water.

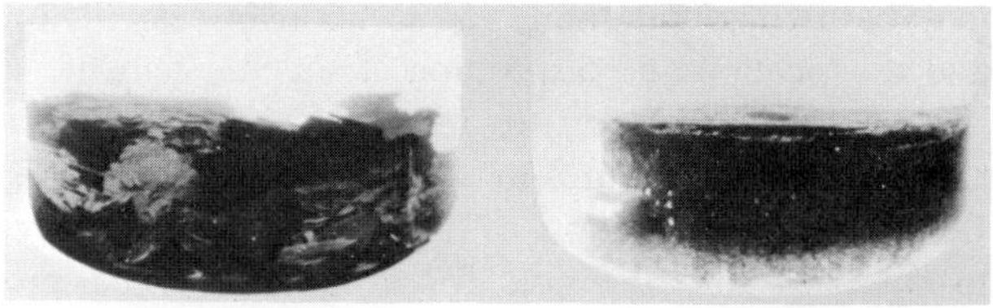

Fig. 3.5 Effect of brine solution on Morrow shale (time of exposure: 1 hour).[24]

TABLE 3.31 — PERMEABILITY OF HYDRATED API CLASS H CEMENT

			Curing Time: 3 Days at 320°F		Curing Time: 28 Days at 320°F	
Silica (percent)	Bentonite (percent)	Hematite (percent)	Compressive Strength (psi)	Permeability (md)	Compressive Strength (psi)	Permeability (md)
0	0	0	2,165	0.031	2,590	4.580
20	0	0	9,590	0.001	5,450	<0.001
30	0	0	8,325	0.001	5,390	<0.001
40	0	0	8,165	<0.001	11,330	<0.001
0	4	0	590	0.548	370	9.720
30	4	0	4,275	<0.001	3,050	<0.001
40	4	0	3,750	<0.001	4,140	<0.001
0	0	28	2,205	0.030	1,600	3.890
30	0	45	9,925	<0.001	7,015	<0.001
4)	0	50	8,525	<0.001	8,450	<0.001

paraformaldehyde and sodium chromate neutralizes certain mud-treating chemicals. It is effective against tannins, lignins, starch, cellulose, lignosulfonate, ferrochrome lignosulfonate, chrome lignin, and chrome lignite.

Mud decontaminants are used primarily in open-hole plug-back jobs and liner jobs, for squeeze cementing, and for tailing out on primary casing jobs.

Silica Flour

Fine silica or silica flour is commonly used in cementing compositions to help prevent loss of strength.[47-49] Research has shown that as temperatures exceed 230°F all manufactured cements lose much of their compressive strength; and the higher the temperature, the greater the loss of strength.[47-49] This loss of strength, which is accompanied by an increase in permeability, is caused by the formation of an alpha hydrate form of calcium silicate in the set cement. Adding a high-water-ratio material such as bentonite accelerates the loss of strength. (See Table 3.31.)

Silica flour can be added to all classes of API cement to prevent the loss of strength that occurs with time and high temperatures. The optimum amount of silica for controlling strength loss is 30 to 40 percent. Silica flour (– 200 mesh) has a water requirement of 40 percent (40 lb, or 4.8 gal, of water per 100 lb of silica flour). Where weighted slurries are required (17 to 20 lb/gal), coarse silica having a particle size range of – 50 to + 150 mesh is frequently used.

Radioactive Tracers

Radioactive tracers are added to cement slurries as markers that can be detected by logging devices. They may be used to determine the location of cement tops and the location and disposition of squeeze cement.

The isotopes commonly used down the hole have half-lives ranging from 8 to 84 days. (See Table 3.32.) By the proper selection of tracer, the time required to get back into the hole for a survey can be programmed.

Radioisotopes are controlled and licensed by the U. S. Atomic Energy Commission and various state agencies, and cannot be used indiscriminately.

Dyes for Cement

Small amounts of indicator dye can be used to identify a cement of a specific API classification or an additive blended in a cementing composition. When the dyes are used down hole, however, dilution and mud contamination may dim and cloud the colors, rendering them ineffective. Table 3.33 lists some materials used as indicators.

Hydrazine

Hydrazine is an additive used to treat the mud column above the cement to minimize corrosion problems in the uncemented portion of the hole.[50] One pound of Hydrazine (2.85 lb of 35 percent solution) is required to remove 1 lb of dissolved oxygen. The California Research Corp. recommends 6 gal of 35 percent Hydrazine solution for 100 bbl of mud. Since Hydrazine is an oxygen scavenger it should be handled

TABLE 3.32 — RADIOISOTOPES COMMONLY USED IN CEMENT

Trade Name	Isotope	Half-Life (days)	Maximum Energy Gammas (mev)
RAC-2*	I 131	8	0.364
RAC-3	Sc 46	84	1.17
Rayfrac®** Sand	Sc 46	84	1.17
Rayfrac®** Walnut Hulls	Sc 46	84	1.17
Rayfrac®** Ucar Props	Sc 46	84	1.17

*Not recommended for squeeze cementing.
**Not recommended for squeeze cementing; primarily used for tracing lost circulation and fracturing.

TABLE 3.33—DYES OR PIGMENTS FOR COLORING CEMENT

Material	Percentage Used	Water/Cement Contact Color	Cement Slurry Color
Indicator Dyes			
Fluorescein	0.1	Green	Green
Phenolphthalein	0.1	Red violet	Purple
Methylene blue	0.1	Blue	Blue
Pigments			
Black oxide	0.1	Faint tannish green	Dark gray with black streaks
Yellow oxide	0.1	Faint tannish yellow	Pale olive green
Red iron oxide	0.1	Faint tannish red	Light brown with orange streaks

with extreme caution. Before a Hydrazine job is performed, a special adaptor should be placed in the suction side of the displacement pump to aid in mixing the Hydrazine with the drilling mud to be pumped ahead of the cement. In determining the amount of Hydrazine to be used, the calculated theoretical volume of mud to be left behind casing should be increased by 20 percent. Hydrazine is expensive, so the quantity should be calculated carefully. However, an excess of it in the mud is not physically detrimental. Upon completion of the job, pumps, lines, and containers should be thoroughly flushed with water.

Fiber in Cement

Synthetic fibrous materials, such as Tuf Fibers*, are frequently added to oilwell cements in concentrations of ⅛ to ½ lb/sk to reduce the effects of shattering or partial destruction from perforation, drill-collar stress, or other down-hole forces.[51] Fibrous materials transmit localized stresses more evenly throughout the cement and thus improve the resistance to impact and shattering (see Fig. 3.6). The most desirable fiber is nylon. It has fiber lengths varying up to 1 in., is resilient, and imparts high shear, impact, and tensile strengths.

The properties achieved with nylon-fiber-reinforced cement on unsupported casing are illustrated in Figs. 3.6 and 3.7.

*A trade mark of Halliburton Services.

Fig. 3.6 Effect of fiber on cement. (Right sample — cement sheath contains no fiber; left — fiber-laden sheath must be broken with wedge.)[51]

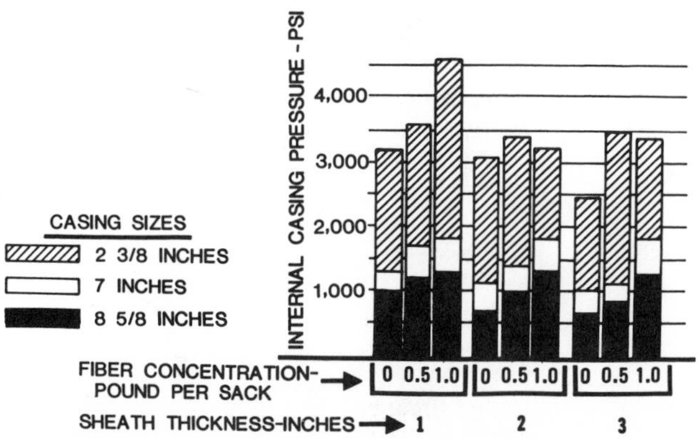

Fig. 3.7 Internal casing pressure required to cause failure of fiber-laden cement.[51]

Gypsum Additives

About 4 to 10 percent gypsum is added to portland cement to achieve (1) flash setting to combat lost circulation, (2) gelling or thixotropic properties, and (3) expansion properties in the set cement.

Adding 30 to 50 percent gypsum to any portland cement will produce a flash set in 12 to 20 minutes even when the slurry is in motion.[52] This has been effectively done to seal off lost-circulation zones in shallow wells where strength is necessary to give stability to the wellbore.

For unconsolidated, highly permeable, fractured, or cavernous formations, 5 to 10 percent gypsum added to a portland cement slurry will cause it to gel rapidly when in a static state.[53,54] This thixotropic property helps the slurry pass permeable formations. The slurry will support its own column weight if circulation is lost and therefore will not fall back into the lost-circu-

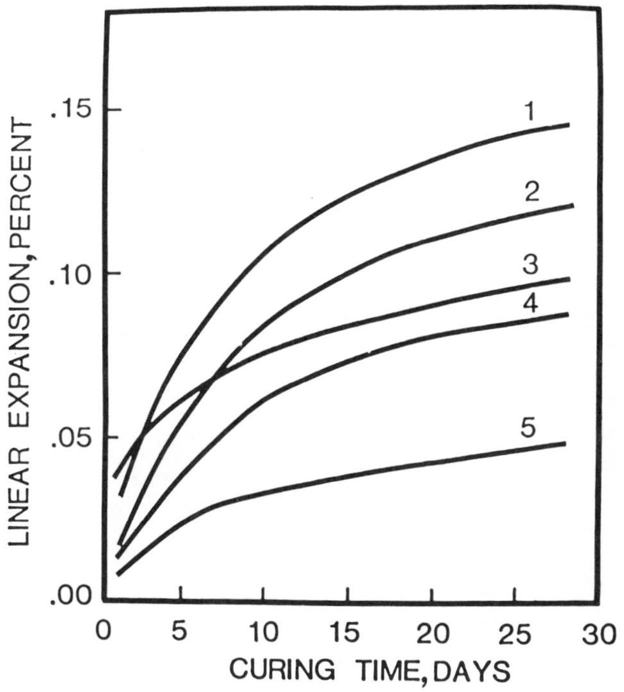

1. API Class A cement, salt-saturated mixing water — 16.1 lb/gal.
2. API Class A cement — 18 percent salt by weight of mixing water — 15.9 lb/gal.
3. Commercial expanding cement — 15.6 lb/gal.
4. API Class A cement densified with 1 percent dispersant — 17.0 lb/gal.
5. Neat API Class A cement — 15.6 lb/gal.

Fig. 3.8 Effect of additives on linear expansion of cements (curing temperature, 80°F; fresh water).

lation zone. Gypsum is used principally in wells less than 6,000 ft deep.

Gypsum added to API Class A, G, or H cement in concentrations of 3 to 6 percent will react with the tricalcium aluminate and expand the set cement. (See Expanding Cements, Section 2.7.) These expanding properties improve the cement bond between pipe and formation, effecting a better seal against gas or fluid annular migration. Typical expansion behavior is illustrated in Fig. 3.8.

3.11 Summary

Table 3.34 is a summary of the most common cementing additives, their uses and benefits, and the cements to which they can be added. Fig. 3.9 shows the minor and major effects of additives on the physical properties of cement.

References

1. "A Basic Oil Well Cement," *API Report 701-61-A*, API (Feb. 1965).
2. Farris, R. F.: "Method for Determining Minimum Waiting-on-Cement Time," *Trans.*, AIME (1946) **165,** 175-188.
3. Davis, S. H., and Faulk, J. H.: "Have Waiting-on-Cement Practices Kept Pace With Technology?" *Drill. and Prod. Prac.*, API (1957) 180.

TABLE 3.34 — SUMMARY OF OILWELL CEMENTING ADDITIVES

Type of Additive	Use	Chemical Composition	Benefit	Type of Cement
Accelerators	Reducing WOC time Setting surface pipe Setting cement plugs Combatting lost circulation	Calcium chloride Sodium chloride Gypsum Sodium silicate Dispersants Sea water	Accelerated setting High early strength	All API Classes Pozzolans Diacel systems
Re arders	Increasing thickening time for placement Reducing slurry viscosity	Lignosulfonates Organic acids CMHEC Modified lignosulfonates	Increased pumping time Better flow properties	API Classes D, E, G, and H Pozzolans Diacel systems
Weight-reducing additives	Reducing weight Combatting lost circulation	Bentonite-attapulgite Gilsonite Diatomaceous earth Perlite Pozzolans	Lighter weight Economy Better fillup Lower density	All API Classes Pozzolans Diacel systems
Heavy-weight additives	Combatting high pressure Increasing slurry weight	Hematite Ilmenite Barite Sand Dispersants	Higher density	API Classes D, E, G, and H
Additives for controlling lost circulation	Bridging Increasing fillup Combatting lost circulation	Gilsonite Walnut hulls Cellophane flakes Gypsum cement Bentonite-diesel oil Nylon fibers	Bridged fractures Lighter fluid columns Squeezed fractured zones Minimized lost circulation	All API Classes Pozzolans Diacel systems
Filtration-control additives	Squeeze cementing Setting long liners Cementing in water-sensitive formations	Polymers Dispersants CMHEC Latex	Reduced dehydration Lower volume of cement Better fillup	All API Classes Pozzolans Diacel systems
Dispersants	Reducing hydraulic horsepower Densifying cement slurries for plugging Improving flow properties	Organic acids Polymers Sodium chloride Lignosulfonates	Thinner slurries Decreased fluid loss Better mud removal Better placement	All API Classes Pozzolans Diacel systems
Special cements or additives				
Salt	Primary cementing	Sodium chloride	Better bonding to salt, shales, sands	All API Classes
Silica flour	High-temperature cementing	Silicon dioxide	Stabilized strength Lower permeability	All API Classes
Mud Kil	Neutralizing mud-treating chemicals	Paraformaldehyde	Better bonding Greater strength	API Classes A, B, C, G, and H
Radioactive tracers	Tracing flow patterns Locating leaks	Sc 46	—	All API Classes
Pozzolan lime	High-temperature cementing	Silica-lime reactions	Lighter weight Economy	—
Silica lime	High-temperature cementing	Silica-lime reactions	Lighter weight	—
Gypsum cement	Dealing with special conditions	Calcium sulfate Hemihydrate	Higher strength Faster setting	—
Hydromite	Dealing with special conditions	Gypsum with resin	Higher strength Faster setting	—
Latex cement	Dealing with special conditions	Liquid or powdered latex	Better bonding Controlled filtration	API Classes A, B, G, and H

4. Bearden, W. G., and Lane, R. D.: "You Can Engineer Cementing Operations To Eliminate Wasteful WOC Time," *Oil and Gas J.* (July 3, 1961) 104.
5. Craft, B. C., and Stephenson, A. H.: "Effect of Calcium Chloride on High Early Strength Cement," *J. Pet. Tech.* (June 1952) 11-12.
6. Montgomery, P. C.: "Cement Accelerators Cut Rig Downtime," *Drilling* (June 1965) **26,** No. 9, 76.
7. Shell, F. J., Hurley, J. R., Bergman, W. E., and Fisher, H. B.: "Low Density Oil Well Cements," *World Oil* (Sept. 1956) 131.
8. Waggoner, H. F.: "Additives Yield Heavy High-Strength Cements with Low Water Ratios," *Oil and Gas J.* (April 13, 1964) 109.
9. Smith, R. C., and Calvert, D. G.: "The Use of Sea Water in Well Cementing," *J. Pet. Tech.* (June 1975) 759-764.
10. Coffer, H. F., Reynolds, J. J., and Clark, R. C., Jr.: "A Ten-Pound Cement Slurry for Oil Wells." *Trans.,* AIME (1954) **201,** 146-148.
11. Dumbauld, G. K., Brooks, F. A., Jr., Morgan, B. E., and Binkley, G. W.: "A Lightweight, Low-Water-Loss, Oil-Emulsion Cement for Use in Oil Wells," *J. Pet. Tech.* (May 1956) 99-104.
12. "Specifications for Oil-Well Cements and Cement Additives," *API Standard 10A,* 20th ed., API, New York (1975).
13. Morgan, B. E., and Dumbauld, G. K.: "A Modified Low-Strength Cement," *Trans.,* AIME (1951) **192,** 165-170.
14. Morgan, B. E., and Dumbauld, G. K.: "Bentonite Cement Proving Successful in Permanent-Type Squeeze Operations," *World Oil* (Nov. 1954) 220.
15. Beach, H. J.: "Improved Bentonite Cements Through Partial Acceleration," *J. Pet. Tech.* (Sept. 1961) 923-926; *Trans.,* AIME, **222.**
16. Smith, D. K.: "Well Cementing Method," U. S. Patent No. 3,227,213 (Jan. 4, 1966).
17. Porter, E. W.: "A Low Water Loss-Low Density Cement," *Drill. and Prod. Prac.,* API (1955) 465 (abstr.); paper 875-9-I presented at API Rocky Mountain Dist. Div. of Production Spring Meeting, Denver, Colo., April 12-15, 1955.
18. "Cement Additive Reduces Weight," *Petroleum Week* (June 13, 1958) **6,** No. 11, 32.
19. Slagle, K. A., and Carter, L. G.: "Gilsonite — A

		Bentonite	Diatomaceous Earth	Pozzolan	Sand	Heavy Weight	Accelerator	Sodium Chloride	Retarder	Friction Reducer	Low-Water-Loss Materials	Lost Circulation Materials
DENSITY	Decrease	(x)	(x)	(x)								
	Increase				(x)	(x)		x				
WATER REQUIRED	Less											
	More	(x)	(x)	x	x					x		x
VISCOSITY	Decreased						x	x	(x)	(x)		
	Increased	x	x	x	x	x						x
THICKENING TIME	Accelerated						(x)	(x)				
	Retarded	x	x					x	(x)	x	x	
EARLY STRENGTH	Decreased	x	x	x					(x)		x	x
	Increased						(x)	(x)		x		
FINAL STRENGTH	Decreased	(x)	(x)	x		x					x	x
	Increased								x	x		
DURABILITY	Decreased	x	x									x
	Increased			(x)								
WATER LOSS	Decreased	(x)							x	x	(x)	x
	Increased		x									

x Denotes minor effects.

(x) Denotes major effects and / or principal purpose for which used.

Fig. 3.9 Effects of additives on the physical properties of cement.[55]

Unique Additive for Oil-Well Cements," *Drill. and Prod. Prac.*, API (1959) 318-328.

20. "Strata-Crete for Lighter Cement Slurries," Great Lakes Carbon Corp., Houston (1951) 10 pp.
21. Bleakley, W. B.: "Cut Lost Circulation While Cementing," *Oil and Gas J.* (Aug. 26, 1963).
22. Underwood, D., Broussard, P., and Walker, W.: "Long Life Cementing Slurries," paper presented at API Southwestern Dist. Div. of Production Spring Meeting, Dallas, March 10-12, 1965.
23. Shell, F. J., and Wynne, R. A.: "Application of Low-Water-Loss Cement Slurries," papèr 875-12-I presented at API Rocky Mountain Dist. Div. of Production Spring Meeting, Denver, Colo., April 1958.
24. Slagle, K. A., and Smith, D. K.: "Salt Cement for Shale and Bentonite Sands," *J. Pet. Tech.* (Feb. 1963) 187-194; *Trans.*, AIME, **228.**
25. Ludwig, N. C.: "Effects of Sodium Chloride on Setting Properties of Oil-Well Cements," *Drill. and Prod. Prac.*, API (1951) 20-27.
26. Scott, P. O., Jr., Lummus, J. L., and Howard, G. C.: "Methods for Sealing Vugular and Cavernous Formations," *Drilling Contractor* (Dec. 1953) 70-74.
27. Goins, W. C., Jr.: "Lost Circulation Problems Whipped with BDO (Bentonite Diesel Oil) Squeeze," *Drilling* (Sept. 1954) **15,** No. 11, 83.
28. Messenger, J. U., and McNeil, J. S., Jr.: "Lost Circulation Corrective: Time Setting Clay Cement," *Trans.*, AIME (1952) **195,** 56-64.
29. Gibbs, M. A.: "Primary and Remedial Cementing in Fractured Formations," paper presented at Southwestern Petroleum Short Course, Texas Technological College, Lubbock, April 22-23, 1965.
30. Einarsen, C. A.: "High Strength Granular Sealing Material Increases Efficiency of Primary Cementing and Squeeze Cementing," *J. Pet. Tech.* (Aug. 1955) 15-18.
31. White, R. J.: "Lost Circulation Materials and Their Evaluation," *Drill. and Prod. Prac.*, API (1956) 352-359.
32. Wieland, D. R., Calvert, D. G., and Spangle, L. B.: "Design of Special Cement Systems for Areas with Low Fracture Gradients," paper presented at API Southwestern Dist. Div. of Production Spring Meeting, Lubbock, Tex., March 12-14, 1969.
33. Stout, C. M., and Wahl, W. W.: "A New Organic Fluid-Loss-Control Additive for Oilwell Cement," *J. Pet. Tech.* (Sept. 1960) 20-24.

33A. Weisend, C. F.: "Method and Composition for Cementing Wells," U.S. Patent 3,132,693 (May 12, 1964) filed Dec. 26, 1961.

34. Beach, H. J., O'Brien, T. B., and Goins, W. C., Jr.: "The Role of Filtration in Cement Squeezing," *Drill. and Prod. Prac.*, API (1961) 27-35.
35. "Latex-Cement Shows High Success Ratio," *Drilling* (19th Annual Exposition-in-Print) (Dec. 2, 1957) 33.

35A. Weisend, C. F.: "Method and Composition for Cementing Wells," U.S. Patent 3,359,225 (Dec. 19, 1967) filed Aug. 26, 1963.

36. Boughton, L. D., Pavlich, J. P., and Wahl, W. W.: "The Use of Dispersants in Cement Slurries To Improve Placement Techniques," paper SPE 412 presented at SPE-AIME 33rd Annual Fall Meeting, Los Angeles, Oct. 8-10, 1962.
37. Howard, G. C., and Clark, J. B.: "Factors To Be Considered in Obtaining Proper Cementing of Casing," *Drill. and Prod. Prac.*, API (1948) 257-272.
38. Slagle, K. A.: "Rheological Design of Cementing Operations," *J. Pet. Tech.* (March 1962) 323-328; *Trans.*, AIME, **225.**
39. Brice, J. W., Jr., and Holmes, B. C.: "Engineered Casing Cementing Programs Using Turbulent Flow Techniques," *J. Pet. Tech.* (May 1964) 503-508.
40. McLean, R. H., Manry, C. W., and Whitaker, W. W.: "Displacement Mechanics in Primary Cementing," *J. Pet. Tech.* (Feb. 1967) 251-260.
41. Parker, P. N., Ladd, B. J., Ross, W. N., and Wahl, W. W.: "An Evaluation of a Primary Cementing Technique Using Low Displacement Rates," paper SPE 1234 presented at SPE-AIME 40th Annual Fall Meeting, Denver, Colo., Oct. 3-6, 1965.
42. Shell, F. J.: "The Effect of Salt on DE Cement," *Cdn. Oil and Gas Ind.* (March 1957) **10,** No. 3, 64-67.
43. Moore, J. E.: "Clay Mineralogy Problems in Oil Recovery," *Pet. Eng.*, Part 1 (Feb. 1960); Part 2 (March 1960).
44. Hewitt, C. H.: "Analytical Techniques for Recognizing Water-Sensitive Reservoir Rocks," *J. Pet. Tech.* (Aug. 1963) 813-818.
45. Cunningham, W. C., and Smith, D. K.: "Effect of Salt Cement Filtrate on Subsurface Formations," *J. Pet. Tech.* (March 1968) 259-264.
46. Beach, H. J., and Goins, W. C., Jr.: "A Method of Protecting Cements Against the Harmful Effects of Mud Contamination," *Trans.*, AIME (1957) **210,** 148-152.
47. Carter, L. G., and Smith, D. K.: "Properties of Cementing Compositions at Elevated Temperatures and Pressures," *J. Pet. Tech.* (Feb. 1958) 20-28.
48. Kalousek, G. L.: Discussion of "The Reactions of Thermochemistry of Cement Hydration at Ordinary Temperatures," *Proc.*, Third Intl. Symposium on the Chemistry of Cement, London (1952) 296-311.
49. Dunlap, I. R., and Patchen, F. D.: "A High-Temperature Oil Well Cement," *Pet. Eng.* (Nov. 1957) B60.
50. Schremp, F. W., Chittum, J. F., and Arczynski, T. S.: "Use of Oxygen Scavengers To Control External Corrosion of Oil-String Casing," *J. Pet. Tech.* (July 1961) 703-711; *Trans.*, AIME, **222.**
51. Carter, L. G., Slagle, K. A., and Smith, D. K.: "Resilient Cement Decreases Perforating Damage," paper presented at API Mid-Continent Dist. Div. of Production Spring Meeting, Amarillo, Tex., April 1968.
52. Clason, C. E.: "Evolution and Use of Gypsum Cement for Oil Wells," *World Oil* (Aug. 1949) 119-126.
53. Boice, D., and Diller, J.: "A Better Way To Squeeze Fractured Carbonates," *Pet. Eng.* (May 1970) 79-82.
54. Goolsby, J. L.: "A Proven Squeeze Cementing Technique in a Dolomite Reservoir," *J. Pet. Tech.* (Oct. 1969) 1341-1346.
55. Dale, O. O.: "The Effects of Some Additives on the Physical Properties of Portland Cement," *Oil Well Cementing Practices in the United States*, API (1959) 62.

Chapter 4

Factors That Influence Cement Slurry Design

4.1 Introduction

Completion depths, well temperatures, hole conditions, and drilling problems must all be considered in designing an oilwell cementing composition. The following factors affect cement slurry design:

1. Well depth
2. Well temperature
3. Mud-column pressure
4. Viscosity and water content of cement slurries
5. Pumping, or thickening, time
6. Strength of cement required to support pipe
7. Quality of available mixing water
8. Type of drilling fluid and drilling fluid additive
9. Slurry density
10. Heat of hydration
11. Permeability of set cement
12. Filtration control
13. Resistance to down-hole brines

Much of the industry's current ability to design a slurry properly has resulted from the standardization of laboratory equipment and testing procedures and from the availabilty of laboratory facilities for testing at simulated down-hole cementing conditions.

4.2 Pressure, Temperature, and Pumping Time

Two basic influences on the down-hole performance of cement slurries are temperature and pressure. They affect how long the slurry will pump and how well it develops the strength necessary to support pipe. Temperature has the more pronounced influence.[1,2] As the formation temperature increases, the cement slurry hydrates and sets faster and develops strength more rapidly. Also, the pumping (or thickening) time is decreased. Fig. 4.1 shows how temperature affects thickening time.

Pressure imposed on a cement slurry by the hydrostatic load of well fluids also reduces the pumpability of cement. In deep wells, hydrostatic pressure plus surface pressure during placement may exceed 20,000 psi (Table 4.1). The influence of pressure upon the pumpability of cements is illustrated in Table 4.2.

Temperature gradients vary in different geographical areas. In West Texas and New Mexico, gradients average about 0.8°F per 100 ft of depth, whereas along the Texas and Louisiana Gulf Coasts they range up to 2.2°F per 100 ft of depth. Estimates of bottom-hole static temperatures may be obtained from surveys run during logging and from drillstem tests. Bottom-hole circulating temperatures are obtained from temperature recording subs run in the drillstring during mud conditioning trips before casing is set. From such data the relationship of bottom-hole static temperatures vs circulating temperatures can be obtained to determine the pumpability of a cement slurry.

Temperature studies conducted along the Gulf Coast of Texas and Louisiana in the early 1950's have formed

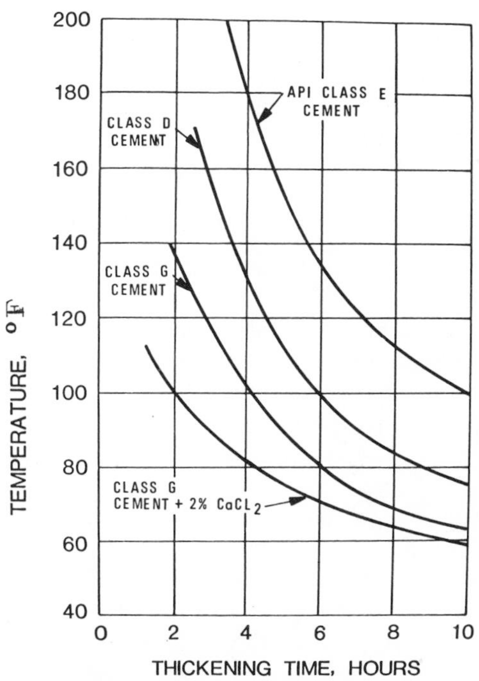

Fig. 4.1 Effect of temperature on thickening time of various API cements at atmospheric pressure.[2]

TABLE 4.1 — HYDROSTATIC PRESSURES ENCOUNTERED IN WELLS

Depth (ft)	Pressure (psi) for Fluid Indicated				
	Water	Drilling Fluid or Cement Slurry			
	8.34 lb/gal	11 lb/gal	13 lb/gal	15 lb/gal	17 lb/gal
1,000	434	572	676	780	884
5,000	2,170	2,860	3,380	3,900	4,420
10,000	4,340	5,720	6,760	7,800	8,840
15,000	6,500	8,580	10,100	11,700	13,300
20,000	8,670	11,400	13,500	15,600	17,700
25,000	10,840	14,260	16,880	19,500	22,120
30,000	13,010	17,120	20,260	23,400	26,540
35,000	15,180	19,980	23,640	27,300	30,960
40,000	17,354	22,840	27,840	27,020	35,380

the basis of API testing schedules and cement specifications for more than 20 years. The schedules are based on bottom-hole temperatures, °F = 80°F + 0.015 × depth in feet. (See Fig. 4.2.) The cooling effect of mud displacement lowers the circulating temperature of the hole considerably during casing cementing. During squeeze cementing, there is less cooling because there is less well fluid preceding the slurry. Thus, a cementing compositon is pumpable longer during casing cementing than during squeeze cementing at the same depth.

The time it takes a cement slurry to reach bottom depends upon casing size and displacement rate. These factors were studied and a survey was made by the API in 1962. As a result, testing schedules were revised to compensate for higher displacement rates in wells of moderate to extreme depths. Tables 4.3, 4.4, and 4.5 list the data used as a basis for the thickening-time specifications and procedures given in current API *Standards 10A* and *10B*.[4]

In designing cement slurries for specific well conditions, the rate of slurry placement per 1,000 ft of depth, as well as horsepower requirements, displacement rates, slurry volumes, and hole and casing size relationships are used as the basis for determining the pumping time to be expected from a given cementing composition. Strength data are based on well temperatures and pressures and indicate the time required for the cement to become strong enough to support the pipe.

4.3 Viscosity and Water Content of Cement Slurries

In primary cementing, the cement slurry should have

TABLE 4.2 — EFFECT OF VARYING PRESSURE ON THE THICKENING TIME OF API CLASS H CEMENT WITH RETARDER

Depth (ft)	Temperature (°F)		Pressure (psi)	Thickening Time (hours:min.)
	Static	Circulating		
10,000	230	144	5,000	2:10
			10,000	1:34
			15,000	1.18
14,000	290	206	10,000	8:35
			15,000	5:19
			20,000	1:14
16,000	320	248	10,000	4:11
			15,000	3:39
			20,000	2:30
			25,000	2:08

a viscosity or consistency that will achieve the most efficient mud displacement and still permit a good bond between the formation and the pipe. (See Chap. 7, Primary Cementing.) To achieve this, most slurries are mixed with that amount of water that will provide a set volume equal to the slurry volume without free water separation. Particle size, surface area, and additives all influence the amount of mixing water required to achieve particular viscosity for a given slurry. There are ranges in viscosity for a given cement slurry and ranges in viscosity that govern how thick a slurry may be and still remain pumpable under a given set of well conditions. These amounts of water are given specific terms, defined as follows.[3]

Maximum Water is that amount of mixing water for any given cementing composition that will give a set volume equal to the slurry volume without more than 1½ percent free water separation. This is measured by a settling test (Fig. 4.3) in a 250-ml graduate after the slurry has been stirred on an atmospheric thickening-time tester. Maximum Water is the amount used for most cementing because the maximum yield or "fillup" is desired from each sack of cement.

Normal Water is the amount of mixing water that will achieve a consistency of 11 Uc (Units of consistency) as measured on an atmospheric thickening-time tester after 20 minutes of stirring. The API uses Units of consistency because the values obtained are not true (poise) viscosity values. Uc's are based on torque or

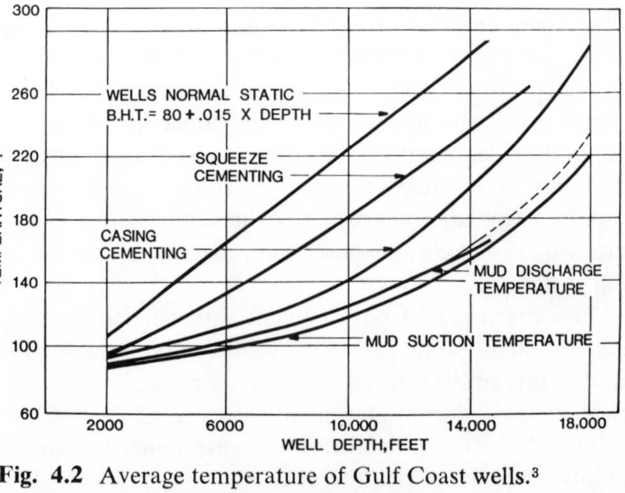

Fig. 4.2 Average temperature of Gulf Coast wells.[3]

drag rather than on water separation. Normal Water is sometimes called Optimum Water, as it provides a good, pumpable slurry.

Minimum Water is the amount of mixing water that will give a consistency of 30 Uc after 20 minutes of stirring. It yields a fairly thick slurry that can be used, for example, to control lost circulation.

Water/cement ratio, slurry volume, and set volume are closely related to the particle size or surface area of a cement (see Table 4.6). For most API Classes, the grind or particle size and the water requirements to achieve certain levels of strength, retardation, pumpability, etc., are specified.[4] API Standards do not list a fineness for Classes G and H cements, but they do specify the amount of mixing water and the allowable free water, which is controllable by the cement fineness.

In a cement column, excess or free water may collect in pockets rather than separating and migrating to the top of the column. Tests[5] performed on a 16-ft cement column using a 1-in. glass annulus showed that a cement

TABLE 4.3 — BASIS FOR CASING-CEMENTING WELL-SIMULATION TEST SCHEDULES[3]

1	2		3		4			5		6		7		8	9
	Depth		Mud Density[1]					Surface Pressure,[2]		Bottom-Hole Circulating Temperature,[3]		Bottom-Hole Pressure,[4]		Time to Reach Bottom,[5]
Schedule No.	ft	m	lb per gal	kg/l	lb per cu ft	psi per 1,000 ft of depth	kg/cm2 per m	psi	kg/cm2	F	C	psi	kg/cm2	min
1	1,000	310	10	1.2	74.8	519	0.120	500	35	80	27	1,020	72	7
4	6,000	1830	10	1.2	74.8	510	0.120	750	53	113	45	3,870	272	20
5	8,000	2440	10	1.2	74.8	519	0.120	1,000	70	125	52	5,160	363	28
6	10,000	3050	12	1.4	89.8	623	0.144	1,250	88	144	62	7,480	526	36
8	14,000	4270	16	1.9	119.7	831	0.192	1,750	123	206	97	13,390	941	52
9	16,000	4880	17	2.0	127.2	883	0.204	2,000	141	248	120	16,140	1135	60

[1]Mud densities obtained from a review of field data.

[2]Surface pressure obtained from a review of field data.

[3]Bottom-hole circulating temperatures averaged from actual field tests run at various depths.

[4]Bottom-hole pressure calculated from surface pressure, mud density and depth as shown in the table.

[5]Time to reach bottom is based on a survey of field operations and reflects conditions as severe as 75 per cent of the jobs surveyed.

TABLE 4.4 — BASIS FOR SQUEEZE AND PLUG-BACK CEMENTING WELL-SIMULATION TEST SCHEDULES[3]

1	2		3		4			5		6		7		8	9
	Depth		Mud Density[1]					Surface Pressure,[2]		Bottom-Hole Circulating Temperature,[3]		Bottom-Hole Pressure,[4]		Time to Apply Final Squeeze Pressure,[5]
Schedule No.	ft	m	lb per gal	kg/l	lb per cu ft	psi per 1,000 ft of depth	kg/cm2 per m	psi	kg/cm2	F	C	psi	kg/cm2	min
12	1,000	310	10	1.2	74.8	519	0.120	500	35	89	32	3,300	232	23
13	2,000	610	10	1.2	74.8	519	0.120	500	35	98	37	4,200	295	25
14	4,000	1220	10	1.2	74.8	519	0.120	500	35	116	46	5,600	394	28
15	6,000	1830	10	1.2	74.8	519	0.120	800	56	136	58	6,700	471	31
16	8,000	2440	10	1.2	74.8	519	0.120	1,000	70	159	71	7,800	548	35
17	10,000	3050	12	1.4	89.8	623	0.144	1,300	91	186	86	9,400	661	38
18	12,000	3660	14	1.7	104.7	727	0.168	1,500	105	213	101	11,800	830	42
19	14,000	4270	16	1.9	119.7	831	0.192	1,800	127	242	117	14,000	984	45
20	16,000	4880	17	2.0	127.2	883	0.204	2,000	141	271	133	16,500	1160	48
21	18,000	5490	18	2.2	134.6	935	0.216	2,200	155	301	149	19,000	1336	51

[1]Mud densities obtained from a review of field data.

[2]Surface pressure obtained from a review of field data.

[3]Bottom-hole circulating temperatures averaged from actual field tests run at various depths.

[4]Bottom-hole pressure calculated from surface pressure, mud density, and slurry density.

[5]Time to apply final squeeze pressure is based on a survey of field operations and reflects conditions as severe as 75 per cent of the jobs surveyed.

TABLE 4.5 — BASIS FOR LINER CEMENTING WELL-SIMULATION TEST SCHEDULES[3]
(Tentative)

1	2		3		4			5		6		7		8	9
	Depth		Mud Density					Surface Pressure,		Bottom-Hole Circulating Temperature,		Bottom-Hole Pressure,		Time to Reach Bottom,
Schedule No.	ft	m	lb per gal	kg/l	lb per cu ft	psi per 1,000 ft of depth	kg/cm2 per m	psi	kg/cm2	F	C	psi	kg/cm2	min
22	1,000	310	10	1.2	74.8	519	0.120	500	35	80	27	1,020	72	3
23	2,000	610	10	1.2	74.8	519	0.120	500	35	91	33	1,540	108	4
24	4,000	1220	10	1.2	74.8	519	0.120	500	35	103	39	2,580	181	7
25	6,000	1830	10	1.2	74.8	519	0.120	750	53	113	45	3,870	272	10
26	8,000	2440	10	1.2	74.8	519	0.120	1,000	70	125	52	5,160	363	15
27	10,000	3050	12	1.4	89.8	623	0.144	1,250	88	144	62	7,480	526	19
28	12,000	3660	14	1.7	104.7	727	0.168	1,500	105	172	78	10,230	719	24
29	14,000	4270	16	1.9	119.7	731	0.192	1,750	123	206	97	13,390	941	29
30	16,000	4880	17	2.0	127.2	883	0.204	2,000	141	248	120	16,140	1135	34
31	18,000	5490	18	2.2	134.6	935	0.216	2,000	141	300	149	18,800	1322	39

with a surface area of 1,500 sq cm/gm, mixed at a slurry density of 15.4 lb/gal, formed a solid cement plug over the entire column. When mixed with more water (15.1 lb/gal) free water separated into horizontal pockets of clear water from ½ to 1½ in. in diameter. The lighter the cement slurry weight, the more water and the greater the number of water pockets. The pockets began to form about 15 minutes after the cement slurry was put in the glass tubing.

It should be emphasized that although increasing the water content will permit longer pumping time and delay the setting of a cement, the water should never be increased unless bentonite or a similar material is blended with the cement to tie up excess water. Excess water always produces a weaker cement with lower resistance to corrosion.

4.4 Thickening Time

Minimum thickening time is the time required to mix and pump the slurry down the hole and up the annulus behind the pipe. Equipment for measuring the thickening time of any cement slurry under laboratory conditions is defined in API Testing Procedures[3] and is available in active drilling areas throughout the world. (See Fig. 4.4.) The thickening-time tester simulates well conditions where static bottom-hole temperatures range up to 500°F and pressures are in excess of 25,000 psi.

As the apparatus applies heat and pressure to the slurry, a continuous consistency measurement is recorded on a strip chart. The limit of pumpability is reached when the torque on the paddle in the slurry cup reaches 100 Uc.

Specific thickening-time recommendations depend largely upon the type of job, the well conditions, and the volume of cement being pumped. When casing is to be cemented at depths of 6,000 to 18,000 ft, a 3- to 3½-hour pumping time is commonly used in designing the slurry. This length of time allows an adequate safety factor, as few cementing jobs, even large ones, require more than 90 minutes for placing the slurry. On deep liner jobs where fairly high temperatures are encountered, the 3- to 3½-hour pumping time is still adequate. For spotting a cement plug, thickening times should not exceed 2 hours, since most jobs are completed in less than 1 hour. In squeeze cementing, the thickening-time requirements may vary for different techniques. Shut-downs during a hesitation squeeze will significantly reduce the pumpability of a cement slurry. Although shut-downs are normally not considered during laboratory testing, they can be a contributing factor to leaving set cement in the tubing before the desired squeeze pressure is obtained. For any critical cement job at depths greater than 12,000 ft, the field water and cement should always be laboratory tested before they are mixed at the job site.

The relationship between various well depths and thickening time is illustrated in Fig. 4.5. Data for simulating standard API casing, squeeze, plugback, and liner cementing conditions are given in Tables 4.3, 4.4, and 4.5.

TABLE 4.6 — INFLUENCE OF VARYING SURFACE AREAS AND WATER RATIOS ON THE VOLUME OF SET CEMENT[5]

Water Content (percent by weight of cement)	Volume of Slurry (cu ft/sk)	Percent Free Water When Set	Volume of Set Cement (cu ft/sk)
Specific surface: 1,890 sq cm/gm*			
40	1.069	0.00	1.069
50	1.220	0.74	1.211
60	1.370	2.34	1.338
70	1.521	4.75	1.449
Specific surface: 1,630 sq cm/gm**			
35	0.994	0.88	0.985
40	1.069	1.33	1.055
50	1.220	7.66	1.114
60	1.370	16.01	1.151
Specific surface: 1,206 sq cm/gm			
35	0.994	3.15	0.963
40	1.069	8.38	0.979
50	1.220	16.20	1.022
60	1.370	22.35	1.064

*Similar to API Class C cement.
**Similar to API Class A, B, and G cements.

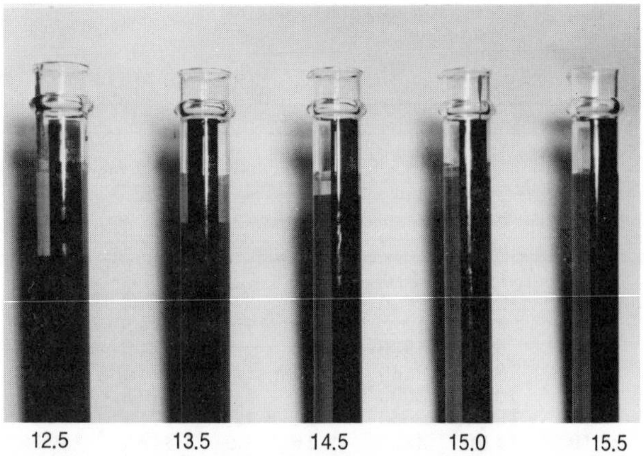

12.5 13.5 14.5 15.0 15.5

SLURRY WEIGHT, LB/GAL

Fig. 4.3 API settling tests with API Class G cement at various cement/mixing-water ratios.

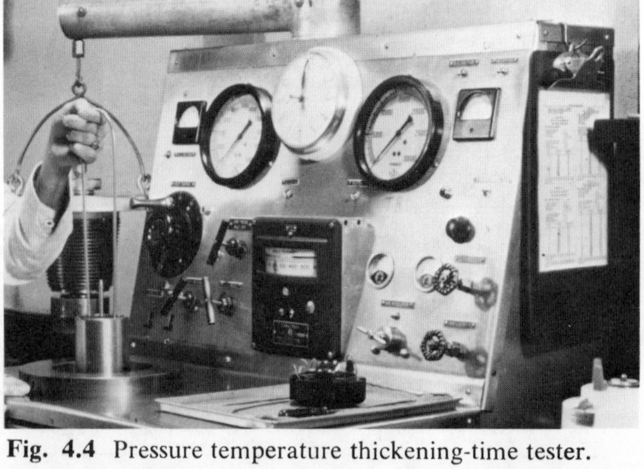

Fig. 4.4 Pressure temperature thickening-time tester.

TABLE 4.7 — LENGTH OF CASING AND SIZE OF DRILL COLLARS SUPPORTED BY A 10-FT-LONG CEMENT COLUMN OF 8-PSI TENSILE STRENGTH[8]

Casing		Drill Collar Size (in.)		Length of Casing (ft)
Size (in.)	Weight (lb/ft)	OD	ID	
7	17.00	4¾	2	94
8⅝	24.00	6¼	2¼	67
10¾	32.75	6¾	2⅞	72
13⅜	48.00	9	3¼	50

4.5 Strength of Cement to Support Pipe

Cement requires very little early strength to support a string of casing.[6-8] Data have shown that a 10-ft annular sheath of cement possessing only 8-psi tensile strength can support more than 200 ft of casing of the lighter-weight sizes, even under rather poor bonding conditions. In setting surface casing, when high bit weights are required for drilling out floating equipment, an additional load must be supported by the casing and cement sheath. Table 4.7 shows the minimum length of casing and the size of drill collar that can theoretically be supported by a 10-ft column of cement of 8-psi tensile strength. Since in cement strength testing (Fig. 4.6) the cement is usually in compression rather than in tension, the values must be converted from compressive strength to tensile strength. As a general rule, compressive strength is approximately 8 to 10 times as great as tensile; that is, the 8-psi tensile strength would be equivalent to 80 to 100 psi compressive strength.

It must be realized that the interval from the time when the cement first sets to the time it develops 100 psi compressive strength can be comparatively short. Field variables — completion procedures, materials, curing conditions — cannot be known or controlled well enough to establish a foolproof curing time. Therefore a reasonable safety factor should be applied. It is generally accepted in the industry and by regulatory bodies that a compressive strength of 500 psi is adequate for most operations, and by using good cementing practices an operator should be able to drill out safely by adhering to this minimum strength requirement. (See Chap. 13, Regulations.)

In deciding how long to wait for cement to set (that is, in selecting a WOC time) it is important (1) to know how strong the cement must be before drilling can begin, and (2) to understand the strength development characteristics of the cements in common use.

It may be observed from the compressive-strength values in Table 4.8 that curing temperature is very significant in strength development. To properly apply laboratory strength information and establish a reasonable WOC time, one must have some knowledge of down-hole curing temperatures. Static bottom-hole temperatures in most geographical areas have been reasonably well defined by using surface isotherm data along with the accepted depth-temperature gradients. The results are verified by temperature surveys conducted in uncased surface holes. In most areas, the formation temperature at surface casing depth equals the mean surface temperature plus 2°F per 100 ft of depth.

The curing temperature of the cement, however, will almost certainly not equal the formation temperature, and, in fact, it does not even have a constant value. It is governed by a complex group of variables, including the temperatures of the drilling mud, cement slurry, and displacement fluid, as well as the heat of hydration of the cement.

The following observations relevant to the strength of cement to support pipe are based on research and field experience.

1. High cement strengths are not always required to support casing during drill-out, and by increasing the slurry density the time required to develop adequate compressive strength is decreased.

2. Densification increases both the strength and the heat of hydration of cement.

3. Cement slurries with excessive water ratios result in weak set cements and so should be avoided around the lower portion of the pipe.

4. By selecting the proper cements and applying

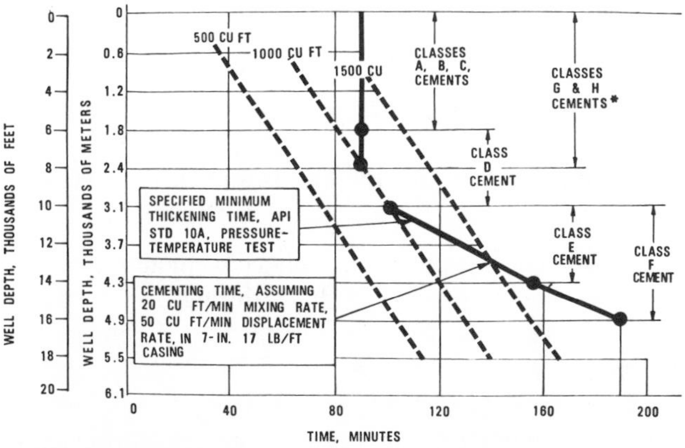

*Specific maximum thickening time: 120 minutes.

Fig. 4.5 Well depth and cementing relationships.[3]

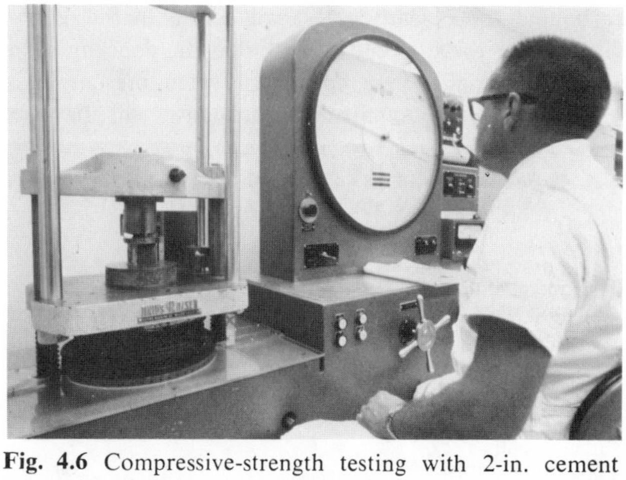

Fig. 4.6 Compressive-strength testing with 2-in. cement cube.

TABLE 4.8 — INFLUENCE OF TIME AND TEMPERATURE ON THE COMPRESSIVE STRENGTH OF API CLASS H CEMENT

Curing Time (hours)	Calcium Chloride (percent)	Compressive Strength (psi) at Curing Temperature and Pressure of			
		95°F 800 psi	110°F 1,600 psi	140°F 3,000 psi	170°F 3,000 psi
6	0	100	350	1,270	1,950
8		500	1,200	2,500	4,000
12		1,090	1,980	3,125	4,700
24		3,000	4,050	5,500	6,700
6	1	900	1,460	2,320	2,500
8		1,600	1,950	2,900	4,100
12		2,200	2,970	3,440	4,450
24		4,100	5,100	6,500	7,000
6	2	1,100	1,700	2,650	2,990
8		1,850	2,600	3,600	4,370
12		2,420	3,380	3,900	5,530
24		4,700	5,600	6,850	7,400

good cementing practices, WOC time for surface casing can be reduced to 3 to 4 hours under summer operating conditions and 6 to 8 hours under winter conditions.

4.6 Mixing Waters

The primary function of water in a cement slurry is to wet the cement solids and carry them down the hole. Many a cementing job has gone awry because of interference from some constituent in the mixing water. Ideally, the water supply for mixing cement should be reasonably clean and free of soluble chemicals, silt, organic matter, alkali, or other contaminants. This is not always practical, so the most readily available source of water must be considered. Water most commonly found in the field or around the rig is obtained from an open pit or reservoir supplied from a shallow water well or lake. Such water is often satisfactory for mixing with cement at well depths less than 5,000 ft, particularly when it is relatively clear and has a total solids content of less than 500 ppm.

Contaminants in mixing waters have been traced to (1) fertilizers dissolved in rainwater runoff from agricultural areas, (2) waste effluents in streams, (3) soluble agricultural products such as sugar cane or sugar beets found in streams during rainy seasons, and (4) soluble chemicals inherent in the soil.

Inorganic materials (chlorides, sulfates, hydroxides, carbonates, and bicarbonates) will accelerate the setting of cement,[9] the rate depending upon the concentration of the material. These chemicals, when present in a mixing water in small concentrations, will have a negligible effect in shallow holes. This same water used for cementing a deep liner at higher temperatures and pressures may cause the cement slurry to set prematurely, particularly if the water contains trace amounts of carbonates or bicarbonates. (See Table 4.9.)

Sea water, because it contains 30,000 to 43,000 ppm solids, accelerates cement. These accelerating chemicals, however, can be neutralized with a retarder so that the water can be used at higher temperatures. (See also Chap. 3 on Cementing Additives, especially Tables 3.7 and 3.8.)

Chloride impurities often cause aeration and foaming during the mixing of cement, which makes it difficult to weigh the slurry accurately.

Natural waters containing organic chemicals from decomposed plant life, waste effluents, or fertilizers will retard the setting of portland cement. A common retarding substance is humic acid, formed by the decaying of plants. Water containing humic acid frequently drains from mountains or high moorland regions and may also be found in lakes or ponds in frozen areas. The retarding properties of organic contaminants are particularly detrimental in cementing surface pipe and shallow holes.

Potable water is always recommended where available. Unless clear water has a noticeably saline or brackish taste, it is usually suitable. Even saline waters may be usable, but the slurries should be laboratory tested first.

To summarize: the water to be used in mixing cement should be the purest available.

TABLE 4.9 — EFFECT OF MIXING WATER ON THE PERFORMANCE OF API CLASS H CEMENT

Type of test: 6,000-ft API Casing Cementing Test.
Curing time: 24 hours.
Curing temperature: 95°F.
Curing pressure: 5,000 psi.

Type of Water	Thickening Time (hours:min.)	Compressive Strength (psi)
Tap water	2:34	2,150
Tap water plus 2,200 ppm carbonates	1:18	2,300
Sea water	1:52	2,610

4.7 Sensitivity to Drilling Fluids and Drilling-Fluid Additives

A significant problem in oilwell cementing is the effective removal of drilling fluids during displacement. Contamination and dilution by mud may damage cementing systems, as may chemicals in the mud and in the filter cake. (See Table 4.10.)

Some contamination of this sort occurs during most jobs, but probably most of it occurs when a cement plug is spotted in a mud system that is highly treated

TABLE 4.10 — EFFECTS OF MUD ADDITIVES ON CEMENT

Additive	Purpose	Effect on Cement
Barium sulfate ($BaSO_4$)	To weight the mud	Increases density, reduces strength
Caustics ($NaOH$, Na_2CO_3, etc.)	To adjust the pH	Accelerate
Calcium compounds (CaO, $Ca(OH)_2$, $CaCl_2$, $CaSO_4 \cdot 2H_2O$)	To condition the hole and control pH	Accelerate set
Hydrocarbons (diesel oil, lease crude oil)	To control fluid loss and to lubricate the hole	Decrease density
Sealants (scrap, cellulose, rubber, etc.)	To seal against leakage to the formation	Retard set
Thinners (tannins, lignosulfonates, quebracho, lignins, etc.)	To disperse mud solids	Retard set
Emulsifiers (lignosulfonates, alkyl ethylene oxide adducts, hydrocarbon sulfonates)	To form oil-in-water or water-in-oil muds	Retard set
Bactericides (substituted phenols, formaldehyde, etc.)	To protect organic additives against bacterial decomposition	Retard set
Fluid-loss-control additives (CMC, starch, guar, polyacrylamides, lignosulfonates)	To reduce loss of fluid from mud to formation	Retard set

with chemicals.[9] The volume of cement in relation to the volume of mud is small, and the degree of mud contamination is never known. Softness of cement as a plug is drilled out is a sign of contamination. (See Chap. 10, Open-Hole Cement Plugs.)

The best way to combat the detrimental effects of drilling mud additives is to use wiper plugs and spacers or flushes. Wiper plugs help eliminate contamination inside the casing, and flushes help to clean the annular space between the casing and the formation. Spacers, or buffer washes, consist of water, solutions of acid, phosphates, cement-water mixtures, and slurries of untreated bentonites and clay in water. For oil or invert-emulsion mud systems, diesel oil flushes — both weighted and unweighted — are effective.

4.8 Slurry Density

The density of a cement slurry should always, except for squeeze jobs, be great enough to maintain well control. There are various ways of controlling density. Table 4.11 shows materials that can be used to achieve a given density. For lower densities — 10.8 to 15.6 lb/gal — materials that require large volumes of mixing water are frequently used. For greater densities — 15.6 to 22 lb/gal — dispersants and weighting materials such as hematite are commonly used.

In field operations, slurry density is customarily monitored with a standard mud balance. (For accuracy, samples are selectively taken from the mixing tub and vibrated to remove the finely entrapped air bubbles from the jet mixer.) Automated weighting devices, however, fitted into the discharge line between the mixing unit and the wellhead, give a more uniform weight record on a strip chart and are being used more widely (see Fig. 4.7).

The fluid density balance (Fig. 4.8), a portable device that is easily assembled on location, measures the cement slurry under sufficient pressure (about 30 psi) to compress the entrained air. This compression yields a more accurate measurement than when the sample is taken directly from the tub during mixing. Table 4.12 illustrates the disparities in values of cement density resulting from various measuring methods.

4.9 Lost Circulation

In selecting and using materials to control lost circulation, two factors must be borne in mind: the material must be of a size that can be handled by the pumping equipment, and the formation openings must be small enough to allow the material to bridge and seal. When formation openings are so large that the sealing agents are relatively ineffective it may become necessary to design semisolid or flash-setting cements. For a more detailed discussion of lost circulation and the materials used to control it, see Section 3.7, "Additives for Controlling Lost Circulation." The effectiveness of those materials has been established not only by laboratory tests, but also by results of field use.[10]

4.10 Heat of Hydration

When cement is slurried with water an exothermic reaction occurs in which considerable heat is liberated.

TABLE 4.11 — ADDITIVES FOR CONTROLLING CEMENT SLURRY DENSITY

Cement Slurry Density (lb/gal)	Approximate Water Content (gal/sk)	Additive	Approximate Concentration of Additive per Sack of Cement (lb)
11.0	25	Diatomaceous earth	40
12.0	13	Bentonite + dispersant	12 + 1
13.0	10.5	Bentonite	8
14.0	6.0	Pozzolans	50
15.0	5.8	None	—
16.0	4.4	None	—
17.0	4.0	Dispersants	1
18.0	4.0	Dispersants + weighting material	1 + 12
19.0	4.0	Dispersants + weighting material	1 + 28
20.0	4.0	Dispersants + weighting material	1 + 46
21.0	4.0	Dispersants + weighting material	1 + 71

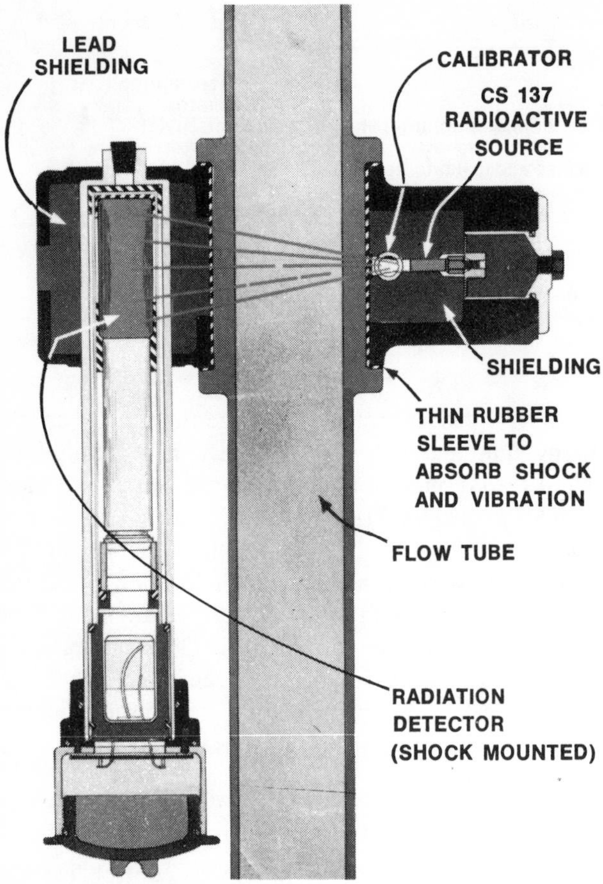

Fig. 4.7 Radioactive densimeter used to weigh drilling fluids and oilfield cements.[16]

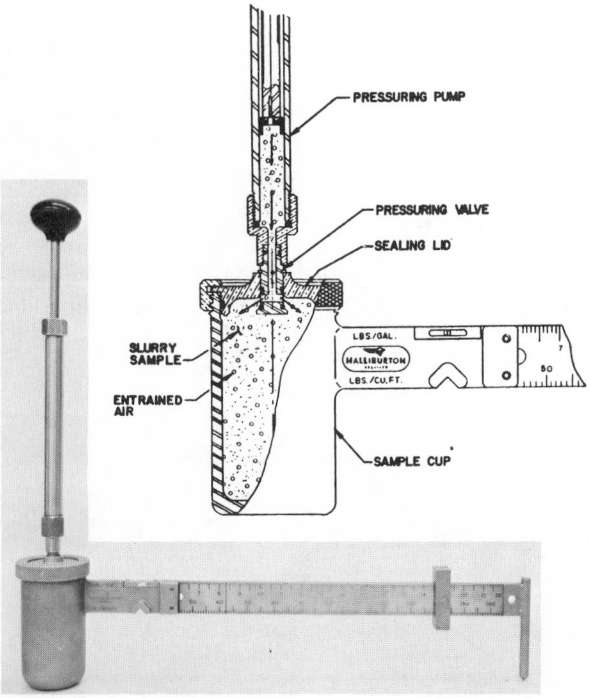

Fig. 4.8 Fluid density balance for weighing cement slurries.[3]

TABLE 4.12 — CEMENT DENSITIES DERIVED BY VARIOUS METHODS OF MEASURING

By Calculation of Slurry Weight (lb/gal)	With Standard Mud Balance (lb/gal)	With Radioactive Densometer (lb/gal)	With Fluid Density Balance (lb/gal)
11.1	9.9	10.9	11.2
13.3	12.8	13.2	13.4
19.0	18.2	18.7	19.1
19.5	18.3	19.2	19.6
19.5	18.5	19.3	19.5

The greater the mass of cement, the greater the evolution of heat. In the laboratory, such heat is usually measured with a calorimeter, an insulated vacuum flask containing a thermocouple connected to a recorder. The increase in temperature is recorded at specific intervals until the maximum temperature is observed. Heat of hydration (sometimes called heat of reaction or heat of solution) is influenced by the fineness and chemical composition of the cement, by additives, and by downhole environment. The higher the formation temperature, the faster the reaction and the more rapid the evolution of heat. (See Fig. 4.9.)

The heat of hydration of pure cement compounds has been studied under controlled laboratory conditions.[11] Some of the results are shown in Table 4.13. Table 4.14 compares the heats of hydration of various cementing compositions.

In most holes the annulus is ½ to 2 in., except in washed-out zones. In a typical surface pipe, the heat of hydration produces a maximum temperature rise of 35° to 45°F (Fig. 4.10).

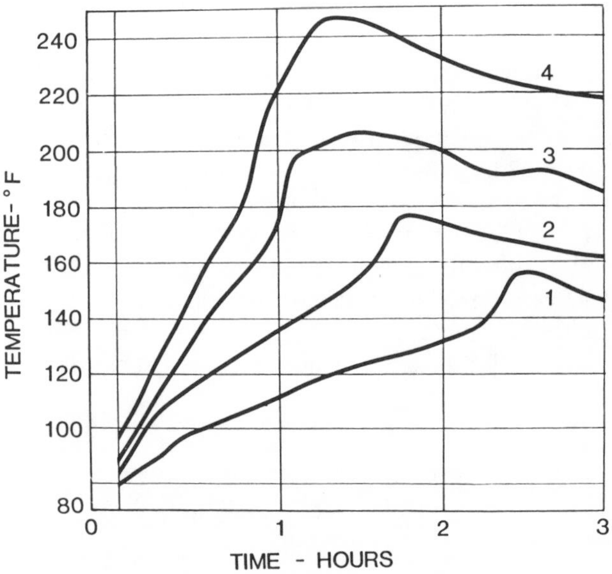

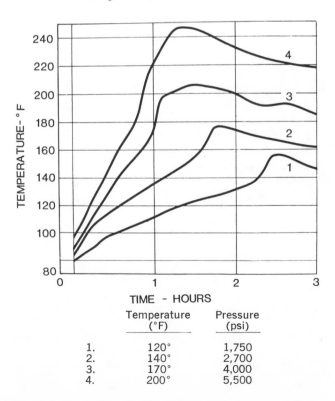

	Temperature (°F)	Pressure (psi)
1.	120°	1,750
2.	140°	2,700
3.	170°	4,000
4.	200°	5,500

Fig. 4.9 Heat of hydration of API Class A cement under varying temperatures and pressures.

TABLE 4.13 — HEATS OF HYDRATION OF PURE CEMENT COMPOUNDS[11]

Components (Abbreviation)	Calories per Gram	Btu's per Pound of Slurry
$3CaO \cdot Al_2O_3$ (C_3A)	207	373
$3CaO \cdot SiO_2$ (C_3S)	120	216
$4CaO \cdot Al_2O_3 \cdot Fe_2O_3$ (C_4AF)	100	180
$2CaO \cdot SiO_2$ (C_2S)	62	112

4.11 Permeability

Although in designing cement slurries only slight emphasis is given to the permeability of set cement, there are means of measuring it for both water and gas. The API has specified a standard system that involves the use of a permeameter.[3]

Set cements have very low permeabilities — much lower, in fact, than those of most producing formations. Data have shown[12] that at temperatures less than 200°F the permeability of cement decreases with age and temperature. After 7 days of curing, the permeability is usually too low to measure.

The permeability of set cement to gas is normally higher than to water, but measurements of the former are less reliable because it is difficult to obtain good representative samples for measuring gas flow.[12] Cements that have set for 3 to 7 days have a gas permeability of less than 0.1 md. Dolomite and limestone have an average of 2 to 3 md and oolitic limestones usually have a very low permeability. Sandstone has a gas permeability ranging from 0.1 to 2,000 md.

For a discussion of the use of silica flour for combatting increases in permeability,[13,14] see Section 3.10.

4.12 Filtration Control

Controlling filtrate in the cement slurry is very important in cementing deep liners and in squeeze cementing. Loss of filtrate through a permeable medium will

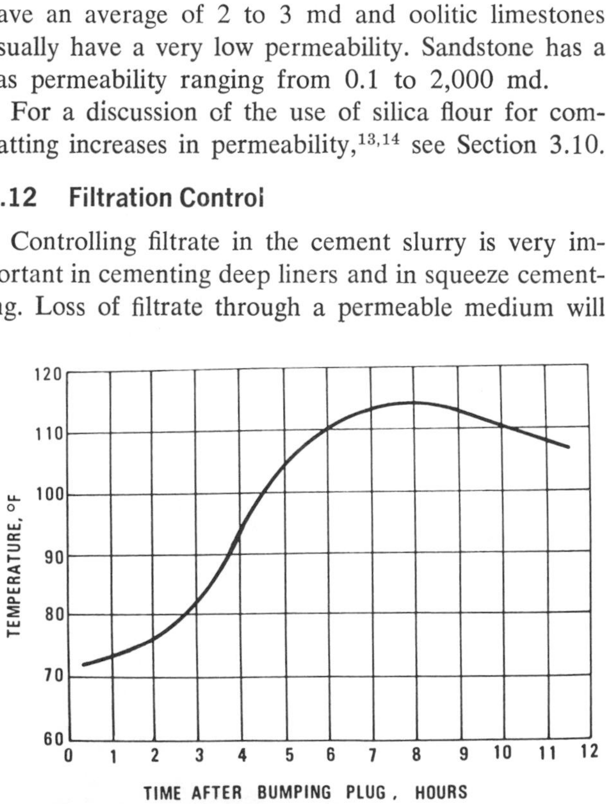

Survey depth: 550 ft.
Slurry weight: 15.4 lb/gal.
Mixing-water temperature: 74°F.
Formation temperature: 65°F.

Fig. 4.10 Temperature/time relationship resulting from heat of hydration of cement slurry used on surface pipe.[8]

TABLE 4.14 — HEAT OF HYDRATION OF OILWELL CEMENT

	Heat Transfer Coefficient, Btu/(hr ft °F)	Heat of Hydration (Btu per lb of slurry)
API Class G cement	0.50	118
API Class H cement	0.50	120
1:1 Class H cement : fly ash plus 2 percent gel	0.49	91.4
1:1 Class H cement : fly ash plus 6 percent gel	0.50	109.1
Refractory cement	0.40	57.0

cause a rise in slurry viscosity and a rapid deposition of filter cake, thus restricting flow. The factors that influence the filter loss of cement slurries are time, pressure, temperature, and permeability. To measure filtration characteristics of cement slurries, the API specifies a standardized 30-minute test at 100 or 1,000 psi.

The API procedure employs a filter assembly (Fig. 4.11) consisting of a frame, a cylinder, and a 325-mesh screen supported on a 60-mesh screen as the filtration medium. A heating jacket makes it possible to simulate formation temperatures. To simulate down-hole placement, slurries may be pumped on a pressure or nonpressure thickening-time tester for a given time before they are removed and poured into the filter cell.

The API filter loss of all cement slurries without additives is high — in excess of 1,000 ml. When all the filtrate is received in the test cell in less than 30 minutes, the following equation is used to calculate the hypothetical 30-minute fluid-loss value.[3]

$$F_{30} = F_t \frac{5.477}{\sqrt{t}},$$

where F_{30} is the quantity of filtrate in 30 minutes and F_t is the quantity of filtrate in t minutes.

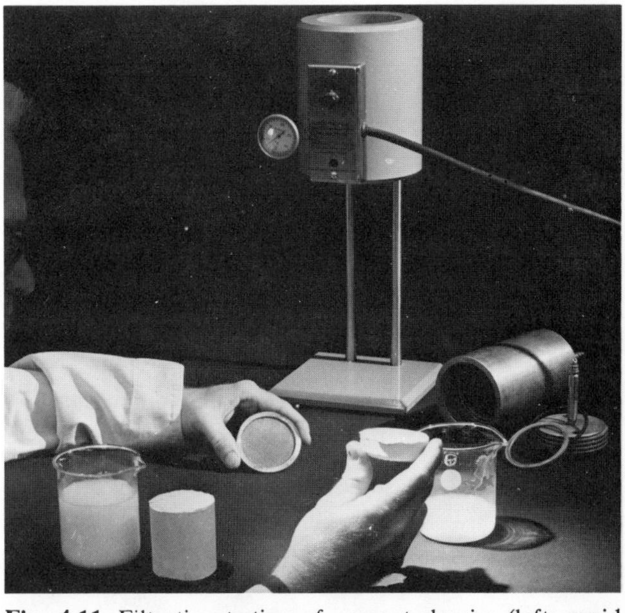

Fig. 4.11 Filtration testing of cement slurries (left: rapid loss of water leaves thick filter cake; right: in 30 minutes, controlled-water-loss slurry leaves thin cake).

Controlled filtration of a cement slurry is normally achieved by the addition of long-chained polymers in concentrations ranging from 0.6 to 1.0 percent by weight of cement. (See Tables 3.23 and 4.15.)

Cement slurries having laboratory fluid-loss values of 50 to 150 ml in 30 minutes are commonly used in squeeze cementing. In cementing deep liners, the API filter loss may be as high as 300 ml.

4.13 Resistance to Down-Hole Brines

The susceptibility of cements to corrosion by formation waters has been the subject of much research.[15] Formation brines containing sodium sulfate, magnesium sulfate, and magnesium chloride are among the most destructive down-hole agents. Such brines are found in West Texas, Kansas, the North Sea, and other oil-producing areas.

Sulfates, generally regarded as the chemicals most corrosive to cement, react with the lime and tricalcium aluminate in cement to form large calcium sulfoaluminate crystals. These crystals require more pore space that the set cement can provide, so they cause excessive expansion and eventual deterioration. Fig. 4.12 shows how this crystal growth has caused expansion at the end of a 12-in. test bar of API Class A cement that has been allowed to cure in a 5-percent sodium sulfate solution.

The studies of corrosive formation waters have particularly emphasized the susceptibility of set cement pastes or concrete. The sodium ion is thought to be more detrimental than the magnesium ion and is often used in laboratory testing.

There appear to be three distinct chemical reactions when sodium sulfate reacts with set cement.

$$Na_2SO_4 + Ca(OH)_2 \rightarrow 2NaOH + CaSO_4 \cdot 2H_2O$$

$$Na_2AO_4 + 3CaO \cdot Al_2O_3 \cdot Aq. \rightarrow 3CaO \cdot Al_2O_3 \; 3CaSO_4 \cdot Aq. + Na_2O \cdot Al_2O_3 + NaOH$$

$$Na_2O \cdot Al_2O_3 + H_2O \rightarrow 2NaOH + 2Al(OH)_3$$

In these reactions, calcium sulfoaluminate and sodium aluminate are formed, and the latter hydrolizes into sodium and aluminum hydroxides. The calcium sulfoaluminate formed at room temperature contains 31 molecules of H_2O. Thus the product is a large molecule and most of the expansion and disintegration is considered to be caused by the deposition of this material in the set cement.

The rate of attack on a hardened cement by solution of sodium sulfate or magnesium sulfate is governed to some extent by the concentration of these salts in the formation water. However, for both compounds there appears to be a limiting concentration beyond which further increases in concentration raise the rate of attack only slightly.

Temperature also influences the sulfate resistance of a hardened cement. From investigations conducted at both low and high temperatures, it was concluded that sulfate attack is most pronounced at temperatures of 80° to 120°F, whereas at 180°F it becomes negligible.[15] This conclusion is supported by the observation that field problems are more common in shallow wells, where temperatures are lower, than in deep wells, where temperatures may exceed 200°F. A cement that is resistant to sulfate attack at low temperatures is likely to perform well at higher temperatures. Lowering the tricalcium aluminate (C_3A) content increases the sulfate resistance of the cement. Therefore, the API classifies types of cement as moderately sulfate resistant (MSR) and highly sulfate resistant (HSR) on the basis of the C_3A content of the cement (MSR = 3 to 6 weight percent C_3A; HSR = 0 to 3 weight percent C_3A. (See also Chap. 2.)

It may be noted that electrolytic corrosion rather than chemical corrosion has been responsible for the weakening and ultimate failure of some casing strings. Most investigations show that a uniform sheath of competently set cement offers excellent protection against electrolytic corrosion of casing. Since a current of 1 amp leaving a pipe carries with it 20 lb of metal per year, the importance of a uniform sheath of permanent cement is quite apparent.[5]

4.14 Conclusion

Many factors must be considered in designing cement slurries for down-hole use. Field laboratories operated by oil companies, service organizations, and cement manufacturers are available throughout the world to aid in gathering the necessary data. On critical wells, test data should be obtained on the same cementing

TABLE 4.15 — EFFECT OF POLYMER FLUID-LOSS ADDITIVE ON API CLASS G CEMENT

Polymer (percent by weight of cement)	Slurry Viscosity (Uc)		API Fluid Loss (cc/30 min.)* at Test Pressure (psi) of		
	Initial	20 min.	100	500	1,000
0.0	3	10	1,000+	1,000+	1,000+
0.6	4	11	96	178	250
0.8	4	11	24	70	100
1.0	5	12	14	32	60
1.2	10	15	10	24	43

*325-mesh screen.

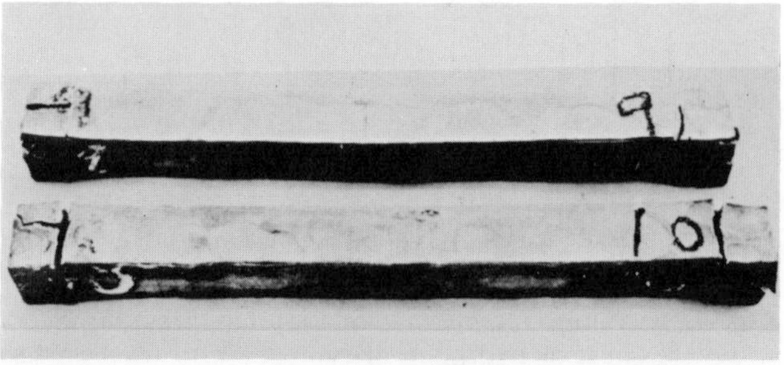

Fig. 4.12 Sulfate attack of set API Class A cement.[15]

materials to be used on the job; otherwise, recommendations are not entirely meaningful.

References

1. Swayze, M. A.: "Effects of High Temperatures and Pressures on Strengths of Oil Well Cements," *Drill. and Prod. Prac.*, API (1954) 72; *Oil and Gas J.* (Aug. 2, 1954) 103-105.
2. Bearden, W. G.: "Effect of Temperature and Pressure on the Physical Properties of Cement," *Oil-Well Cementing Practices in the United States*, API, New York (1959) 56.
3. "Recommended Practice for Testing Oil Well Cements and Cement Additives," *API RP 10B*, 20th ed., API Div. of Production, Dallas (1974).
4. "Specifications for Oil Well Cements and Cement Additives," *API Standards 10A*, 19th ed., API, New York (1974).
5. Craft, B. C., Holden, W. R., and Graves, E. D., Jr.: *Well Design: Drilling and Production*, Prentice-Hall, Inc., Englewood Cliffs, N. J. (1962) 43-48, 55-79, 212-213.
6. Farris, R. F.: "Method for Determining Minimum Waiting-on-Cement Time," *Trans.*, AIME (1946) **165**, 175-188.
7. Davis, S. H., and Faulk, J. H.: "Have Waiting-on-Cement Practices Kept Pace with Technology?" *Drill. and Prod. Prac.*, API (1957) 180.
8. Maier, L. F.: "Understanding Surface Casing Waiting-on-Cement Time," paper presented at CIM 16th Annual Tech. Meeting, Calgary, Alta., Canada, May 1965.
9. Anderson, F. M.: "Effects of Mud-Treating Chemicals on Oil-Well Cements," *Oil and Gas J.* (Sept. 29, 1952) 283-284.
10. Lummus, J. L.: "A New Look at Lost Circulation," *Pet. Eng.* (Nov. 1967) 69-73.
11. Lea, F. M., and Desch, C. H.: *The Chemistry of Cement and Concrete*, Arnold & Co., London (1935; reprinted 1937, 1940).
12. Goode, J. M.: "Gas and Water Permeability Data for Some Common Oilwell Cements," *J. Pet. Tech.* (Aug. 1962) 851-854.
13. Saunders, C. D., and Walker, W. A.: "Strength of Oilwell Cements and Additives Under High Temperature Well Conditions," paper 390-G presented at SPE-AIME 29th Annual Fall Meeting, San Antonio, Tex., Oct. 17-20, 1954.
14. Ludwig, N. C., and Pence, S. A.: "Properties of Portland Cement Pastes Cured at Elevated Temperatures and Pressures," *Proc.*, American Concrete Institute (1956) **52**, 673-687.
15. "Report on Cooperative Tests on Sulfate Resistance of Cements and Additives," API Mid-Continent Dist. Study Committee on Cementing Practices and Testing of Oil-Well Cements (1955).
16. Guest, R. J., and Zimmerman, C. W.: "Compensated Gamma Ray Densimeter Measures Slurry Densities in Flow," *Pet. Eng.* (Sept. 1973).

Chapter 5

Hole and Casing Considerations

5.1 Introduction

Before the correct casing string for a well can be selected, certain information must be obtained: (1) the setting depth; (2) the size of the hole and of the casing in which the string is to be run and of the hole to be drilled below the casing; (3) the mud-column and reservoir pressures; (4) what type of well it is (for example, a high-pressure gas well, a wildcat well, or a well in an established producing area); (5) the formation conditions; (6) the drilling objectives.

The casing string must be designed so that it will not fail in tension, will not collapse or burst, and will resist down-hole corrosion.[1-3] Table 5.1 lists the types and functions of well casing and liner.

5.2 Casing String Design

Casing strings should be designed to withstand internal and external pressures as well as lateral loads from down-hole formations. To achieve these objectives, they are frequently composed of casing with different weights and grades, especially where hole conditions are critical. The strength of the casing string must be considered during running, landing, and cementing. The strength and conditions of open-hole formation and factors such as breakdown pressure must also be considered.

To assure an adequate margin of safety, most casing strings are designed with a safety factor of 1.5 to 1.8 for tensile stress; 1.0 to 1.25 for external pressure stress (collapse); and 1.1 to 1.33 for internal pressure stress (burst).

Tensile force is significant in all strings except the conductor string. The greatest stress is imposed in the upper portion of each string or, if the strings are tapered, in the upper portion of each section. Tensile-

TABLE 5.1 — TYPES AND FUNCTIONS OF WELL CASING AND LINER*

Type	Size (in. OD)	Setting Depth (ft)	Function
Conductor casing	16 to 30	40 to 1,500	Stabilizes collar and protects rig foundation. Restrains unconsolidated formations. Confines circulating fluids. Helps prevent water flow and loss of circulation.
Surface casing	7 to 16	To 4,500	Helps prevent contamination of fresh-water zones. Connects blowout preventer and wellhead. Supports deeper casing and tubing string. Confines shallow zones and prevents loss of circulation.
Intermediate casing	7 to 11¾	Varies with hole conditions	Helps prevent sloughing and enlargement of hole during deeper drilling. Protects production string from corrosion. Helps to resist high formation pressure. Protects against loss of drillstring in key-seated or "sticky" holes. Helps prevent loss of circulation.
Production casing	2⅜ to 9⅝	Through producing zone	Protects hole. Isolates fluids and prevents fluid migration. Helps provide well control if tubing fails. Protects down-hole equipment. Allows selective production of oil and gas.
Liner	5 to 7	Through producing zone	Functions like production casing. Limits need for full string of casing.

*After Ref. 4.

strength calculations are based primarily on load per unit of cross-sectional area for the grade of steel used.[5] (See Fig. 5.1.)

External yield (collapse) pressure should be considered in choosing all strings except the conductor string. Maximum external pressure is exerted at the bottom of each string, or, in the case of tapered strings, at the bottom of each of the sections. Calculations are based upon the maximum weight of the column of fluid on the outside of the pipe minus the weight of the column of fluid on the inside of the pipe.[5] (See Fig. 5.2.)

Design for internal yield (burst) pressure is most likely to be important where wellhead pressures are relatively high. Where they are not high, pipe that will withstand the tensile and collapse forces will be adequate to withstand the possible burst forces. (See Fig. 5.3.)

The casing string should always be designed to resist burst failure, particularly where gradients range from 0.6 to 1.0 psi per foot of depth.[5]

Buckling force should be considered only if a well is to be drilled in water or where the surface bed is incompetent, such as in a marsh. Buckling force must be allowed for in designing only the conductor and surface strings. The other strings are generally resistant to buckling forces, except where the land is shifting and causing shearing of the casing string (for example, along fault planes).

The Performance Properties of Casing and Tubing (in *API Standards 5A*) define the physical properties of steel from which casing is produced. (See Table 5.2.) The numbers used to define these types represent the minimum yield strength of the steel in thousands of pounds per square inch. Specifications control the minimum tensile strength and minimum elongation of the material in each grade. Each grade is available in weights per foot that vary according to wall thickness and coupling length.[6]

The casing for any well should be carefully studied to insure that it is properly designed for the requirements of the hole. Most casing manufacturers have computerized programs or design charts that can be used to design casing strings, taking into account well stress, mud weights, and safety factors.[8,9]

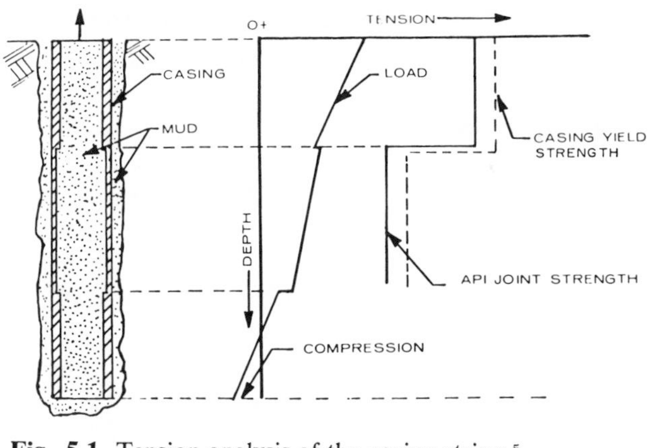

Fig. 5.1 Tension analysis of the casing string.[5]

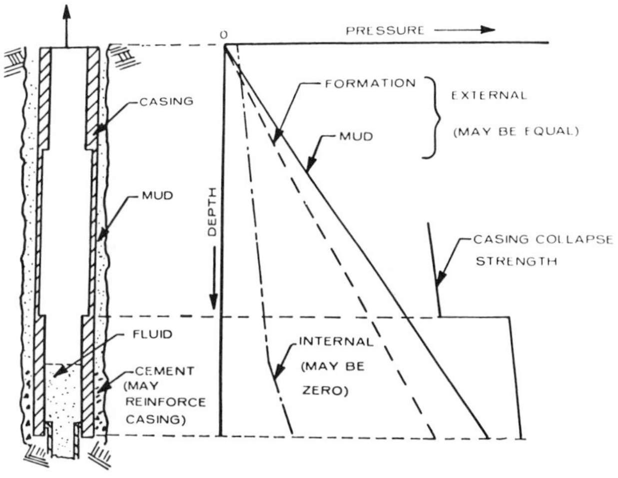

Fig. 5.2 Collapse analysis of the casing string.[5]

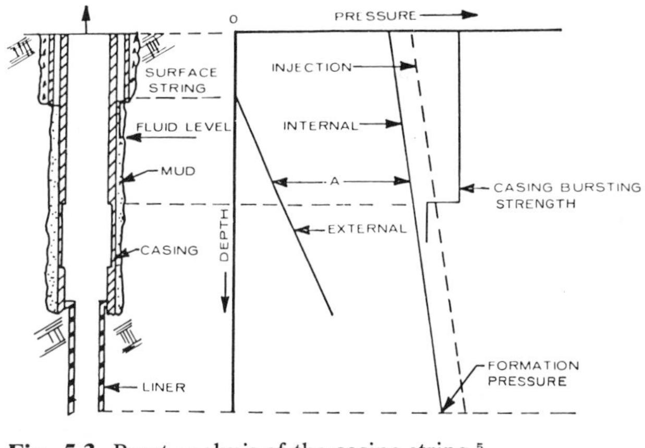

Fig. 5.3 Burst analysis of the casing string.[5]

TABLE 5.2 — PHYSICAL PROPERTIES OF TUBULAR GOODS[7]

Grade of Casing and Tubing	Yield Strength (psi)		Tensile Strength (psi)		Minimum Elongation of 2-in. Strip Specimens* (percent)
	Minimum	Maximum	Minimum	Maximum	
H-40	40,000	—	60,000	—	29.5
J-55	55,000	80,000	75,000	—	—
K-55	55,000	80,000	95,000	—	19.5
C-75	75,000	90,000	95,000	—	19.5
N-80	80,000	110,000	100,000	—	18.5
C-95	95,000	110,000	105,000	—	—
P-110	110,000	140,000	125,000	—	15.0

*0.75 sq in. area and greater.

5.3 Wellbore Conditioning and Running Casing

If a primary cementing job is to be successful, before the casing is run the hole must be properly conditioned. Drilling mud should be circulated to condition and clean the hole at a pumping rate equal to or greater than the drilling circulating rate. Mud properties that tend to cause poor cement jobs are high gel strength, high viscosity, high density, and excessive chemical content. The plastic viscosity and the yield strength of the mud should be as low as possible.[9,10] In cementing gas wells, circulation should last long enough to bring up any gas entrained in the drilling mud.

It is a good practice to "break" (start) circulation every 1,000 to 3,000 ft while running the casing to remove the filter cake that has collected around the collars, centralizers, and scratchers. Some scratchers tend to accumulate an abnormal amount of filter cake as the pipe goes into the hole, and the cake may prevent circulation when the pipe reaches bottom.

The condition of the hole — the presence or absence of bridges, the straightness of the hole, and the clearance between the pipe and the hole — determines the speed at which casing can be run. Under ideal conditions, casing can be run from 1,000 ft/hr on the Gulf Coast to 2,000 ft/hr in hard-rock country.

Fig. 5.4 shows a bottom-hole pressure chart on a typical casing and cementing job as it would appear on a pressure recorder at the bottom of a well containing an 11.8 lb/gal mud.[12] Pressures are expressed as equivalent mud weights required to equal the normal static mud weight plus the surge pressure. If the equivalent pressure falls below 10.5 lb/gal, formation fluids will enter the wellbore; or if the weight exceeds 15.4 lb/gal, the formation will break down and circulation will be lost. Surges begin to increase as the annular path of the return mud increases with the running of casing.

As wall cake is removed, mud weight increases, depending on the volume removed in scratching. After the plug container or cementing head is hooked up, there is a swabbing upward of the casing before circulation is begun. During mixing, pumpng pressure — hence circulation rate — is usually lower because cement trucks have only one pump on the well. After mixing, pumping and reciprocation are stopped to release the cementing plug and change over to the rig pump. The heavier cement column causes an upward movement of the down-hole annular fluid, maintaining a displacing pressure against the formation. When pumping and reciprocation are resumed, the top wiper plug catches up with the cement and the displacement rate returns to normal. The displacement pressure continues to increase as additional cement enters the annulus until the plug bumps at the float collar.

5.4 Casing-Landing Procedures

Casing should be landed in the hole so that future operating conditions imposed on the casing will not cause excessive loading and lead to casing failure. Damage to landed and cemented casing can result from excessive formation temperatures, pressures, and rarely, earth movements.[11,13,14] Formation pressure and temperature can usually be calculated fairly accurately on the basis of estimates made before the casing was set. Temperature increases resulting from the production of well fluids will reduce the tensile stresses imposed on the casing during its landing. If the temperature of the casing drops below that which prevailed when it was set, the casing will shrink and the tensile loading will increase. Stresses induced by earth movements are rare and are difficult to evaluate.

The most common causes of casing failure are extreme hole deviations (doglegs), which cause severe bending stresses; corrosion, either external or internal; internal casing wear; and changes in well conditions that increase the stresses on the casing (Fig. 5.5).[13]

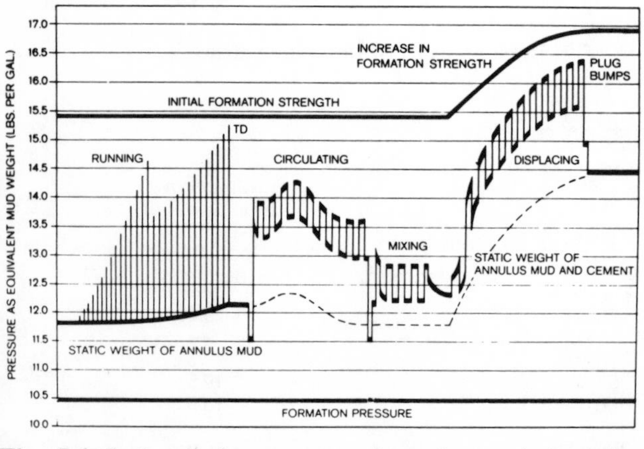

Fig. 5.4 Bottom-hole pressure chart for typical casing and cementing job.[12]

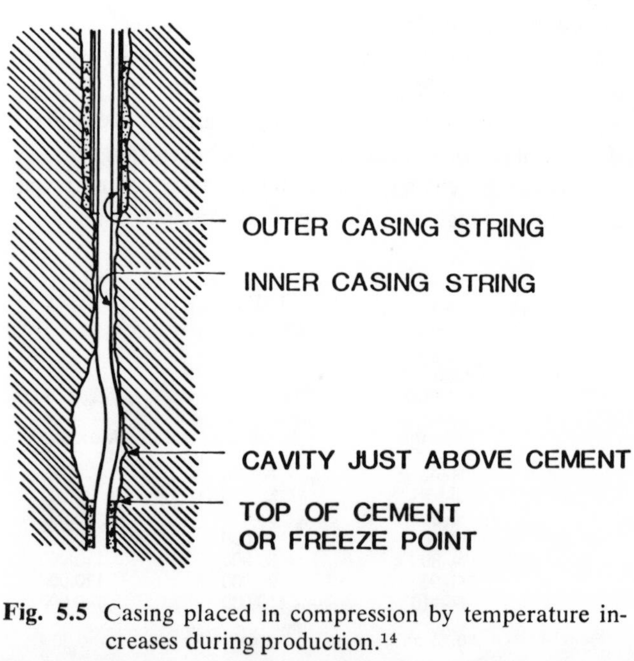

Fig. 5.5 Casing placed in compression by temperature increases during production.[14]

Casing does not usually fail until some time after it has been run, when well pressures and conditions have changed.[11,14,15] An example of this may be found in salt formations of North Dakota and Montana and in the North Sea.[15] Where salt formations are enlarged during drilling it is impossible to confine the enlarged area with cement. As a result the salt sections, which are always capable of flowing, eventually move, causing casing failures (Fig. 5.6). The solutions to this problem are to use heavier pipe through the salt formations and to centralize and move (rotate or reciprocate) the pipe. In designing casing, one should take into account not only collapse, tensile, and burst forces, but also bending stresses, reciprocation loads, and squeeze cementing effects.[16]

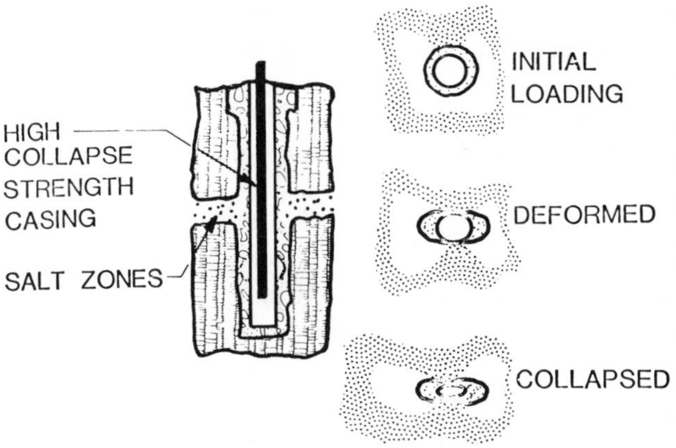

Fig. 5.6 Casing failure in salt formations.[15]

5.5 Loss of Casing Down Hole

Improperly cemented casing is vulnerable to the shocks and vibrations caused by prolonged drilling and tripping of the drill string. High repair costs or even the loss of the hole may result if because of a poor cement job the bottom joints of casing unscrew or break off.

Failures in the bottom joints of surface and intermediate casing strings are common in some areas.[17] Such failures are not normally recognized until the well is logged. Then it may be found that one or more joints have parted from the casing string and dropped down the hole. The parted section of casing may have uncovered a lost-circulation zone or may have shifted laterally, restricting the passage of drilling equipment. Remedial work is required to realign the parted casing and seal the exposed formations. Fig. 5.7 illustrates various ways that casing can be lost down the hole.

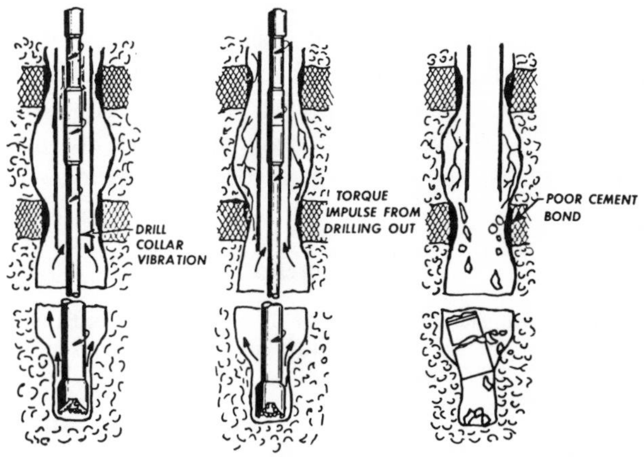

Fig. 5.7 Loss of casing down hole, caused by (1) drill-collar vibration; (2) torque impulses from drilling out; (3) poor cement bond resulting from contamination by mud.[17]

Failures can be eliminated by cementing all casing strings with two plugs and by strengthening the coupling joints at the bottom of surface or intermediate casing strings.

Casing strings composed of steel of K-55 grade or stronger should have the bottom three or four joints coated with thread-locking compound and tightened to the recommended make-up torque. Strings of lower-strength casing may be strengthened with a weld around the lower casing joint. Floating equipment should always be strengthened with a thread-locking compound because it is manufactured from N-80 grade steel. Rotary speeds should be limited to provide a safety factor of 2 between the torsional strength of the thread-locked joints and the maximum torque impulses of the bottom collars.[17] (Fig. 5.8 gives the maximum rotary speeds for drilling out cement.)

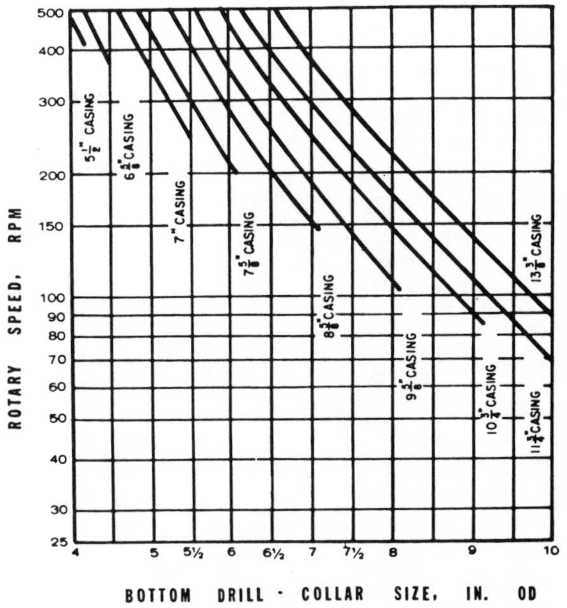

Fig. 5.8 Recommended maximum rotary speed for drilling out cement and cementing equipment (safety factor of 2). (All grades of casing strengthened with thread-locking compound; H-40 grade casing strengthened also with full-circumference weld.)[17]

5.6 Summary

In running the casing string, one must provide safety factors: 1.5 to 1.8 for tensile force (influenced by longitudinal loading); 1.0 to 1.25 for collapse pressure (influenced by unbalanced external pressure); and 1.1

to 1.33 for burst pressure (influenced by unbalanced internal pressure). Corrosion by down-hole formation fluids must also be taken into account.

Concerning the wellbore, drilling fluid should be circulated before cementing; mud cake and gas pockets should be removed; and cement channelling must be minimized. Casing should be run at 1,000 ft/hr — more slowly where there is a lost circulation or restricted annular clearance. Pressure surges should be avoided when the pipe is being run.

In landing the casing, excessive loading should be avoided. To minimize the risk of casing failure, tensile and compressive forces should be allowed for in setting and cementing the casing. Measures should be taken to protect against excessive pressures, excessive temperatures of produced fluids, and earth movements.

To avoid losing joints of casing down the hole in drilling out, the following precautions should be taken: (1) use top and bottom plugs; (2) clean the casing threads; (3) use thread-locking compounds; (4) limit rotary speeds; and (5) pick a competent casing seat.

References

1. Lubinski, A.: "Influence of Tension and Compression on Straightness and Buckling of Tubular Goods in Oil Wells," *API Proc. (Prod. Bull. 237)* (1951).
2. Bowers, C. N.: "Design of Casing Strings," paper 514-G presented at SPE-AIME 30th Annual Fall Meeting, New Orleans, La., Oct. 2-5, 1955.
3. "Casing String Design Factors," paper 851-29-I, report of API Mid-Continent Dist. Study Committee on Casing Programs (March 1955).
4. Hendrickson, J. F.: "How To Design and Run Casing Strings," *Pet. Eng.* (July 1961).
5. Hills, J. O.: "A Review of Casing-String Design Principles and Practices," *Drill. and Prod. Prac.,* API (1951) 91.
6. "Casing, Tubing, Drill Pipe and Line Pipe Properties," *API Bull. 5C3,* 2nd ed. (Nov. 1974).
7. "Specifications for Casing, Tubing, and Drill Pipe," *API Standards 5A,* 32nd ed. (March 1973).
8. Holmquist, J. L., and Nadai, A.: "A Theoretical and Experimental Approach to the Problem of Collapse of Deep-Well Casing," *Drill. and Prod. Prac.,* API (1939) 392.
9. Moore, P. L., and Cole, F. W.: *Drilling Operations Manual,* Petroleum Publishing Co., Tulsa (1965).
10. Burkhardt, J. A.: "Wellbore Pressure Surges Produced by Pipe Movement," *J. Pet. Tech.* (June 1961) 595-605; *Trans.,* AIME, **222.**
11. "Casing-Landing Recommendations," *API Bull. D7,* API, Dallas (June 1955).
12. Clark, E. H., Jr.: "A Graphic View of Pressure Surges and Lost Circulation," *Drill. and Prod. Prac.,* API (1956) 424-438.
13. Peret, J. W.: "Casing String Design, Handling, and Usage," *Fundamentals of Oil and Gas Production,* 3rd ed., Petroleum Engineer Publishing Co., Dallas (June 1959).
14. Lubinski, A., and Blenkarn, K. A.: "Buckling of Tubing in Pumping Wells, Its Effects and Means for Controlling It," *Trans.,* AIME (1957) **210,** 73-78.
15. Cheatham, J. B., Jr., and McEver, J. W.: "Behavior of Casing Subjected to Salt Loading," *J. Pet. Tech.* (Sept. 1964) 1069-1076.
16. Texter, H. G.: Three papers on oilwell casing and tubing troubles, variously titled, *Oil and Gas J.* (July 4, Aug. 1, and Aug. 29, 1955).
17. Schuh, F. J.: "Failures in the Bottom Joints of Surface and Intermediate Casing String," *J. Pet. Tech.* (Jan. 1968) 93-101; *Trans.,* AIME, **243.**

Chapter 6

Surface and Subsurface Casing Equipment

6.1 Introduction

Floating equipment, cementing plugs, stage tools, centralizers, and scratchers are mechanical devices commonly used in running pipe and in placing cement around casing.[1-3] Specifications covering such equipment are limited and variable, and standards are primarily the responsibility of the manufacturer. This chapter provides a general description of those mechanical aids and discusses applications.

6.2 Floating and Guiding Equipment

Floating equipment (Fig. 6.1) is commonly used on the lower sections of casing to reduce derrick stress by allowing the casing to be floated into place. The guide shoe directs the casing away from ledges and minimizes sidewall caving as the casing passes through deviated sections of the hole. Some basic types of floating and guiding equipment are (1) the guide shoe, with or without a hole through the guide nose, (2) the float shoe containing a float valve and a guide nose, and (3) the float shoe and float collar containing an automatic fillup valve.

The simplest guide shoe is an open-end collar, with or without a molded nose. It is run on the first joint of casing and simply guides the casing past irregularities in the hole. Circulation is established down the casing and out the open end of the guide shoe, or through side ports designed to create more agitation as the cement slurry is circulated up the annulus.[4,5] If the casing rests on bottom or is plugged with cuttings, circulation can be achieved through the side ports.

The combination guide or float shoe usually incorporates a ball or spring-loaded backpressure valve. The outside body is made of steel, of the same strength as that of the casing. The backpressure valve is enclosed in plastic and high-strength concrete. The valve, which is closed by a spring or by hydrostatic pressure from the fluid column in the well, prevents fluids from entering the casing while pipe is lowered into the hole. Suspension is controlled by the volume of fluid placed inside the casing by filling either from the surface or through a fillup device in the float shoe. Once the casing has been run to the desired depth, circulation is established through the casing and float valve and up through the annulus. When the cement job is completed, the backpressure valve prevents cement from flowing back into the casing.

For shallow wells where it is not necessary to guide or float the casing to bottom, a simple casing shoe is used.

Float collars are normally placed one to three joints above the float or guide shoe in the casing string and serve the same basic functions as the float shoe.[4,5] (See Fig. 6.2.) They contain a backpressure valve similar to the one in the float shoe and provide a smooth surface or latching device for the cementing plugs. The space betwen the float collar and the guide shoe serves

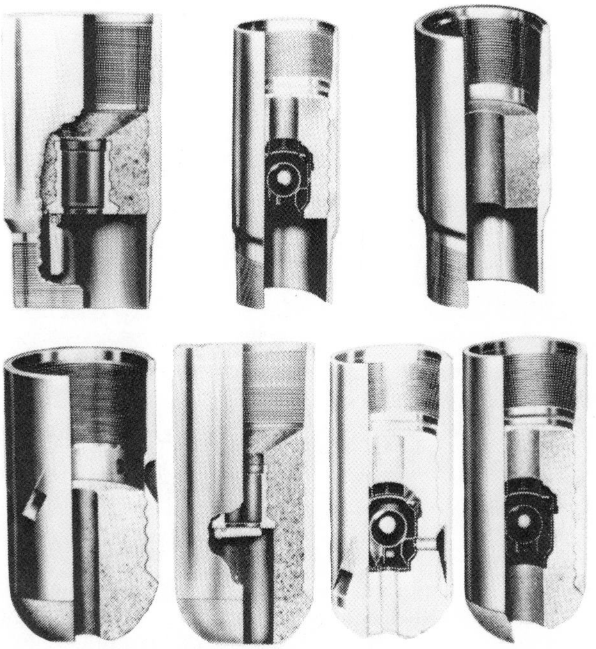

Fig. 6.1 Types of floating equipment. Top row — float collars; bottom row — float shoes.

as a trap for contaminated cement or mud that may accumulate from the wiping action of the top cementing plug. The contaminated cement is thus kept away from the shoe, where the best bond is required.

When the cement plug seats at the float collar, it shuts off fluid flow and prevents overpumping of the cement. A pressure buildup at the surface indicates when cement placement is complete. For larger casing, float collars or shoes may be obtained with a special stab-in device that allows the cement to be pumped through tubing or drillpipe. (This method of placement is often called inner-string cementing.) Such a device eliminates the need for large cementing plugs and oversize plug containers. It also reduces the volume of cement that can be lost if the job must be terminated before all the slurry is mixed and pumped. (See Fig. 6.3.)

For reasons of economy, a simple insert flapper valve and seat may be installed in the casing string one or two joints above the guide shoe. It is strong enough to satisfy most of the pressure requirements of a casing job that may not demand the standard floating equipment. The insert flapper valve, like the float collar, provides a space for isolating contaminated cement. It also provides a surface for landing the cement plug.[4,5] (See Fig. 6.4.)

Differential fillup and automatic fillup float collars and float shoes are devices that permit a controlled amount of fluid to enter the bottom of the casing while the casing is being run in the hole (Figs. 6.5 and 6.6). They operate on the principle that hydrostatic pressure in the annulus will tend to proportionally balance the hydrostatic pressure inside the casing. A restricted area allows a controlled amount of fluid to enter the casing through the bottom of the float shoe while the casing is being run, thereby shortening running time and re-

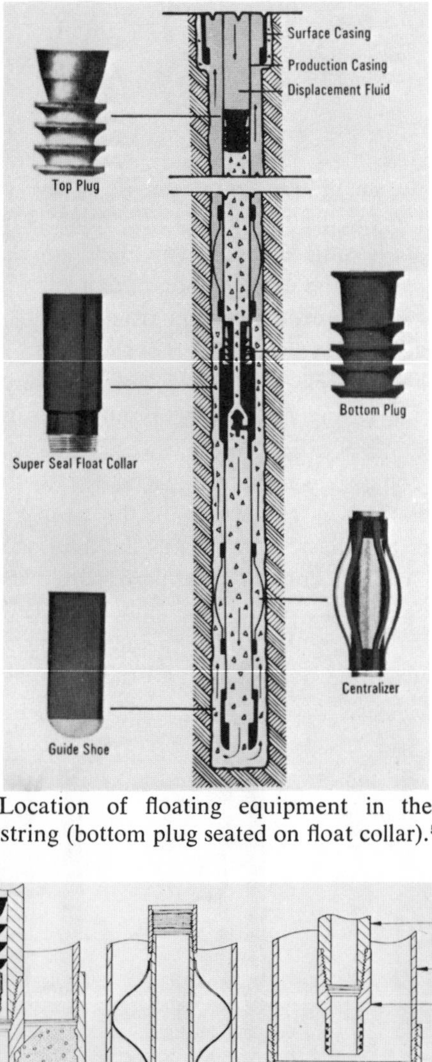

Fig. 6.2 Location of floating equipment in the casing string (bottom plug seated on float collar).[5]

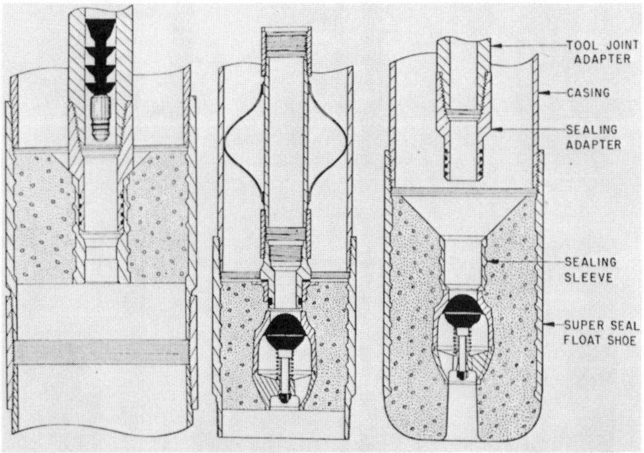

Fig. 6.3 Floating equipment for cementing through tubing and drillpipe (left — collar with latch-down plug and sealing sleeve; center and right — float collar and shoe for stabbing in tubing string).[5]

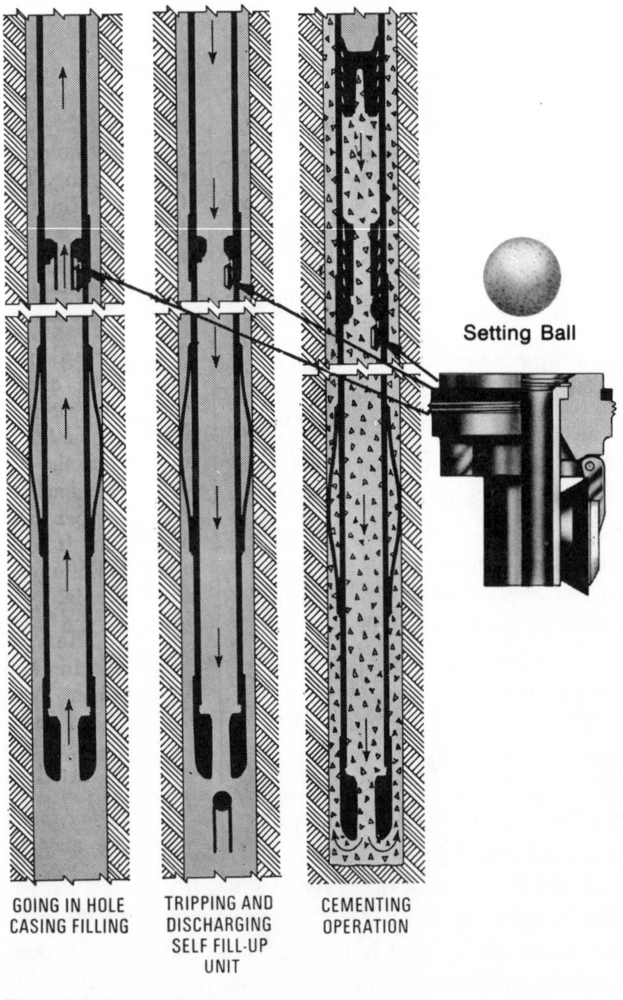

Fig. 6.4 Insert flapper valve.[5]

ducing pressure surges against the formation. The back-pressure valve in automatic fillup equipment is held out of service until it is released by a predetermined pump pressure or flow rate applied from the surface.[4,5] The rate of flow into the casing is usually low enough to hold the fluid level within 10 to 300 ft of the surface.

In purchasing floating equipment it is important to specify the outside diameter, the threading, the material grade, and the pipe weight.

6.3 Formation Packer Collars and Shoes

Formation packer shoes (and formation packer collars) are floating equipment containing expanding packers. They are used when hole conditions require the casing to be set above the producing formation or where a formation is to be isolated at a specific point below the tools. The packer, when set, packs off the open hole below the casing to help protect the formation against breakdown and cement contamination. One variation of this tool contains a float valve to aid in lowering the casing to bottom and to keep cement slurry from backing up into the casing. Another type (Fig. 6.7) uses a flapper valve and latch-down plug. Both types of packer shoes allow fluids to circulate through the bottom of the shoe until the packer is expanded and set.

The formation packer collar is similar to the packer shoe but is installed in the casing above the shoe joint. It may be used to set liners and to set perforated pipe or screen in disposal wells. It is also used in waterflood wells where full-hole cementing is desirable above the injection zone.

6.4 Stage-Cementing Tools

When it is desirable to cement two or three separate sections behind the same casing string or to cement a long section in two or three stages, multiple-stage cementing tools are used. Stage cementing usually reduces mud contamination and lessens the possibility of formation breakdown, which is often a cause of lost circulation.[4,5]

Stage tools are installed at a specific point in the casing string as casing is being run into the hole. After the cement has been placed around the bottom of the casing (the first stage) the tool can be opened hy-

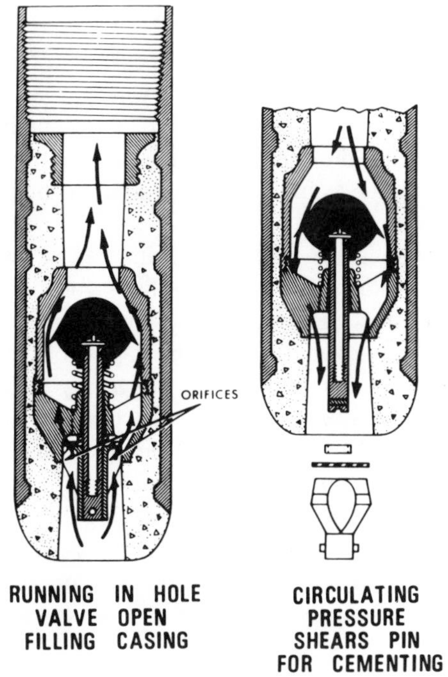

Fig. 6.5 Automatic fillup float shoes.[5]

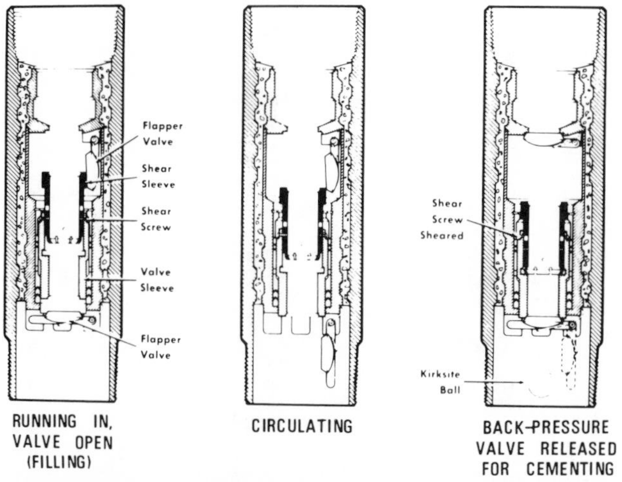

Fig. 6.6 Differential fillup float collar.[4]

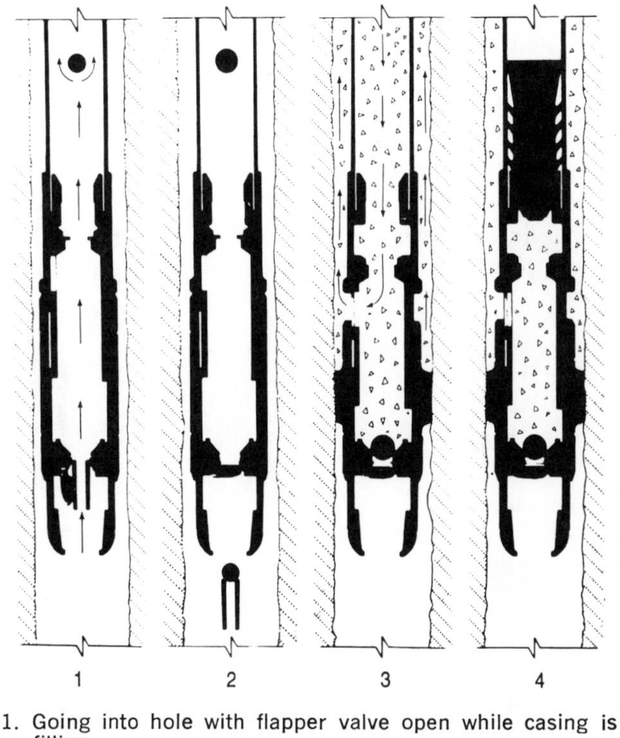

1. Going into hole with flapper valve open while casing is filling.
2. Casing in position. Flapper valve closed. Fillup tube and ball pumped out. Setting ball falling.
3. Packer set. Setting ball on valve seat. Cementing under way.
4. Job completed. Latch-down plug latched in baffle.

Fig. 6.7 Formation packer shoe — flapper-valve type.[5]

draulically either with a free-falling opening plug dropped down the casing or with a plug pumped down the casing. When the tool is opened, fluid, such as cement, can be circulated through its outside ports. When all the cement slurry has been placed, a closing plug closes a sleeve over the side port.

The free-fall stage-cementing method is used when the first-stage cement is not required to fill the annulus from the bottom of the casing all the way to the stage tool or when the distance between the tool and the casing shoe is fairly long. The primary advantage of this method is that the shutoff plug used in the first stage prevents overdisplacement of the first-stage cement.[5] (See Fig. 6.8a.)

The displacement stage-cementing method is used when the cement is to be placed in the entire annulus from the bottom of the casing up to or above the stage tool (Fig. 6.8b). The displacement method is often used in deep or deviated holes in which too much time is needed for a free-falling plug to reach the tool. Fluid volumes must be accurately calculated and carefully measured to prevent overdisplacement or underdisplacement of the first stage.

Two-stage cementing is the most widely used multiple-stage cementing technique. However, when a cement slurry must be distributed over a long column and hole conditions will not allow circulation in one or two stages, a three-stage method can be used. The same steps are involved as in the two-stage methods, except that there is an additional stage.

6.5 Plug Containers and Cementing Plugs

Plug Containers

Plug containers hold the top and bottom cementing plugs that are released ahead of and behind the cement slurry. There are two types of containers. One of them is the quick-change plug container, through which cement plugs may be inserted directly into the casing before and after the cementing operation. The other is the continuous cementing head, which holds one or two plugs that may be loaded before the cement slurry is mixed. During the cementing operation, plugs can be released from the container as required without interrupting the pumping.

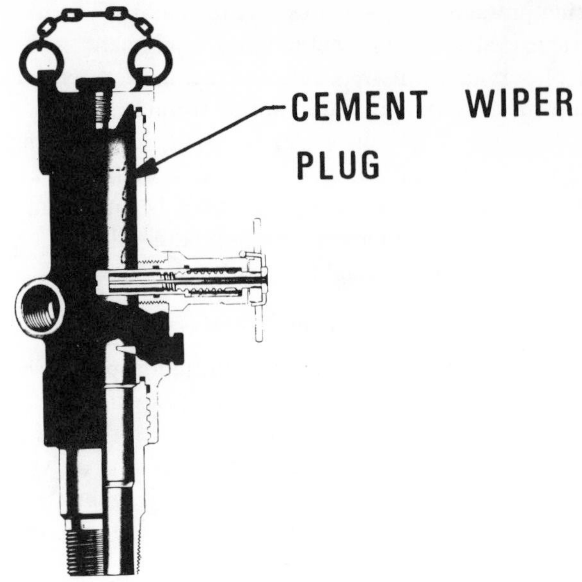

Fig. 6.9 Continuous-head plug container.[5]

Plug containers are equipped with valves and connections for attaching cementing lines for circulation and displacement. For rotation, swivels between the collar and the plug container make it possible to rotate while the casing is suspended by the rotary table slips. Unions permit fast connection of the plug container to the casing when the last joint is landed so that circulation can be started immediately.

For ease of operation, the cementing head should be as near the level of the rig floor as possible. A typical plug container (Fig. 6.9) allows a bottom plug to be inserted through the container into the casing ahead

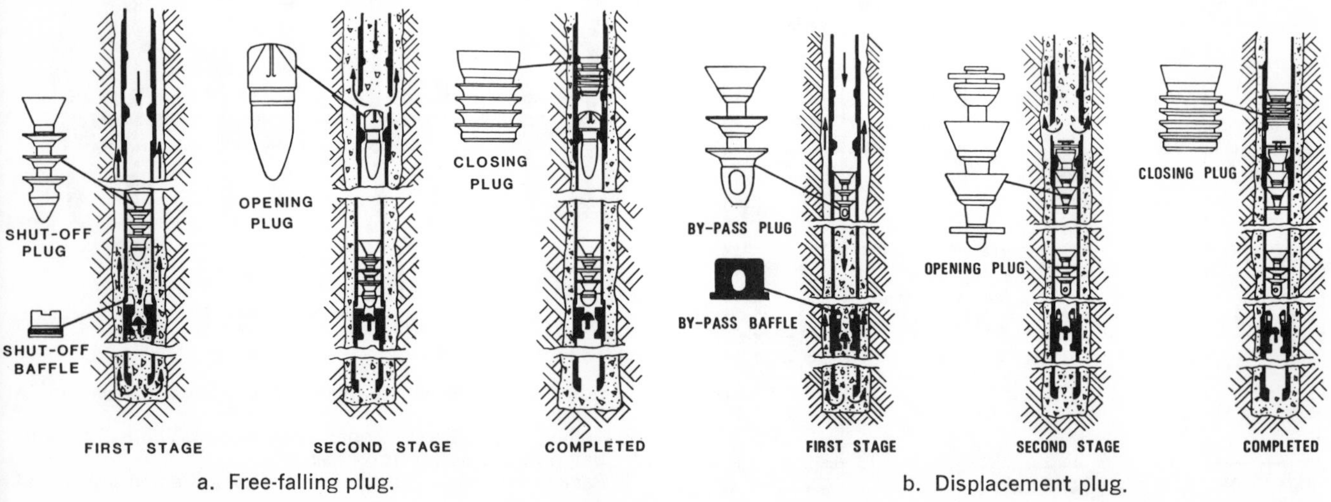

Fig. 6.8 Multiple-stage cementing tool with free-falling plug (a) and displacement plug (b) for cementing a hole in two stages.[5]

of the cement slurry. The top plug is loaded into the plug container, where it rests on a support bar. It is released by retracting the support bar after the cement is mixed. A lever on some types of plug containers indicates the passage of the plug as it leaves the container.

Cementing Plugs

Unless a well is drilled with air or gas, the casing and hole are usually filled with drilling fluid before cementing. To minimize contamination of the interface between the mud and the cement in the casing, a bottom plug is pumped ahead of the cement slurry. This plug wipes the mud from the casing wall as it moves down the pipe. When it reaches the float collar, differential pressure ruptures a diaphragm on top of the plug, allowing the cement slurry to flow through the plug and the floating equipment and up the annular space between the pipe and the hole (see Fig. 6.2). The top cementing plug is landed at the float collar or float shoe. It keeps the displacement fluid from channelling with and contaminating the cement slurry and causes a pressure buildup in the pipe.

Although top and bottom cementing plugs are similar in outward appearance (Fig. 6.10), their internal structures are different. The top plug, with its drillable insert and rubber wipers, is built to withstand the landing force of the cement column and displacement fluid and to provide dependable sealing or shutoff.[6] For cable tool operations, plugs are made with plastic inserts to reduce drilling time.

Although the conventional wiper plugs are the most widely used (Fig. 6.10), there are other designs available for primary cementing: balls, wooden plugs, subsea plugs, and tear-drop or latch-down devices such as those shown in Fig. 6.11.[5,7] The latch-down casing plug (Fig. 6.3) and baffle may be used with most conventional floating equipment but commonly they are used in small-diameter tubing for inner-string cementing. This type of plug system, supplementing the float valve, prevents fluid from re-entering the casing string. When the cement has all been pumped, the latch-down plug permits surface pressure to be released immediately, and also prevents the cement and plug from being backed up into the casing by air compressed below the plug. If completions are made fairly close to the float collar, the latch-down plug system eliminates the need to drill out the cement.

6.6 Casing Centralizers

The uniformity of the cement sheath around the pipe determines to a great extent the effectiveness of the seal between the wellbore and the casing. Since holes are rarely straight, the pipe will generally be in contact with the wall of the hole at several places.[2,8-10] (See Fig. 6.12.) Hole deviation may vary from zero to — in offshore directional holes — as much as 70° to 80°. Such severe deviation will greatly influence the number and spacing of centralizers. (See Fig. 6.13.)

A great deal of effort has been expended to determine

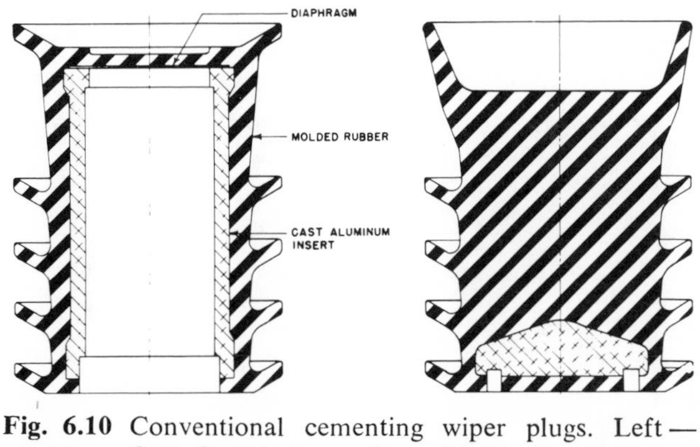

Fig. 6.10 Conventional cementing wiper plugs. Left — five-wiper bottom plug; right — five-wiper top plug.[5]

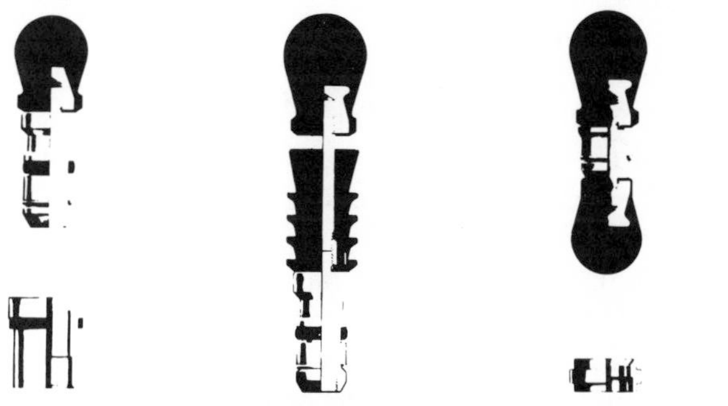

Fig. 6.11 Latch-down tubing plugs.[5]

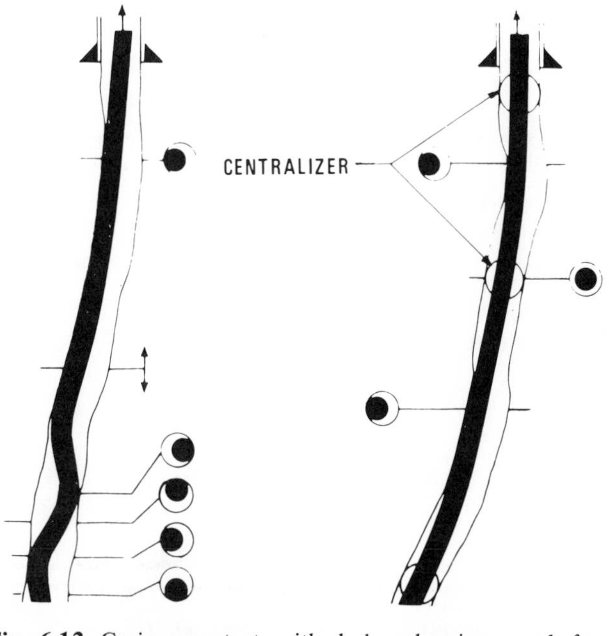

Fig. 6.12 Casing contact with hole, showing need for centralizers.[10]

TABLE 6.1 — API CASING CENTRALIZER SPECIFICATIONS

1. Minimum Centralizer Restoring Force:

Casing Size (in.)	Weight of Medium-Weight Casing (lb/ft)	Minimum Restoring Force (lb_f)
4½	11.6	464
5	13.0	520
5½	15.5	620
6⅝	24.0	960
7	26.0	1,040
7⅝	26.4	1,056
8⅝	36.0	1,440
9⅝	40.0	1,600
10¾	51.0	1,020
13⅜	61.0	1,220
16	65.0	1,300
20	94.0	1,800

2. Starting Force: maximum force should be less than the weight of a joint of 40-ft casing between centralizers.

the relative success of running casing strings with and without centralizers. Although authors differ about the proper approach to an ideal cement job, they agree unanimously that success hinges on the proper centralizing of casing.[2,3,10] Centralizers are one of the few mechanical aids covered by API specifications.[12]

When properly installed in gauge sections of a hole, centralizers (1) prevent drag while running pipe into the hole, (2) center the casing in the wellbore, (3) minimize differential sticking, and thus help to equalize hydrostatic pressure in the annulus, and (4) reduce channelling and aid in mud removal.

Most service companies, as well as the API, publish tables on the proper placement of centralizers, based on casing load, hole size, casing size, and hole deviation.[4,5,11,12]

One of the little-emphasized benefits of centralizers is that they reduce pipe sticking caused by pressure differentials.[9] The force holding casing against a permeable section in the hole is proportional to the pressure differential across the pipe and to the area of the pipe in contact with the wellbore isolated from the hydrostatic pressure by thickened mud cake.

The design of centralizers varies considerably, depending on the purpose and the vendor (Fig. 6.14).[4,5,11] For this reason, the API specifications, as defined in *API Standards 10D*,[12] insure minimum strength requirements. These requirements are based on a starting force and a restoring force. The starting force is that force required to start the centralizer into the previously run casing. The restoring force is the force exerted by a centralizer against the borehole to keep the pipe away from the borehole wall. The minimum strengths for varying pipe sizes are given in Table 6.1.

6.7 Casing Scratchers

Scratchers, or wall cleaners, are devices that attach to the casing to remove loose filter cake from the wellbore. They are most effective when used while the cement is being pumped. Scratchers, like centralizers, help to distribute the cement around the casing. There are two general types of scratchers — those that are used when the casing is rotated (Fig. 6.15), and those that are used when the casing is reciprocated (Fig. 6.16).

The rotating scratcher is either welded to the casing or attached with limit clamps.[4,5,11] The scratcher claws are high-strength steel wires with angled ends that cut and remove the mud cake during rotation. The claws may have a coil spring at the base to reduce breaking or bending when the casing is run into the hole. When the pipe must be set at a precise depth, rotating scratchers should be used, but there must be assurance

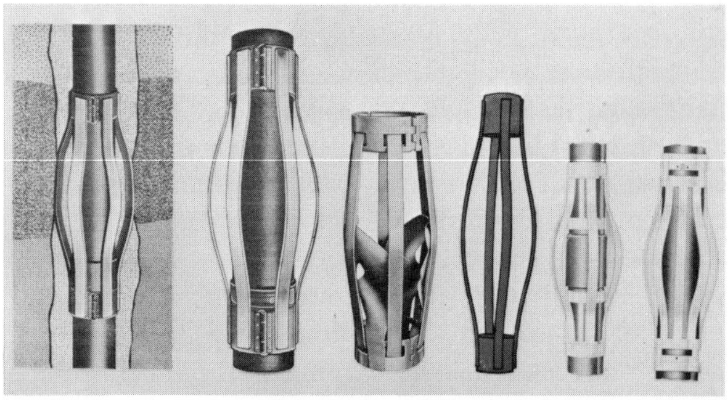

Fig. 6.14 Types of casing centralizers.

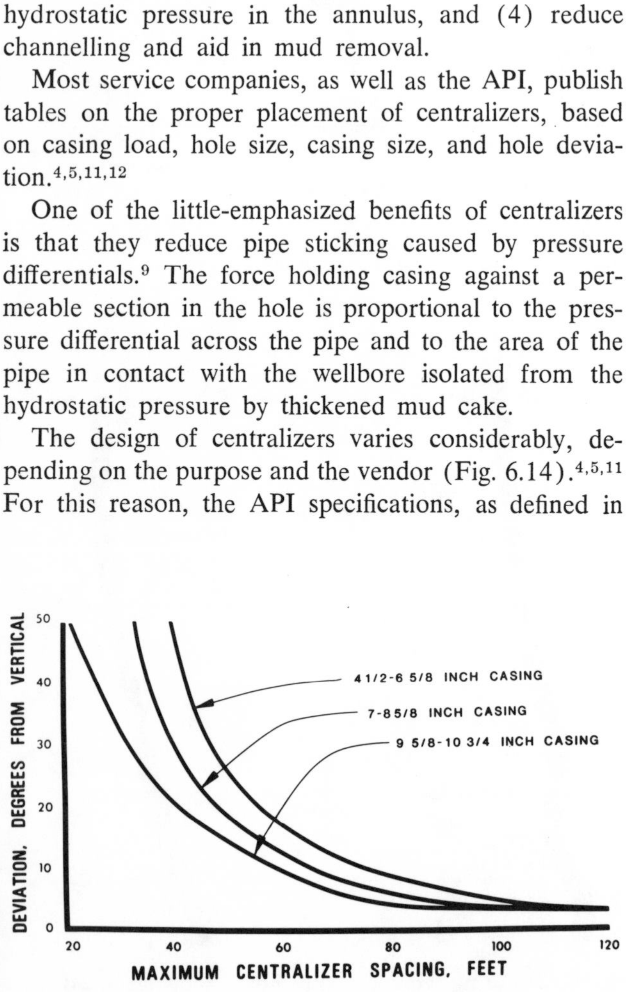

Fig. 6.13 Spacing of centralizers in deviated holes.[13]

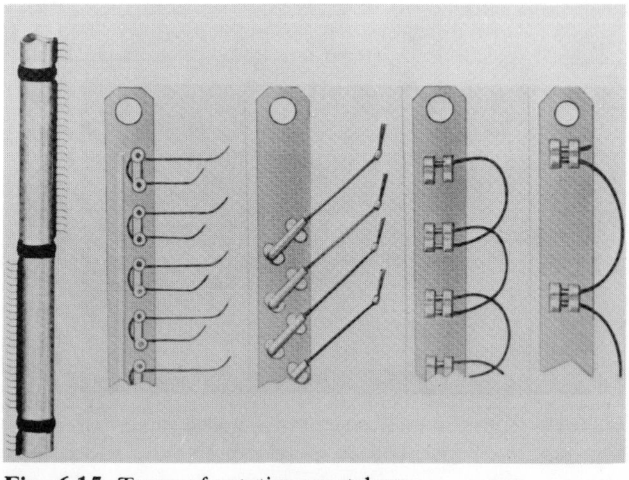

Fig. 6.15 Types of rotating scratchers.

that the pipe can be freely rotated. Because rotating scratchers are damaged by excessive torque on the casing, they are not generally used where the risk of excessive torque is high, such as in deep or deviated wells.

Reciprocating scratchers, also constructed of steel wires or cables, are installed on the casing with either an integral or a separate clamping device. When the desired depth is reached, reciprocating the casing (that is, working it up and down) cleans the wellbore on the upstroke by removing mud and filter cake. Reciprocating scratchers are the more effective kind where there is no depth limitation in setting casing and where the pipe can be either rotated or reciprocated after it is landed.

6.8 Special Equipment

Bridge plugs are devices that are set in open hole or casing as temporary, retrievable plugs or permanent, drillable plugs. They cannot be pumped through and are used to prevent fluid or gas from moving in the wellbore. Bridge plugs are also used to (1) isolate a lower zone while an upper section is being tested; (2) establish a bridge above or below a perforated section that is to be squeezed, cemented, or fractured; (3) provide a pressure seal for casing that is to be tested or for wells that are to be abandoned; (4) seal off zones to be abandoned to allow the upper casing to be recovered; and (5) plug casing while surface equipment is being repaired.

Cement baskets and external packers (Figs. 6.17 and 6.18) are used with casing or liner at points where porous or weak formations require help in supporting the cement column until it takes its initial set. Baskets may be installed by slipping them over the casing, using either the collars or limit clamps to hold them in place.[5] External packers are placed in the casing string as it is run in the well. They are expanded before cementing begins.[4]

6.9 Summary

Table 6.2 summarizes the surface and subsurface casing accessories used in cementing.

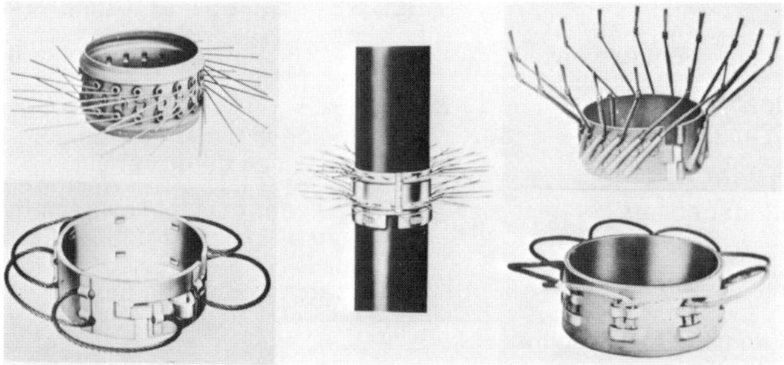

Fig. 6.16 Types of reciprocating scratchers.

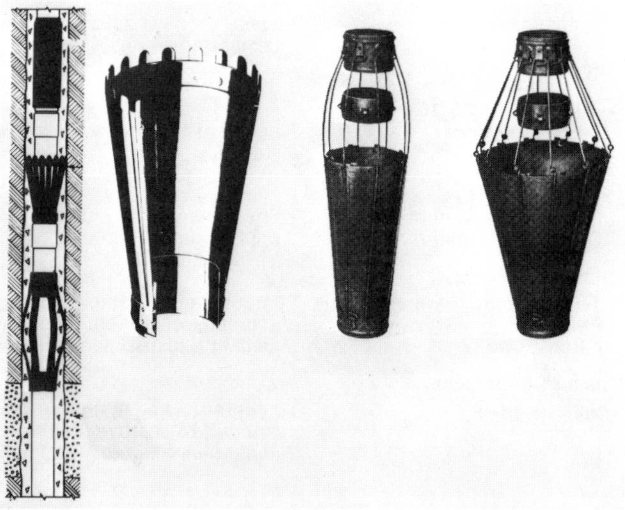

Fig. 6.17 Cementing baskets.

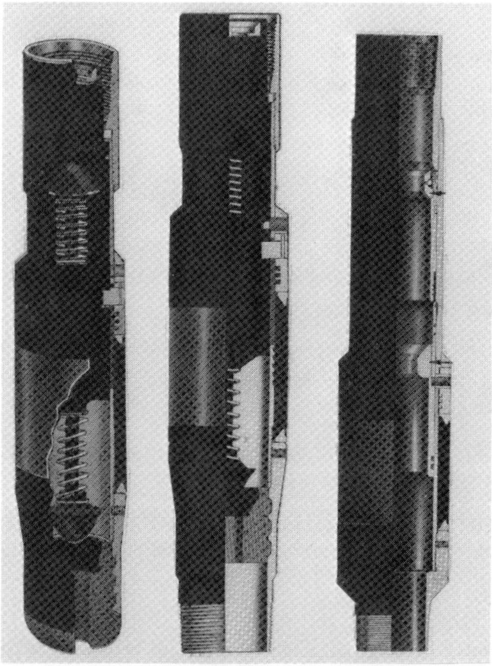

Fig. 6.18 External casing packer.

References

1. Cannon, G. E.: "Improvements in Cementing Practices and the Need for Uniform Cementing Regulations," *Drill. and Prod. Prac.*, API (1948) 126-133; *Pet. Eng.* (May 1949) B42.

2. Teplitz, A. J., and Hassebroek, W. E.: "An Investigation of Oil Well Cementing," *Drill. and Prod. Prac.*, API (1946) 76-101; *Pet. Eng. Annual* (1946) 444.

3. Hilton, A. G.: "Mechanical Aids and Practices for the Improvement of Primary Cementing," *Oil-Well Cementing Practices in the United States*, API, New York (1959) 123.

4. *Technical Sales Catalog*, Baker Oil Tools, Inc., Houston (1974).

TABLE 6.2 — DIGEST OF CEMENTING EQUIPMENT AND MECHANICAL AIDS

Equipment	Function or Application	Placement
Floating Equipment		
Guide shoes	To guide casing into well. To minimize derrick strain.	First joint of casing.
Float collars	To prevent cement flowback. To create pressure differentials to improve bond. To catch cementing plugs.	One joint above shoe in wells less than 6,000 ft deep; two to three joints above shoe in wells deeper than 6,000 ft.
Automatic Fillup Equipment		
Automatic fillup float shoes and collars; Differential fillup float shoes and collars	Same as those of ordinary float shoes and collars; also to control hydrostatic pressure in annulus while casing is being run.	Same as for ordinary float shoes and collars.
Formation Packer Tools		
Formation packer shoes	To protect lower zones by expanding during cementing.	First joint of casing.
Formation packer collars		As hole requirements dictate.
Stage-Cementing Tools		
Two-stage tools; Three-stage tools	To cement two or more sections in separate stages.	Based on critical zones and formation fracture gradients.
Plug Containers		
Quick-opening containers; Continuous cementing heads	To hold cementing plugs in casing string until plugs are released.	To joint of casing at surface of well.
Cementing Plugs		
Top and bottom wiper plugs; Ball plugs; Latch-down plugs	To act as a mechanical spacer between mud and cement (bottom plug) and between cement and displacement fluid (top plug).	Between well fluids and cement.
Casing Centralizers		
Various types	To center casing in hole or provide minimum standoff to improve distribution of cement in annulus and prevent differential sticking.	In a straight hole: one per joint through and 200 ft above and below pay zones; one per every three joints in open hole to be cemented. In a crooked hole: variable, depending on hole deviation.
Scratchers (Wall Cleaners)		
Rotating scratchers	To remove mud cake and circulatable mud from wellbore.	Through producing formations and 50 to 100 ft above. (Pipe should be rotated 15 to 20 rpm.)
Reciprocating scratchers	To aid in creating turbulence. To improve cement bond.	Same as for rotating scratchers. (Pipe should be reciprocated 10 to 15 ft off bottom.)
Bridge Plugs		
Wireline bridge plugs; Tubing bridge plugs	To plug permanently or temporarily in open hole or casing.	In well on wireline, on tubing, or below retrievable squeeze packers.
Special Equipment		
Cementing baskets and external packers	In setting casing or liner, to help weak formations support the cement column until it sets.	Below stage tools or where weak formations exist down hole.

5. *Technical Service Catalog No. 37,* Halliburton Services, Duncan, Okla. (1974).

6. Owsley, W. D.: "Improved Casing Cementing Practices in the United States," *Oil and Gas J.* (Dec. 15, 1949) 76.

7. *Technical Sales Catalog,* BJ Services, Arlington, Tex. (1974).

8. Craft, B. C., and Hawkins, M. F.: *Applied Petroleum Reservoir Engineering,* Prentice-Hall, Inc., Englewood Cliffs, N. J. (1959) 319.

9. Melmick, W. E., and Longley, A. J.: "Pressure-Differential Sticking of Drill Pipe and How It Can Be Avoided or Relieved," *Drill. and Prod. Prac.,* API (1957) 55-61.

10. Goins, W. C., Jr.: "Selected Items of Interest in Drilling Technology," *J. Pet. Tech.* (July 1971) 857-862.

11. *Well Completion Service and Equipment Catalog,* B&W Incorporated, Houston (1974).

12. "Casing Centralizers," *API Spec. 10D,* API, Dallas, 1st ed. (April 1971); 2nd ed. (Feb. 1973).

13. Carter, L. G., Cook, C., and Snelson, L.: "Cementing Research in Directional Gas Well Completions," paper SPE 4313 presented at SPE-AIME European Spring Meeting, London, April 2-3, 1973.

Chapter 7

Primary Cementing

7.1 Introduction

The engineering and economic considerations of a primary cementing job cannot be overemphasized. A poor cementing job can result in a failure to isolate zones and can be very costly in the productive life of any well. Failure to isolate between producing zones can lead to (1) ineffective stimulation treatments, (2) improper reservoir evaluation, (3) annular communication with unwanted well fluids, (4) lifting of excessive well fluids, and (5) accumulation of gas in the annulus.

In some wells, corrosion failures, which do not show up until the later stages of production, are a result of improper cementing. Corrosion holes in the pipe can be repaired only by expensive and possibly damaging workovers.

A number of factors contribute to cement failures. Table 7.1 lists some practices followed in the displacement period that are generally associated with poor cement jobs.[1,2]

7.2 Considerations in Planning a Cementing Job

Many factors determine the success or failure of a primary cementing operation. Even a simple casing job can become complex, so it should be properly planned. Items to consider are listed in Table 7.2.

Selection of Cement To Suit Well Requirements

Cement manufactured to API depth and temperature requirements may be purchased in most oil-producing areas of the world. Table 2.5 and Figs. 4.5 and 7.1 show the conditions to which the various API Classes of cement apply.

In 1948 and 1965 the API conducted surveys to establish the basis of cement testing conditions required for 90 percent of primary cementing operations.[3] The survey reflects the volume of cement, the mixing rate, the displacement rate, and the cementing time required to complete most primary cementing jobs (see Table 7.3).

The volume of cement required for a specific fillup on a casing job should be based on field experience and regulatory requirements. In the absence of specific guides a volume equal to 1.5 times the caliper survey volume should be used. Caliper logs may be necessary to determine hole enlargements and the proper location of centralizers or scratchers to obtain maximum mud displacement. Although regulations or hole conditions may dictate the fillup requirements for a given cement-

TABLE 7.1 — FACTORS THAT CONTRIBUTE TO CEMENTING FAILURES

Type of Failure	Contributing Factor
Premature Setting in Casing	
	Contaminants in mixing water
	Incorrect temperature estimate
	Dehydration of cement in annulus
	Use of improper cement
	Plugged cement shoe or collar
	Insufficient retarder
Failure To Bump Plug	
	Lodging of plug in head
	Running of top plug on bottom
	No allowance for compression
	Incorrect displacement calculations
Incomplete Mixing	
	Mechanical failure
	Insufficient water or pressure
	Failure of bulk system
Gas Leakage in Annulus	
	Insufficient hydrostatic head
	Gelation at cement/mud interface
	Failure of cement to cover gas sands
	Cement dehydration
Channeling	
	Contact of pipe with formation
	Poor mud properties (high plastic viscosity and high yield point)
	Failure to move pipe
	Low displacement rates
	Hole enlargement
Too-Rapid Setting of Cement	
	Improper water ratio
	Incorrect temperature assumption
	Mechanical failures
	Wrong cement or additives for well conditions
	Hot mixing water
	Slurry allowed to remain static to perform rig operation
	Improper choice of mud/cement spacers

TABLE 7.2 — ITEMS TO CONSIDER IN PLANNING FOR PRIMARY CEMENTING

Area	Factors of Influence
Wellbore	Diameter, depth, temperature, deviation, formation properties
Drilling Fluid	Type, properties, weight, compatibility with cement
Casing	Design, size of thread, setting depth, floating equipment, centralizers, scratchers, stage tools
Rig Operations	Time and rate of placing casing, circulation time before cementing
Cementing Composition	Type, volume, weight, properties, additives, mixing, pretesting of well blend with field water
Mixing and Pumping Units	Type of mixer, cementing head, plugs, spacers, movement during cementing, displacing fluids
Personnel	Responsibilities of involved parties

TABLE 7.3 — RESULTS OF 1948 and 1965 API SURVEYS OF CASING CEMENTING CONDITIONS

	At Well Depth of 4,000 ft		At Well Depth of 10,000 ft	
	1948	1965	1948	1965
Slurry volume, cu ft	330	704	330	1,186
Mixing rate, cu ft/min	18.3	32	18.3	33
Displacement rate, cu ft/min	50	31	50	57
Cementing time, minutes	37	41	65	93

ing operation, it is always desirable to have a minimum of 300 to 500 ft of fillup behind the intermediate and production strings of casing. It is better to use too much cement than too little, especially where there is a possibility of mud contamination or dilution.

Once the desired fillup has been ascertained, the volume of cement slurry may be calculated from data similar to that in Table 7.4. Cement manufactured to API requirements will give a recommended slurry yield in terms of cubic feet per sack based on the amount of mixing water.[3] The weight of a sack of cement is 94 lb, except in Canada, where the standard oilfield sack weighs 80 lb, and in certain European countries, where a sack of cement weighs 50 kg, or 110 lb.

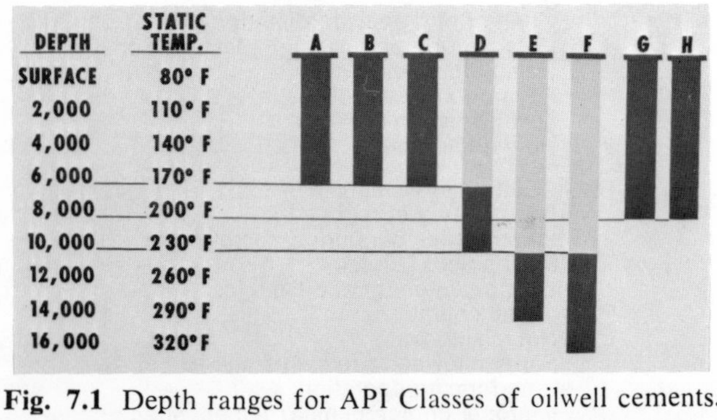

Fig. 7.1 Depth ranges for API Classes of oilwell cements.

TABLE 7.4 — YIELD OF TYPICAL CEMENT SLURRY

Weight of cement: 94 lb/sk.
Curing time: 24 hours.
Curing temperature: 100°F.

API Cement	Water (gal/sk)	Slurry Weight (lb/gal)	Yield (cu ft/sk)	Compressive Strength (psi)
Class A	5.2	15.6	1.18	2,610
Class C	6.3	14.8	1.32	2,705
Class D-E	4.3	16.4	1.06	—
Class G	5.0	15.8	1.15	2,410
Class H	4.3	16.4	1.06	2,200
Class G				
4% gel	7.8	14.1	1.55	1,015
8% gel	10.4	13.1	1.92	510
12% gel	12.3	12.6	2.19	395
Class C*				
0% gel	7.6	14.1	1.50	1,150
4% gel	10.3	13.1	1.88	665
8% gel	12.4	12.5	2.19	495
12% gel	14.6	12.1	2.51	315
Class A				
5% salt**	5.2	15.7	1.19	4,350
10% salt**	5.2	15.8	1.20	4,730
15% salt**	5.2	15.9	1.21	4,480
20% salt**	5.2	16.0	1.22	3,495
Salt-saturated	5.2	16.1	1.27	1,485

*High fineness.
**By weight of water.

Specifications and costs for some example cementing jobs are given in Appendix A.

Bulk Handling and Storage

In most areas of the world, oilwell cement is handled in bulk. With such a system, compositions can be mixed to suit any well condition.

At the bulk blending stations, cement is moved pneumatically at 30 to 40 psi air pressure into weather-tight bins or tanks. For a specific cementing job the dry ingredients are blended and loaded into bulk transport units of about 300-cu-ft capacity. Water-borne vessels using the pneumatic pressure system may be equipped with their own weighing and blending plants, or they may obtain weighed and blended materials from a support vessel or from nearby shore stations. (See Figs. 7.2 through 7.5.)

Bulk handling has greatly enhanced both the economics and the technology of cementing. Special compositions can be blended more quickly and easily, and large jobs are practical that would be impossibly time-consuming with sack cement.

For high-volume jobs, several field storage bins may be required. These bins may be located at the well and filled ahead of time.

Offshore, supplies and pumping equipment must always be on hand in case of emergency. Bulk materials are moved pneumatically from supply/service vessels to containers aboard the rig. The rig system can then transfer these materials back to a supply boat if necessary. Most rigs use pressurized bulk storage systems that are not dependent on gravity feed. They can be placed almost anywhere, and in any position, on the rig, and dry bulk materials can be piped to the cementing equipment at any location.

Discharge pipes from the pressurized containers lead to a surge tank at the cement mixer, where the dry materials are separated from the compressed air and fed into a mixing hopper. The surge tank may be equipped with scales. Most pressurized storage containers are vertical cylinders with domed heads and conical bottoms. A standard container is 10 ft in diameter, 16 ft high, and has a capacity of 820 cu ft. Bulk mud materials are commonly handled by the same type of system.

Water Sources and Supply

On any cementing job site there must be an "essential volume" of water available. An essential volume is that amount required to properly mix the slurry at the desired water/cement ratio, plus an allowance for pump priming, line testing, and pump cleanup. The rate at which water is supplied to the mixing and pumping units will affect and control the rate of mixing of the cement slurry. When these rates are too slow, the time of pumping may encroach on the time the slurry starts to set.[4]

The water normally available at a drilling well may vary in quality from reasonably clean and free of soluble chemicals to badly contaminated with silt, organic matter, alkali, salts, or other impurities. (See Section 4.6.) To emphasize what has already been said, the water for mixing with cement should be the purest available.

The rate of water feed should be based on the number of sacks of cement to be mixed at the highest expected rate. If the cementing unit can handle approximately 50 cu ft/min. of cement requiring 5.0 gal of water per sack, then 250 gal/min. must be the lowest feed rate from the water supply. Thus a 1,000-sk job must have 5,000 gal of water plus an additional 500-gal allowance per cementing unit. An additional 10 percent should be allowed for differences in approach from operator to operator.

In warm weather, the temperature of the mixing water is important. Water above 100°F usually results in a viscous slurry with a shorter pumping time. Water temperature must be considered in winter, also — primarily because lines may freeze, but also because very cold water can significantly retard the set of a slurry. Ideally, the cement and water should produce a cement slurry temperature between 60° and 90°F as it goes into the well (see Fig. 7.6).

The water source should be fitted with an outlet to which the pumping equipment can be connected so water will be quickly available for cementing. Although a pressured source is best, high tanks with good gravity feed are preferred to pits from which the cementing truck must run a suction line.

7.3 Considerations During Cementing

Pumping Equipment

Cement pumping units may be mounted on a truck,

Fig. 7.2 Bulk cement rail transports.

Fig. 7.4 Aircraft bulk units for use in remote areas.

Fig. 7.3 Land-based bulk storage and blending plant.

Fig. 7.5 Marine bulk cementing and pumping units.

trailer, skid, or water-borne vessel (see Figs. 7.7 through 7.10). They are usually powered either by internal-combustion engines or by electric motors, and are operated intermittently at high pressures and at varying rates. Pumping units must be capable of high horsepower input and output over wide torque limits, and yet possess the lowest practical weight-to-horsepower ratio to facilitate transportation.

Cementing units may be equipped with two or three pumps. On a high-pressure system, one pump mixes while the other displaces. On a low-pressure system, a centrifugal pump mixes and two positive-displacement pumps displace. For recirculating mixing, one pump supplies water to the mixing jet, and one or more pick up the slurry and pump it into the well. The pumps must be readily controlled so that pressure and volume can be adjusted as well conditions demand. For pumps driven by internal-combustion engines, this control capability imposes severe design restrictions and, because of the high torque and pressures required at low pump input speeds, almost precludes a direct engine-to-pump drive.

Nearly all cementing pumps are positive-displacement, and are either duplex double-acting piston pumps or single-acting triplex plunger pumps. Either is satisfactory within its design limits. For heavy-duty pumping, plunger pumps discharge more smoothly and can usually handle higher horsepower and greater pressure than duplex pumps.

Most cementing work involves a maximum pressure of less than 5,000 psi, but pressures as high as 20,000 psi are not uncommon. Because of widely varying operating conditions, the cementing pump and its power

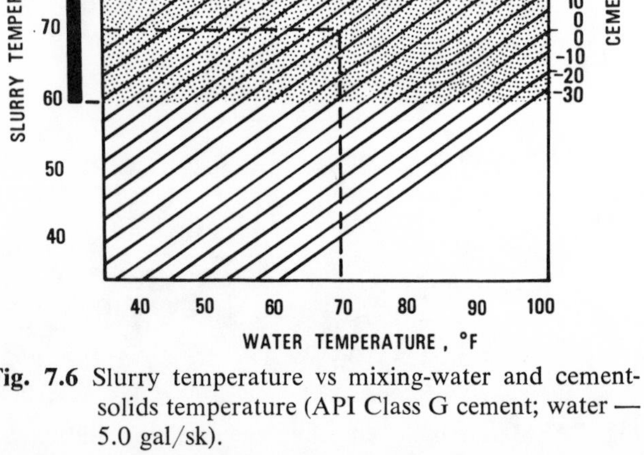

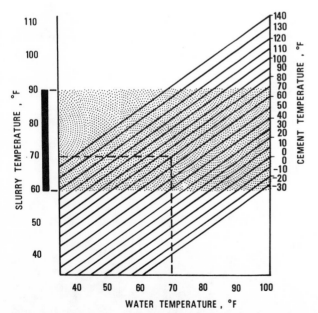

Fig. 7.6 Slurry temperature vs mixing-water and cement-solids temperature (API Class G cement; water — 5.0 gal/sk).

Fig. 7.7 Truck-mounted cementing unit with jet system for mixing and pumping.

Fig. 7.8 Marine cementing unit carrying bulk cement.

Fig. 7.9 Skid-mounted cementing unit.

Fig. 7.10 Offshore platform cementing system.

train are designed for the maximum rather than the average expected pressures.

For a given job, the number of trucks used to mix cement will depend upon the volume of cement, the well depth, and the expected pressures. For surface and conductor strings one truck is usually adequate, whereas for intermediate or production casing as many as three units may be required. On jobs requiring more than 1,000 sk or where high pressures are expected, two and sometimes three mixing trucks are used. A separate mixing hopper is used for each truck, with each unit tied to a common pumping manifold with the mud pumps. If the pipe is to be reciprocated, the mixing trucks are tied in to a temporary standpipe, which supports a flexible line leading to the cementing head.

Field slurries are usually mixed and pumped into the casing at the highest feasible rate, which varies from 20 to 50 sk/min., depending upon the capacity of each mixing unit. As a result, the first sack of cement on a primary cement job reaches bottom-hole conditions rather quickly.[3] (See Table 7.5.)

Cement Mixing

The mixing system proportions and blends the dry cementing composition with the carrier fluid (water), supplying to the wellhead a cementing slurry with predictable properties.

The most widely used mixer is the jet mixer (Fig. 7.11). It consists of a funnel-shaped hopper, a mixer bowl, a discharge line, a mixing tub, and water-supply lines. The mixer forces a stream of water through a jet and into the bowl, where it mixes with cement from the hopper to form a slurry. The slurry is forced into the discharge line, then into the mixing tub, from which it is taken away by the cementing pumps. Mixers of this type can produce a normal slurry at a rate of 50 cu ft/min.

Mixing speed is controlled by the volume of water forced through the jet, and by the amount of cement fed into the hopper during mixing. To lower the slurry weight (increase the water/cement ratio) extra water can be supplied to the bowl discharge line through a bypass line. For some cementing compositions, to control the volume and rate of water going through the jet, the size of the jet outlet may have to be changed.

Water is supplied to the mixer jet from a pump and storage tank near the drilling rig. Mixing pressures may range from 150 psi (low-pressure system) to 500 psi (high-pressure system), depending on the rate of material feed and the water ratio required.

The recirculating mixer, designed for mixing heavy densified slurries, is a pressurized jet mixer with a large tub capacity (see Figs. 7.12 and 7.13). It uses recirculated slurry and mixing water to partially mix and discharge the slurry into the tub. Additional shear is provided by the recirculating pump and agitation jets, and an in-tub eductor provides additional energy and improves mixing. The result is a more uniform cement slurry, with a density as high as 22 lb/gal, that can be pumped as slowly as 0.5 bbl/min.

Table 7.6 compares mixing rates and slurry densities for jet mixers and recirculating mixers.

Batch mixing is used to blend a cement slurry at the surface before pumping it into the well. It is a rather simple way to prepare a specific volume of slurry to exacting physical requirements before it is pumped down the hole. Figs. 7.14 and 7.15 and Fig. 7.16 show two types of batch mixers.

Density Control

Slurry density should be monitored and recorded to insure that the correct water/solids ratio is maintained. To avoid the effect of aeration, samples for weighing

TABLE 7.5 — BOTTOM-HOLE CIRCULATING CONDITIONS DURING SQUEEZE AND CASING CEMENTING
(Basis for API well-simulation test schedules.)

Well Depth (ft)	Bottom-Hole Static Temperature (°F)	Bottom-Hole Circulating Temperature (°F)		
		Casing	Squeeze	Liner
2,000	110	91 (9)*	98 (4)*	91 (4)*
6,000	170	113 (20)	136 (10)	113 (10)
8,000	200	125 (28)	159 (15)	125 (15)
12,000	260	172 (44)	213 (24)	172 (24)
16,000	320	248 (60)	271 (34)	248 (34)
20,000	380	340 (75)		

*Values in parentheses indicate time in minutes for the first sack of cement to reach bottom-hole conditions.

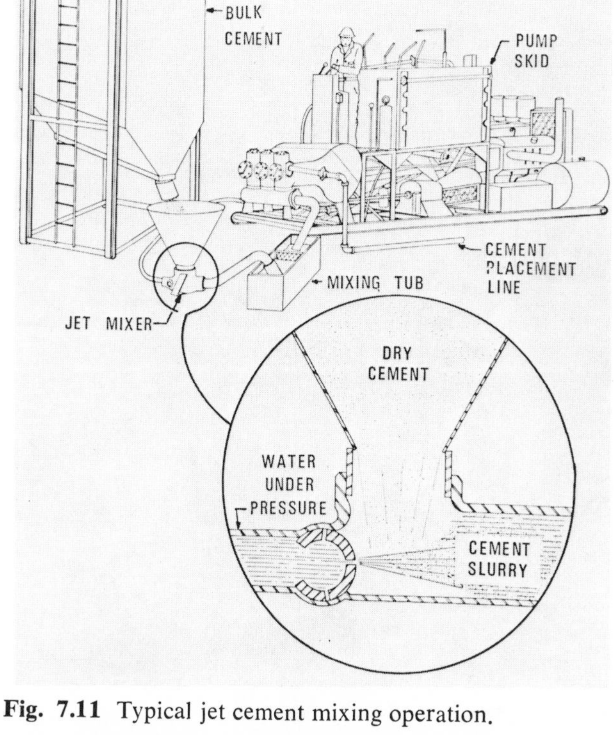

Fig. 7.11 Typical jet cement mixing operation.

Fig. 7.12 Recirculating mixer mounted on truck.

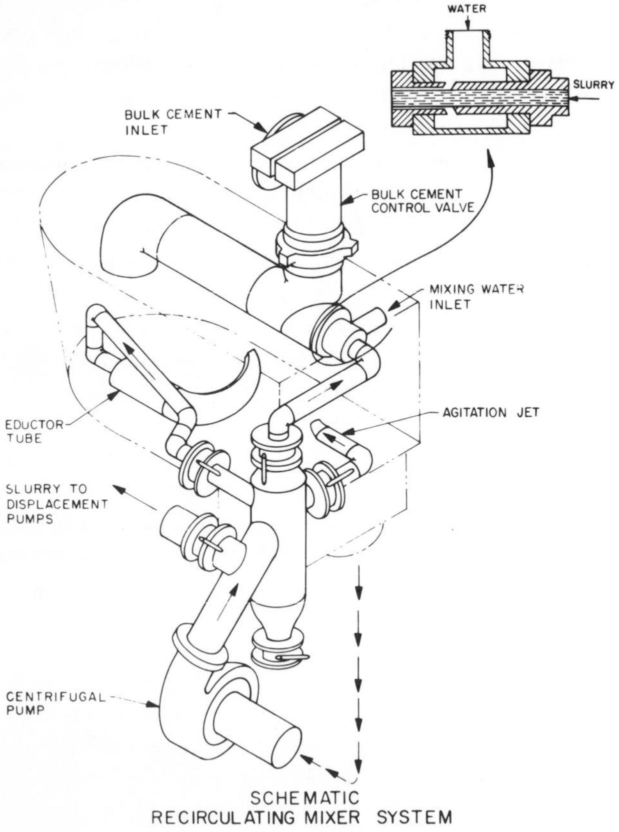

Fig. 7.13 Schematic of recirculating mixer.

Fig. 7.14 Pneumatic batch mixer.

TABLE 7.6 — RANGES OF MIXING RATES AND DENSITIES FOR VARIOUS SLURRIES

	Jet Mixer		Recirculating Mixer	
	Mixing Rate (bbl/min)	Slurry Density (lb/gal)	Mixing Rate (bbl/min)	Slurry Density (lb/gal)
Densified and weighted slurries	2 to 5	16 to 20	0 to 6	16 to 22
Neat slurries	2 to 8	15 to 17	0 to 8	14 to 18
High-water-ratio slurries	2 to 14	11 to 15	0 to 12	11 to 15

should be obtained from a special manifold on the discharge side of the displacement pump rather than from the mixing tub. (See Section 4.8.)

Cement slurries are usually mixed with less water toward the end of the job to achieve better strength. This is especially important with the last volume mixed, since it is placed around the shoe joint.

For mixing densified or heavy-weight slurries to be pumped at rates of less than 5 bbl/min., a recirculating mixer produces a more uniform slurry than a standard jet mixer.

Preflushing

Preflushes, functioning as spacers, minimize mixing and interfacial gelation in the annulus. They have various characteristics, depending on the mud system,

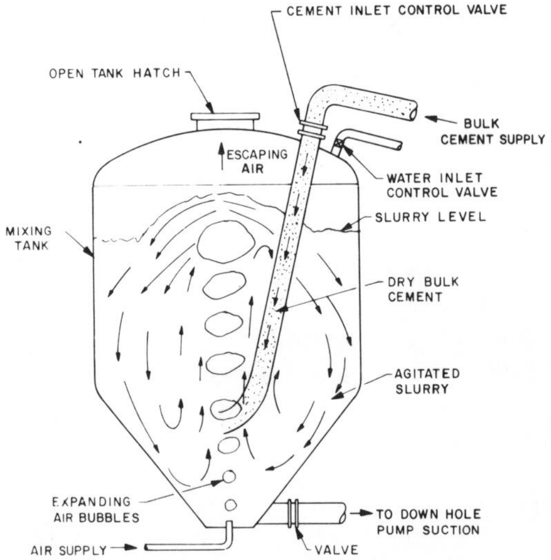

Fig. 7.15 Schematic of pneumatic batch mixer.

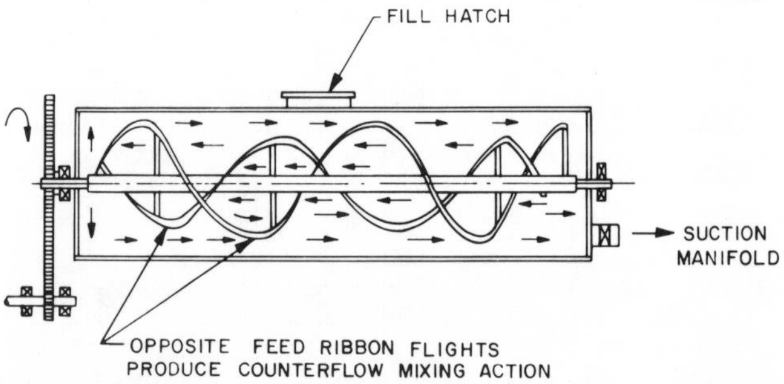

Fig. 7.16 Ribbon batch mixer.

TABLE 7.7 — TYPES OF CEMENT AND MUD PREFLUSHES

Preflush	Function	Recommended Volume	Recommended Flow
Chemical dispersants (acid-phosphates or emulsions, diesel oil for oil muds)	To thin mud	Enough to obtain 300- to 500-ft annular fill	Turbulent
Thin cement slurries (excessive water with neat cement)	To scavenge or scour	30 to 50 bbl slurry; 50 to 150 sk cement	Turbulent
Guar or HEC added to water or cement	To increase viscosity	Enough to obtain 300- to 500-ft annular fill	Plug or laminar

and various functions. Some contain additives to thin the mud and to penetrate and loosen wall cake; some contain abrasive materials to scour the hole; and some have a high apparent viscosity to remove drilling mud by buoyancy. Table 7.7 lists some preflushes and the recommended volumes to be used.

For simple water-base muds, water in sufficient volume is an excellent wash since it is cheap, easy to put into turbulence, and has little effect on the setting of cement. Salt water may lessen the tendency of shales to swell and slough, but may have a detrimental effect on a fresh-water mud. Some mud thinners (quebracho, lignosulfonates) added to water will retard the setting of cement and should be avoided. Fifty barrels of spacer, or 300 to 500 ft of annular fill, should be used, except where the hydrostatic head is reduced excessively in high-pressure zones. Table 7.8 shows the volumes that are required to achieve annular-fill columns of 300, 500, and 1,000 ft for varying casing and hole sizes.

Diluted portland or pozzolan cement is an excellent scavenging preflush since it is easy to put into turbulence and its solid particles erode gelled mud and filter cake.

Contact Time

Contact time[5] is the period of time that a cement slurry flows past a particular point in the annular space during displacement. Studies indicate that when turbulent flow is attained, a contact time of 10 minutes or longer provides excellent mud removal.[5] The volume of fluid needed to provide a specific contact time is

$$V_t = (t_c)(q_d)(5.615),$$

where

V_t = volume of fluid (turbulent flow), cu ft,

t_c = contact time, minutes,

q_d = displacement rate, bbl/min, and

5.615 = cu ft/bbl.

The calculation is simple, since only two factors are required, and those are readily available, and the calculation is independent of casing and hole size. The equation holds as long as all of the fluid passes the point of interest.

Cementing Wiper Plugs

Cementing plugs are highly recommended to separate mud, cement, and displacing fluid.[4] (See Section 6.5.) A bottom plug is used first to wipe mud from the inner surface of the casing ahead of the cement, and to separate mud and cement. A top plug separates mud and cement, and provides shutoff when the cement is in place. If the bottom plug is omitted, mud film wiped by the top plug accumulates ahead of the top plug as shown in Fig. 7.17.

There are times, however, when a bottom plug should not be used; for example, when the cement contains large amounts of lost-circulation material, or when the casing being used is badly scaled. Under such conditions, a bottom plug could cause bridging and plugging of the casing.

A cementing manifold is commonly used with a discharge line to the pit for flushing the cement truck. It is assembled so as to permit pumping the plug out of the cementing head with the displacing fluid.

When the top plug is to be displaced by mud or

TABLE 7.8 — VOLUMES REQUIRED FOR VARIOUS HOLE AND CASING SIZES

Casing Size (in.)	Hole Size (in.)	Volume per Linear Foot (gal)	Volume (bbl) To Fill Annulus to Height of 300 ft	500 ft	1,000 ft
13⅜	17½	5.2	37	62	124
10¾	12¼	1.4	10	17	33
9⅝	12½	2.6	18	31	62
7⅝	9⅞	1.6	11	19	38
7⅝	8½	0.6	4	7	14
7	8¾	1.1	8	14	27
5½	7⅞	1.3	9	16	31
5	6½	0.7	5	9	17
4½	6¼	0.8	5	9	18

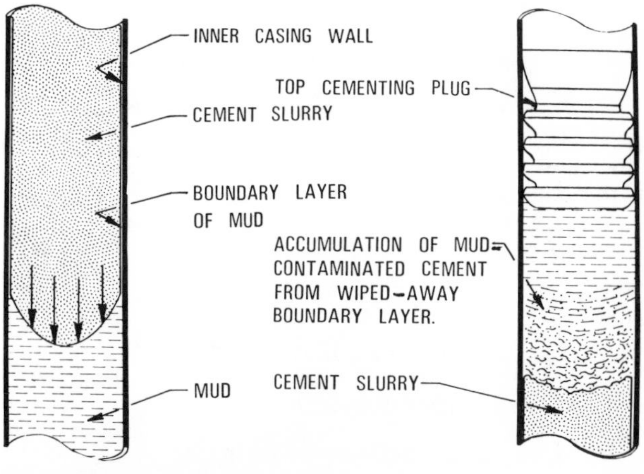

Fig. 7.17 Mud contamination caused by omission of bottom wiper plug.[4]

water, the volume of the displacing fluid should be measured at the cement pumps and compared with the volume measured in the water or mud tanks. Where there is a flow meter, it can be used to cross check. If the top plug does not "bump" (i.e., seat at the float collar, causing a pressure increase) at the calculated displacement volume, the pumping should be stopped so as not to overdisplace cement slurry out of the casing.

If casing movement is employed, it should be continued throughout the mixing cycle. Frequently, movement is continued while plugs are being released and until the top plug seats (bumps), although it is not uncommon to stop while either or both plugs are being inserted.

Displacing Fluid Behind Top Plug

Mud is normally used as the displacing fluid on surface or intermediate casing, although fresh water may be more desirable in deeper wells where density is not critical. Other commonly used displacing fluids are fresh water, salt water or sea water, and sometimes weak acid solutions, depending on the completion program. The selection should be aimed at minimizing formation damage and completion time. Diesel oil may be used to reduce swabbing time. Water containing sugar or other retarding additives is sometimes placed immediately above the top plug in small-diameter casings to inhibit the setting of any cement that may have bypassed the top plug.

Cement slurry should be in turbulent flow, provided the hole is drilled to gauge and there is no danger of excessive pressure that would break down the formation.[1,5-7] Lower flow rates, approaching plug flow, should be used where weak formations are exposed or hole sections are washed out, or in cementing large pipe.[8]

The velocity of cement slurry inside drillpipe, casing, or tubing, is calculated from

$$v_i = 3.056 \frac{q}{d_i^2}$$

where

v_i = velocity inside pipe, ft/sec,

d_i = inside diameter of pipe, in., and

q = flow rate, cu ft/min.

The equation for calculating the velocity of the slurry in the annulus is as follows:

$$v_a = \frac{3.056\, q}{d_h^2 - d_o^2}$$

where

v_a = velocity in annulus, ft/sec,

d_h = diameter of hole, in., and

d_o = outside diameter of casing, in.

The constant 3.056 changes to 17.157 if q is in bbl/min.

Casing Movement During Cementing

Casing movement significantly affects the successful displacement of mud.[1,7,9,10] Ideally, casing should be either reciprocated or rotated until the top plug reaches bottom. However, frictional drag, the weight of the pipe, and differential sticking can restrict or prevent casing movement. Differential sticking — often the most critical factor — can be attributed to (1) the contact area between pipe and mud filter cake, (2) differential pressure between mud column and formation, (3) the available pulling force, and (4) the drilling mud properties.

Sticking coefficient is a function of the mud properties (primarily water loss) and the time that the casing remains stationary against a permeable formation. Differential sticking can often be eliminated by reducing "stationary time," using a fast cementing-head hookup as soon as the casing is on bottom.

During cement pumping, casing movement (reciprocation or rotation) should be slow as the cement reaches bottom and faster when the cement is in the annulus and the top plug bumps. The reciprocation should be on a 2-minute cycle over 15- to 20-ft intervals. Reciprocation causes high pressure and turbulence on the downstroke of the casing. If the pipe shows any tendency to stick, it should be moved near the desired landing point. The likelihood of sticking is indicated by an increase in the *difference* between the weight on the upstroke and that on the downstroke rather than by a weight increase alone.

Whenever possible, the top of the casing should be landed just above the rotary table to save time in attaching the cementing head and to improve safety.

Flotation Factor in Big Pipe

In running large-diameter pipe, flotation can become a problem. When pumping is begun, the pipe will start coming out of the hole if the pressure exceeds a certain level. To deal with the problem, the pipe should be unchained and the elevators attached to the pipe when the pump is started. As the pipe comes out of the hole, the pump should be slowed down (this can be done safely, since circulation is usually established by the time the pipe has risen a few feet). Pump pressure should be increased gradually to clean the hole; the pipe should settle down then to its original position. If the pipe does not go down the hole, it can be raised and lowered during pumping and worked into place. If this fails, alternatives are to (1) cement the pipe where it is, (2) mix mud to stop the cavings and circulate them from the hole, or (3) pull the pipe and run drillpipe and bit to bottom to clean and condition the hole. The last practice is often unsuccessful unless the second is used also. Another common practice is to drill the small hole considerably beyond the casing point to form a pocket into which the cavings can fall.

7.4 Placement Techniques

Most primary cement jobs are performed by pumping the slurry down the casing and up the annulus; however, there are modified techniques for special situations. Fig. 7.18 illustrates the following methods of placement:

1. Cementing through casing (normal displacement technique).
2. Stage cementing (for wells having critical fracture gradients).
3. Inner-string cementing through drillpipe (for large-diameter pipe).
4. Outside or annulus cementing through tubing (for surface pipe or large casing).
5. Reverse circulation cementing (for critical formations).[11]
6. Delayed-set cementing (for critical formations and to improve placement).
7. Multiple-string cementing (for small-diameter tubing).

Cementing Through Pipe and Casing

Conductor, surface, protection, and production strings are usually cemented by the single-stage method,

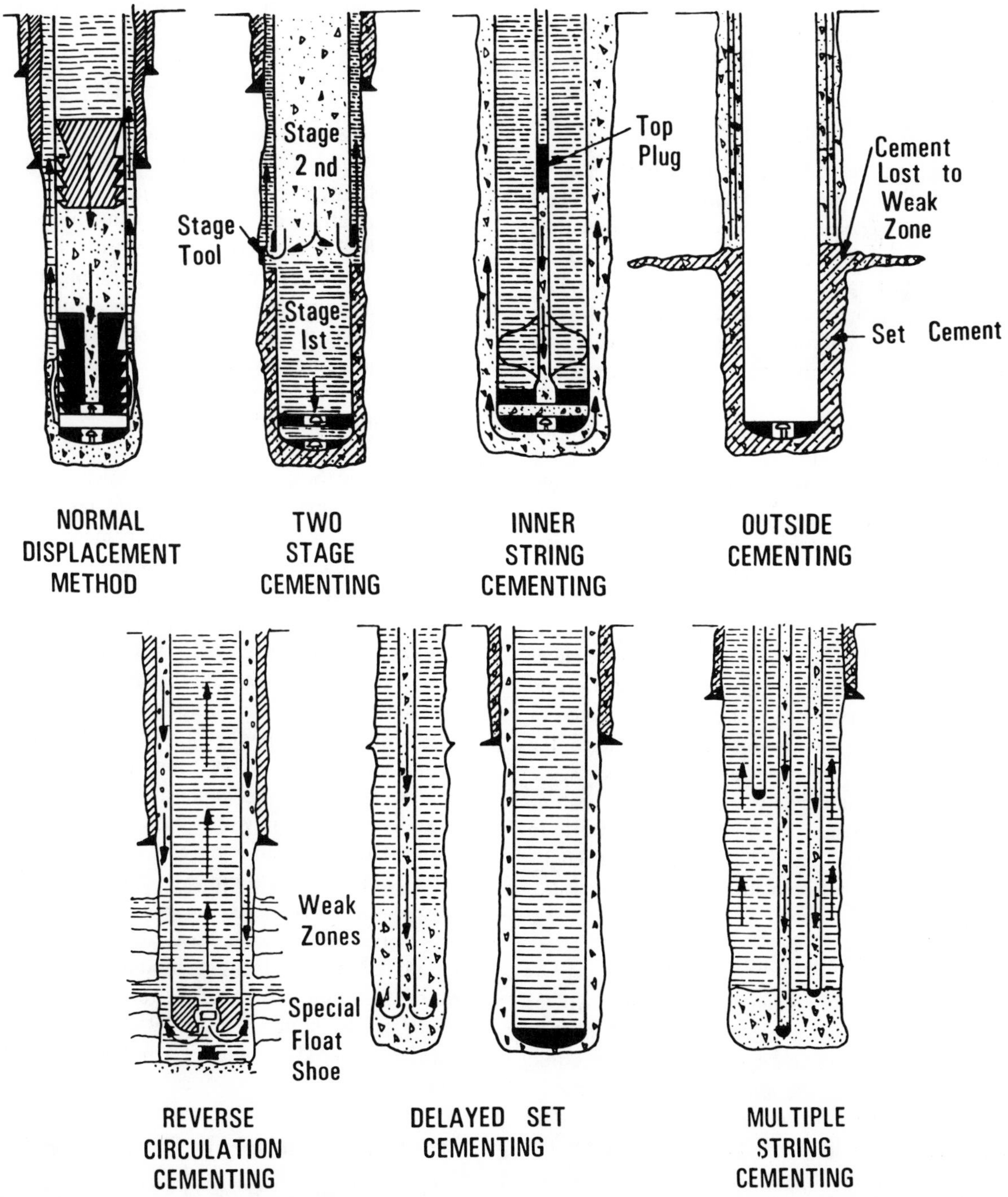

Fig. 7.18 Methods of placing cement down hole.

which is performed by pumping cement slurry through the casing shoe, and using top and bottom plugs. There are various types of heads for continuous cementing, as well as special adaptors for rotating or reciprocating casing.

Stage Cementing

Stage cementing is primary cementing performed in two or three parts or stages. (For more information, see Section 6.4.) It is commonly used in wells that require a long column of cement and where weak formations are exposed that will not support the hydrostatic head during cementing.

One disadvantage of stage cementing is that the casing cannot be moved (either rotated or reciprocated) after the first stage. This increases the possibility of channeling and decreases the probability of removing all of the circulatable mud.

Inner-String Cementing

When large-diameter pipe is cemented, tubing or drillpipe is commonly used as an inner string to place the cement. This procedure reduces the cementing time and the volume of cement required to pump the plug. It avoids having to drill out the great amount of cement that a large casing would hold if it were cemented in the conventional manner. The technique uses modified float, guiding, or baffle equipment, together with sealing adaptors attached to small-diameter pipe. Cementing through the inner string permits the use of small-diameter cementing plugs. And where the casing is equipped with a backpressure valve or latch-down baffle, as soon as the plug is seated the inner string can be disengaged and withdrawn from the casing while preparations are made to drill deeper.

Outside or Annulus Cementing

Pumping cement through tubing or small-diameter pipe run between casings or between the casing and the hole is a method commonly used on conductor or surface casing to bring the top of the cement to the surface. This is sometimes used for remedial work.

Casing can suffer damage when gas sands become charged with high pressure from surrounding wells. In such instances, the casing may have to be repaired by cementing the annulus between strings through a casing-head connection.

Reverse Circulation Cementing

The reverse circulation cementing technique involves pumping the slurry down the annulus and displacing the mud back up through the casing.[11] It requires that the floating equipment, the differential fillup equipment, and the wellhead assembly all be modified. The method is used when it is not possible to pump the cement slurry in turbulent flow without breaking down the weak zones above the casing shoe. It allows for a wider range in slurry compositions, so that heavier or more retarded cement can be placed at the lower portion of casing and lighter or accelerated cement can be placed at the top of the annulus. A drawback to this method of cementing is that the end of the cement displacement period cannot be detected from pump pressure. This difficulty leads to errors in the calculation of annular volume, required amounts of cement slurry, and volume of mud necessary to achieve complete cement placement. To be certain that the shoe is properly cemented, an excess of at least 300 ft of cement must be tolerated in the casing above the shoe. Caliper surveys should be made before the casing is run to accurately determine the volume of cement plus excess to be used. During cement placement, it is absolutely essential that accurate volumes be known if overplacement is to be minimized.

Delayed-Set Cementing

Delayed-set cementing is a way to obtain a more uniform sheath of cement around the casing than may be possible with conventional methods.[12] It involves placing a retarded cement slurry containing a filtration-control additive in a wellbore before running the casing. The cement is placed by pumping it down the drillpipe and up the annulus. The drillpipe is then removed from the well, and casing or liner is sealed at the bottom and lowered into the unset cement slurry. After the cement slurry is set, the well can be completed using conventional methods.

This technique has been used in tubingless-completion wells by placing the slurry down one string and lowering multiple tubing strings into the unset cement.

When the casing is run into the cement slurry, drilling mud left in the annulus mixes with the cement slurry. While this is not highly desirable, it is better than leaving the drilling mud in the annulus as a channel or mud pocket. The delayed-set cement slurry allows protracted reciprocation of the casing string, which is more likely to assure a uniform cement sheath.

A disadvantage is that the cement slurry requires a somewhat longer WOC time than do conventional slurries. This could be expensive if a drilling rig must be kept on location while the cement sets and gains strength. If the drilling rig can be moved off location and a workover rig employed to complete the well, the cost can be reduced.

Cements used in the delayed-set technique usually contain 6 to 8 percent bentonite and dispersant to control filtration and sufficient retarder to delay setting for 18 to 36 hours. Table 7.9 gives the amounts of retarder required to achieve various fluidity times for delayed-set cements.

7.5 Displacement — The Critical Period*

The predominant cause of cementing failure appears

*See also Chap. 11 — Flow Calculations.

TABLE 7.9 — TYPICAL DELAYED-SET CEMENT SLURRIES

Cement: API Class G.
Water: 10.2 gal/sk.
Bentonite: 8 percent.
Slurry weight: 13.1 lb/gal.

Filtration-Control Agent (percent)	Static Temperature (°F)	Percent Retarder To Achieve Minimum Fluidity Time of			
		12 hours	18 hours	24 hours	30 hours
1.0	110	0.6 (1,090)*	0.8 (1,030)	1.0 (215)	1.2 (N.S.)**
1.0	125	0.6 (1,030)	0.8 (935)	0.8 (420)	1.0 (N.S.)
1.0	140	0.8 (860)	0.8 (1,035)	0.8 (610)	1.0 (N.S.)
1.0	170	0.6 (1,220)	0.8 (1,350)	0.8 —	0.8 —
1.0	200	0.8 (1,315)	0.8 (1,325)	1.0 (1,220)	1.25 (1,125)

*Values in parentheses denote strength (psi) after 3 days of curing.
**N.S. = Not set.

to be channels of gelled mud remaining in the annulus after the cement is in place. If mud channels are eliminated, any number of cementing compositions will provide an effective seal.

In evaluating factors that affect the displacement of mud, it is necessary to consider the flow pattern in an eccentric annulus; that is, where the pipe is closer to one side of the hole than the other. Flow velocity in an eccentric annulus is not uniform and the highest velocity occurs in the side of the hole with the largest clearance, as shown in Fig. 7.19.

If the casing is close to the wall of the hole (as in the right-hand diagram of Fig. 7.20), it may not be possible to pump the cement at a rate high enough to develop uniform flow throughout the entire annulus.

Field studies have shown that the length of time cement moves past a point in the annulus in turbulent flow is important.[5] If a mud channel is put in motion, even though its velocity is much lower than that of cement flowing on the wide side of the annulus, given enough time the mud channel may move above the critical productive zone. Contact time is not important when the cement is in laminar flow because apparently the cement does not exert enough drag stress on the mud to start the mud channel moving.

It is significant that essentially all recently published laboratory and field work credits the same factors with the success of primary cementing during this critical displacement period. There is general agreement that each of the italicized factors in the following statements has a bearing on mud removal.[2,7,13-15]

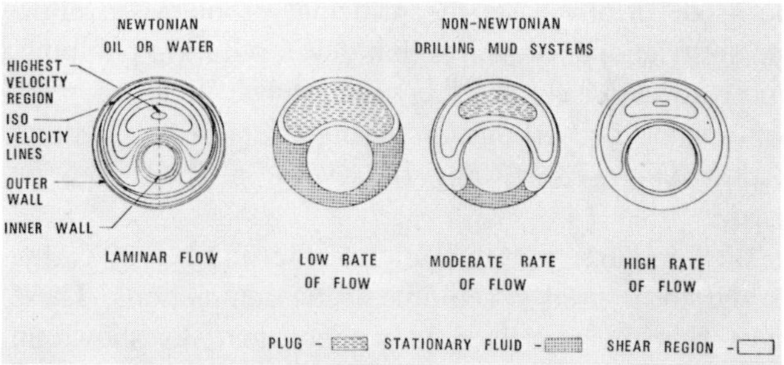

Fig. 7.19 Flow of Newonian and non-Newtonian systems in an eccentric annulus.[15]

1. *Pipe centralization* significantly aids mud displacement.

2. *Pipe movement,* either rotation or reciprocation, is a major driving force for mud removal. Pipe motion with scratchers substantially improves mud displacement where holes are enlarged.

3. *A well conditioned mud* (low plastic viscosity and low yield point) greatly increases mud displacement efficiency.

4. *High displacement rates* promote mud removal. At equal displacement rates a thin cement slurry in turbulent flow is more effective than a thick slurry in laminar flow.

5. *Contact time* (at least 10 minutes) aids in mud removal if cement is in turbulent flow in some part of the annulus.

6. *Buoyant force* due to the density difference between cement and mud is a factor in mud removal,

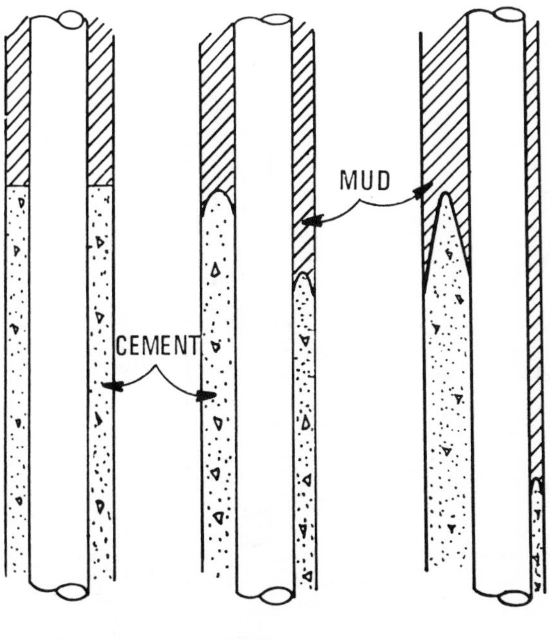

Fig. 7.20 Effect of centralization on uniformity of cement placement.[13]

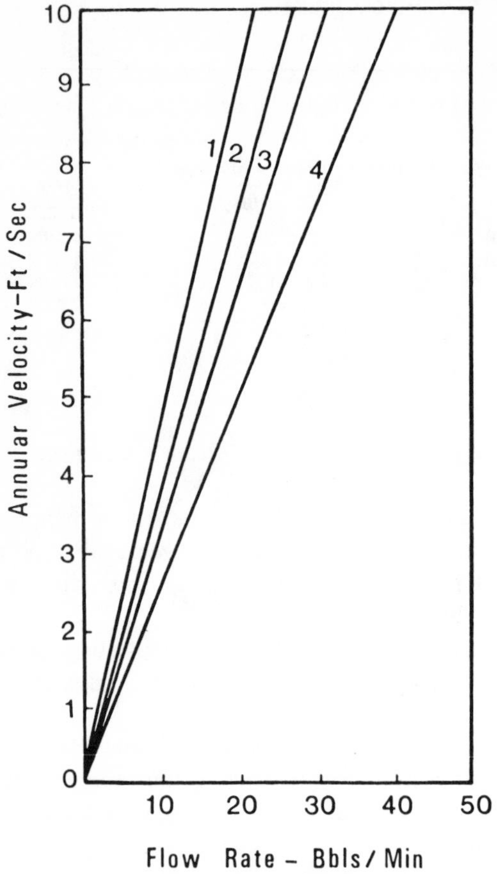

1. 2⅞-in. tubing in 6¾-in. casing.
2. 2⅞-in. tubing in 7⅞-in casing.
3. 2⅞-in. tubing in 8¾-in. casing.
4. 2⅞-in. tubing in 9⅞-in. casing.

Fig. 7.21 Relation between flow rate and fluid velocity in multiple-string cementing.[24]

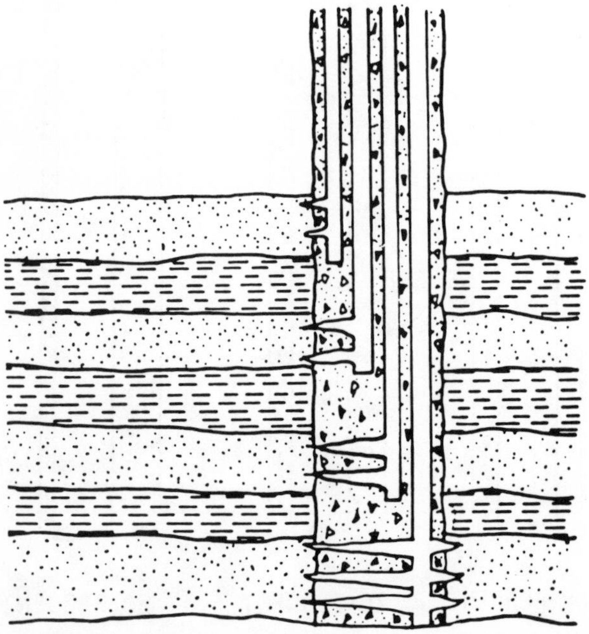

Fig. 7.22 Tubingless completion — four strings of 2⅞-in. casing cemented to varying depths in 9⅝-in. open hole.

though a relatively minor one.

In a given situation it may not be possible, or even necessary, to maximize each factor. To some extent, one may compensate for another. In any case, the most important of them are centralization, pipe movement, and mud conditioning.[16]

7.6 Cementing Multiple Strings

Multiple-casing completions are used when single or conventional completions may not be economically attractive.[17-19] When multiple strings are placed in a well each string is usually run independently, and the longest string is landed first.[20,21] The first string is set in the hanger and usually circulated before the second string is run. In the Gulf Coast area it is the practice to circulate down the first string while running the second string. After the second string is landed in the hanger, it is circulated while the third string is run. In areas where lost circulation is known to be a problem, cement may be placed through the longest casing string. Once the cement fillup has been established, the remainder of the hole is filled with cement slurry through a shorter string.

Centralizers are frequently used, one per joint from 100 ft above to 100 ft below productive zones. Other casing equipment in these small-diameter holes includes landing collars for cement wiper plugs, full-opening guide shoes, and limited-rotating scratchers for single completions. Care must be taken that all float equipment, centralizers, and scratchers will pass the hanger assembly in the casinghead.

Other factors considered in designing the cement slurry are little different from those considered in designing the slurry for a single string of pipe. Normally the cement is pumped down the two longest strings simultaneously, although this is not mandatory. The idle strings may be pressured to 1,000 to 2,000 psi during cementing to safeguard against leakage, thermal buckling, or collapse.

If it seems possible that cement pumped through the longer strings will not come up around the shortest casing string, all the strings may be used to place the cement.

The pumping rate required to achieve a desired annular velocity (Fig. 7.21), the friction losses, the setting depths of the strings, and other factors pertaining to a given job may dictate the cementing procedure to be used. In any case, the principal problems to allow for are frictional pressures that develop during pumping, and restricted clearance between strings. When as many as four or five strings of casing are run to different depths (Fig. 7.22), any of several methods can be used.[20,21]

Well requirements will dictate the number of casing strings to be used in multiple-string completions. These have been standardized to some extent, as shown in Table 7.10.

Following are some general rules for cementing

multiple strings.[22,24]

1. Preflushes or scavenger slurries should be designed to effectively thin and flush the mud ahead of the cement.

2. The cement slurry should usually be designed to the same specification as for other primary cementing jobs in the vicinity.

3. As many strings as practical should be run to the bottom to achieve flexibility for future completions and to aid mud displacement at lower pumping rates.

4. One or more casing strings should be reciprocated during cementing.

5. In small-diameter casing, cement should be placed through at least two casing strings to avoid high friction pressures.

6. Plug-landing collars and latch-in plugs should be used in setting each casing string. Landing collars can be installed 15 to 25 ft below the expected producing interval.

7.7 Cementing Directional Holes

Every well drilled deviates from the theoretical path, forcing the casing string against the wall in many places. In deliberately deviated or directionally drilled holes, this will occur throughout the entire interval. Even with an optimum centralizer program it is probable that the casing will be off center in many parts of the wellbore, creating severe lateral forces in the eccentric annulus. (See Fig. 7.23.) On the upper side, the resistance to flow is less, so the mud displacement is more efficient; but on the lower side, channels of mud are formed.[16,25,26] Another complication is that at some time during cement placement the casing contains a rather heavy cement slurry, the annulus contains a lighter mud, and the total load on the centralizers becomes greater than the simple weight of the casing.[27] After the top plug is placed, loading forces are reversed — that is, the heavier cement slurry is in the annulus and the lighter displacing fluid is in the casing — thus reducing the actual weight of the casing and causing it to float against the upper portion of the hole. The heavier net casing weight causes (1) a more eccentric annulus with static mud on the under side of the casing, (2) greater contact of the casing with the mud cake, and (3) the possibility that, in a fairly soft formation, the centralizer bows will be buried. Any of these conditions will make it more difficult or even impossible for the centralizers to recenter the pipe as the cement slurry moves up the annulus and the net weight is reduced. Although it might seem wise to use a lighter slurry, it would actually be better to use a heavier slurry in conjunction with a much lighter displacing fluid so that the casing can float in the cemented part of the hole and be centralized off the upper side. To achieve this, for example, a 9⅝-in., 40-lb/ft casing would require at least a 16.6 lb/gal cement slurry in the annulus if diesel fuel were used to pump the top plug into place. The mixing and pumping rates after equalization should be as high as possible so that the maximum amount of mud in the annulus can continue to circulate and thus not gel.

Another procedure for cementing in deviated holes is to use a delayed-set slurry. Here buoyancy helps to center the casing, since the heavier slurry is already in the hole when the casing is run; the net casing weight is controlled by filling the casing with a lighter fluid.

TABLE 7.10 — TYPICAL COMBINATIONS OF CASING AND HOLE SIZE IN THE U.S.

Type of Completion	Size of Surface Casing (in.)	Size of Hole Below Surface Casing (in.)	Size of Production Casing (in.)
Single	7 to 8⅝	6¼ to 7⅞	2 or 2½
Dual	8⅝	7⅞	2 or 2½
Dual offshore*	—	11½	3½
Triple	9⅝	8⅝	2 or 2½
Quadruple	10¾	9⅝	2 or 2½

*Deviated wells.

7.8 Gas Leakage After Cementing

Gas communication through a cemented annulus (Fig. 7.24) was first recognized in gas storage wells. It was found that even when the leakage rate is small, excessive pressure can build up if the annulus is not vented.[16,28] In some cases, gas leakage rate is great enough to warrant connecting the annulus to a gathering line. In deep holes, leakage resulting from dehydration, gelling, or bridging of cement or mud in the annulus can cause a pressure buildup behind the production and intermediate casings or behind the liners.

The following paragraphs describe various conditions that are directly related to gas migration in a wellbore.

Ineffective Hydrostatic Head

The density of wellbore fluids (cement, mud, etc.) determines the hydrostatic pressure exerted at any particular depth. In completing gas wells, the hydrostatic pressure in the annulus must always exceed the formation pressure to prevent gas from entering the annulus.

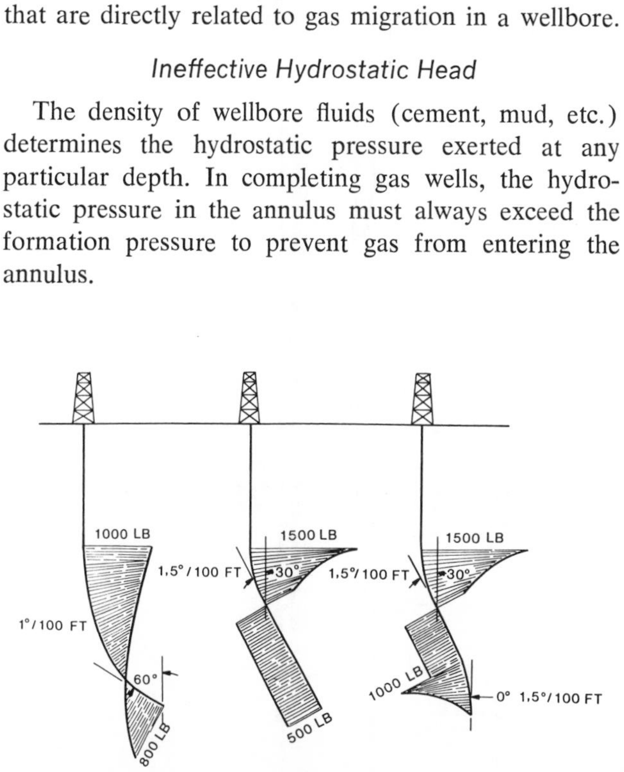

Fig. 7.23 Distribution of lateral forces in directionally drilled holes.[27]

One cubic foot of gas migrating to the surface from a depth of 12,000 ft expands to more than 350 cu ft.

Before it is cemented, a gas well should be circulated long enough to condition the mud and remove any gas bubbles trapped in the annulus. After circulation and before cementing, the pumps should be shut down briefly and then the hole should be circulated again. This is to determine if the mud is free of microscopic gas bubbles that may have been trapped in the washout areas or may have adhered to the walls of the hole. Such gas is a potential second gas kick, and if it is not removed before the cement is placed it may lower the density of the fluid column. During cementing, the cement slurry should remain as heavy as possible to resist being cut by gas.

Cement Dehydration

Controlling cement dehydration — a primary contributor to gas migration in a wellbore — depends principally upon controlling differential pressure and cement filtration rate.[29]

For cement dehydration to occur in a wellbore and permit gas leakage there must first be a permeable formation above the gas interval. Cement particles (filter cake) bridging against this permeable zone will begin to support the cement column above the zone and reduce the effective hydrostatic pressure. After the pressure becomes equal to or less than the pressure of the highest-pressured gas reservoir, it takes only a short time for gas to start entering the wellbore, since the hydrostatic head has been reduced.

The gas, migrating up the unset cement column, further lowers the hydrostatic pressure, which in turn increases the rate of gas channeling. If, when the gas reaches the point up the hole where the dehydrated cement forms the first bridge, the cement is not solidly set, the accumulating gas can build up to a pressure that will cause channeling either through the weak cement column or into the permeable zone on which the bridge, or filter cake, is built. If the gas enters and pressures the permeable zone and the cement is not set or bonded in the early state, the gas can bypass the bridge and re-enter the unset cement column at the top of the permeable zone. The result can be total gas cutting or channeling in the annulus, which will show up at the surface.

Fig. 7.25A, a Scanning Electron Microscope photograph, shows the pore space in a dehydrated cement filter cake immediately after it loses water or filtrate. This cake has permeability for a short time until it hydrates more completely into an impermeable seal. If a filtration-control material is added to the slurry (Fig. 7.25B), it acts as a bridging agent in the pore space of the cement; thus water is retained until the cement sets without going through a change in permeability.

Bridging

Bridging of particles during cementing may restrict

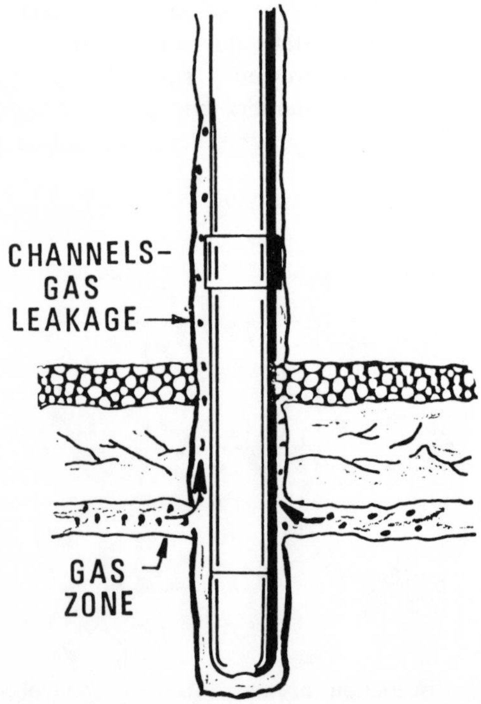

Fig. 7.24 Gas migration in a cemented annulus.[28]

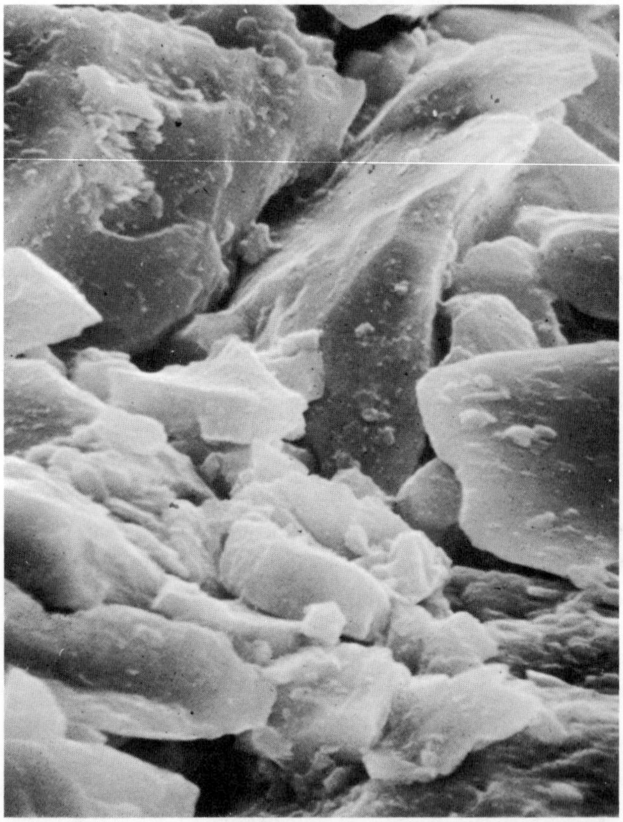

Fig. 7.25A Scanning Electron Microscope photograph of neat cement slurry. Cement slurry, once it dehydrates, goes through a transition stage having some permeability before setting. (Magnification 5.0 × 10^3.)[29]

the effective hydrostatic pressure.[29,30] The bridging can occur anywhere in the wellbore, and in a liner job it can occur at the top of the hanger. Bridging can be attributed primarily to sloughing formations or to mud and cement filter cake.

Annulus bridging up the hole contributes to gas leakage in much the same way that dehydration does. If bridging occurs during primary cementing before displacement is complete, there is an increased possibility of losing returns and, in turn, of losing hydrostatic pressure. Such bridging can also leave cement inside the casing string that must be drilled out. (Fig. 7.26 shows some effects of dehydration and bridging on primary cementing.)

Gelation

Gelation at the cement and drilling mud interface up the hole is another factor to consider with regard to lowering the hydrostatic head to prevent leakage. When cement is mixed with water, a chemical reaction is started and the slurry changes from a pumpable composition to a set material. During this change, gelation, or a significant increase in viscosity, may occur. The length of time the cement remains a gel depends upon temperature, cementing composition, pressure, and water/cement ratio.

Recognizing that a potential leakage problem exists in any gas well completion, it is important to prevent gas leakage during primary cementing. The three factors most important to control at all times are fluid column density, cement filtration, and cement setting speed.

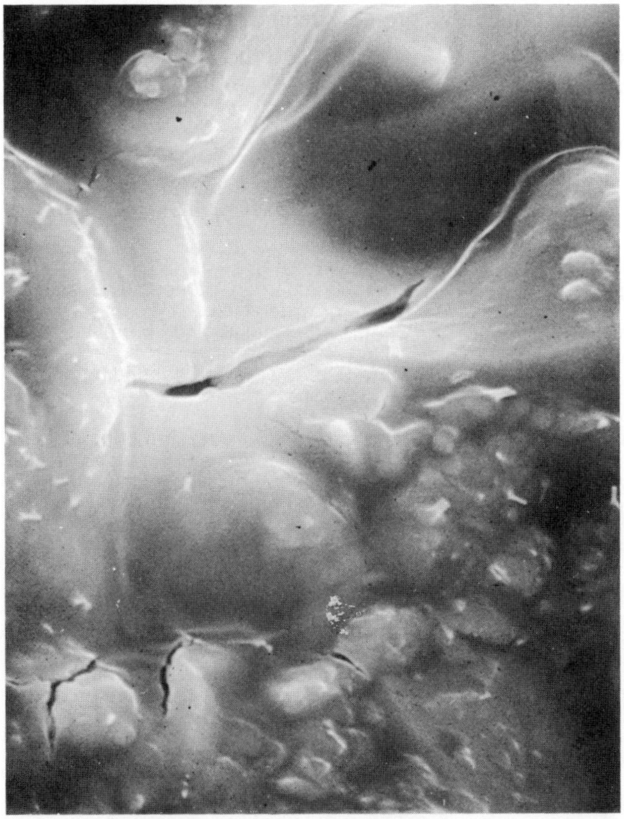

Fig. 7.25B Scanning Electron Microscope photograph of cement slurry with fluid-loss additive, which gives the slurry a very low permeability during the transition from a slurry to a solid. (Magnification 4.8×10^3.)[29]

7.9 Cementing Through Soluble Formations

In setting casing through salt formations, it is imperative to obtain a good bond to the soluble zones.[31] This fact was first recognized along the Gulf Coast and in the Williston Basin area of North Dakota and Montana. Casing failures (collapse) through salt stringers prompted a number of studies dealing with the design of casing subjected to salt loading.[32] Salt deforms easily even under low confining pressures. In triaxial loading tests, samples have tolerated strains in excess of 30 percent without fracturing. In cementing through salt or potash deposits, the casing must withstand these plastic flow forces if collapsing is to be avoided. The forces can approach the overburden stress of 1 psi per foot of depth.

An open, unconfined wellbore through salt zones at depths greater than 3,000 ft will tend to close unless properly stabilized with cement.

The pressure on casing subjected to uniform loading by salt can increase until it eventually equals the overburden pressure. Casing grade and weight can be specified for uniform salt pressure in the same manner as for fluid pressure, even though modes of casing failure may be different.

For uniform loading of casing, it is necessary for the cement to fill the annulus completely and uniformly. If the casing is too weak to withstand the uniform salt pressure, it is helpful either to cement a liner inside the casing or to apply internal pressure to the casing. If the cement job is faulty, the casing may be subjected to nonuniform loading, which can bend or collapse the pipe, as shown in Fig. 7.27. Internal pressure is of little help in resisting initial failure from this type of loading, but it may prove of value in preventing complete collapse. Although a liner increases the load-carrying capacity, it may not lend enough strength to withstand the most severe cases of nonuniform loading.[32]

Oil-base or salt-saturated drilling fluids, which minimize washouts, should be used in addition to good

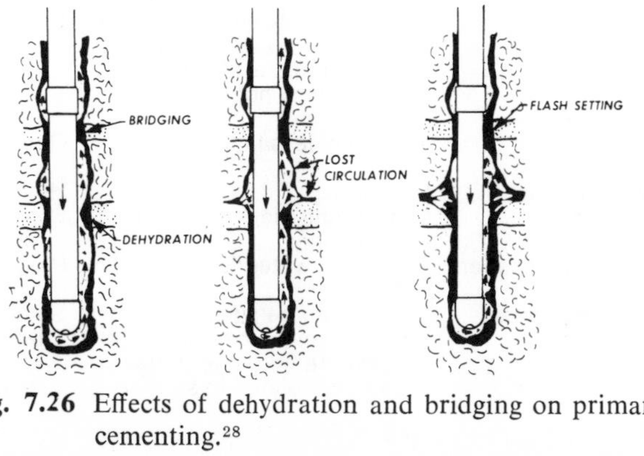

Fig. 7.26 Effects of dehydration and bridging on primary cementing.[28]

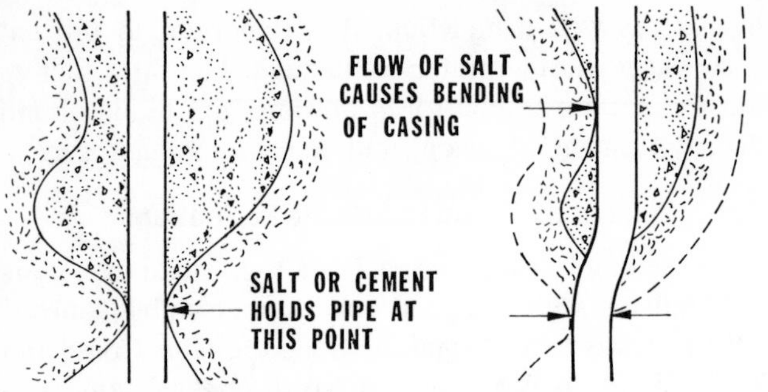
FLOW OF SALT CAUSES BENDING OF CASING
SALT OR CEMENT HOLDS PIPE AT THIS POINT

Fig. 7.27 Bending of casing by salt flow.[33]

cementing practices to prevent casing failures in salt zones.

7.10 Considerations After Cementing

To prevent casing expansion and improve bonding, internal casing pressure should be released when it has been established that the float-collar and guide-shoe backpressure valves are holding. Early practice was to hold pressure on the casing until the cement set. However, research on bonding and bond logging has shown that this can create a microannulus between the casing and the cement, and thus provide a potential avenue of communication. In small-diameter casing (used in tubingless competions), it may be desirable to hold pressure on the casing. This creates additional tension to prevent the casing from buckling before the cement takes its initial set. Bonding may be adversely affected, but expansion due to differential pressure is minor in small-diameter pipe.

Waiting-on-cement time is variable, and state and federal regulations should be checked in local operating areas. Where no rule exists, a reasonable WOC time should permit cement to attain sufficient strength to anchor the pipe and withstand the shock of subsequent operations, to seal permeable zones, and to confine fracture pressures. If densified cements (API Class A, G, or H) and 2 to 3 percent calcium chloride are used, WOC time can be as short as 4 to 6 hours in warm weather and 6 to 8 hours in cold weather. The WOC time depends on the class of cement, the additives in the cement, the time required for placement, and well temperatures and pressures.

Before drilling out, the minimum desirable compressive strength is 500 psi. The recommended time to run a cement bond log and perforate for production or stimulation is 24 to 72 hours, or after the cement achieves a compressive strength of 2,000 psi.

Some states require a pressure test of the casing (Texas) or a production test (California) to define by the GOR or the water content whether a cementing job is a success or a failure.

Testing Primary Cementing Job Performance

A temperature survey is an excellent method for locating cementing tops. A temperature log run 6 to 12 hours after the plug is bumped identifies the top of the cement by noting anomalies in the temperature gradient. Where cement is present behind the pipe more heat is liberated because of the exothermic reaction from the cement setting process. When the cement top is

TABLE 7.11 — DIGEST OF CONDITIONS AND RECOMMENDED PRACTICE INVOLVED IN PRIMARY CEMENTING

	Conductor Pipe	Surface Pipe	Intermediate and/or Production String
Casing size, in.	20 to 30	7 to 20	4½ to 11¾
Setting depth, ft	30 to 1,500	40 to 4,500	1,000 to 15,000
Hole conditions	Probably enlarged	Probably enlarged	Probably enlarged (particularly in salt)
Mud	Native	Native	Native or water-base
Mud properties	Viscous, thick cake	Viscous, thick cake	Controlled viscosity and controlled fluid loss
Cement pumped to	Surface	Surface	Surface pipe or lower, depending on conditions
Cement, API Class	A, G, H	A, G, H	A, C, G, H
Additives normally used with cement	2 to 3 percent CaCl	Bentonite or pozzolan	High gel, filter, or pozzolan bentonite; dispersant + retarder if needed; salt, for cementing through salt sections
Tail-in slurry	Like initial slurry, densified (ready-mix concrete may be dumped into annulus)	Densified for high strength (deep well may use high-strength slurry for entire job)	Densified for high strength over lower 500 to 1,000 ft
Technique	Through drillpipe using small plugs and sealing sleeve, or down casing with large plugs, or into annulus; float collar may or may not be used.	Same as for conductor pipe. Centralize lower casing.	Down casing with plugs (top and bottom), or in stages, depending on fracture gradient. If string is very heavy, it may be set on bottom and cemented through ports. Use float collar and guide shoe; centralize in critical areas.
Placement time	Generally less than 30 min.	Generally less than 45 min.	Variable, depending on cement volume—45 min. to 2½ hours.
Placement rate	Low or high	High	High
WOC time	6 to 8 hours, depending on regulations	6 to 12 hours, depending on regulations	6 to 12 hours, depending on regulations
Mud/cement spacers	Plugs and water flush	Plugs and water or thin cement	Plugs and thin cement, or spacer compatible with mud
Cementing hazards	Casing can be pumped out of hole; cement may fall back down the hole after it has been circulated to surface.	Same as for conductor. Lower joints may be lost down the hole with deeper drilling; casing can easily stick.	There may be both weak and high-pressure zones, requiring variable-weight cement slurries. Prolonged drilling may damage casing. Wells may be hot, necessitating measurement of bottom-hole temperatures.

not at the desired height, it can usually be assumed that channeling has occurred, provided the correct volume of slurry has been used.

Cement bond logs are also frequently used to ascertain how successfully the cement has been placed between the pipe and the formation. A good bond indicates a satisfactory cement job; a poor bond indicates a questionable job.

A radioactive tracer survey may be used to identify cement tops but requires the addition of a radioactive material to the cement slurry to allow detection behind the pipe. The procedure is not widely used because it requires special preparation and planning.

7.11 Summary

Table 7.11 summarizes recommended practices related to primary cementing. Flow calculations for example primary cementing jobs appear in Appendix D.

References

1. Howard, G. C., and Clark, J. B.: "Factors To Be Considered in Obtaining Proper Cementing of Casing," *Drill. and Prod. Prac.*, API (1948) 257-272.
2. Goins, W. C., Jr.: "Selected Items of Interest in Drilling Technology," *J. Pet. Tech.* (July 1971) 857-862.
3. "Recommended Practice for Testing Oil Well Cements and Cement Additives," *API RP 10B*, 20th ed., API Div. of Production, Dallas (1974).
4. Owsley, W. D.: "Improved Casing Cementing Practices in the United States," *Oil and Gas J.* (Dec. 15, 1949) 76.
5. Brice, J. W., Jr., and Holmes, R. C.: "Engineered Casing Cementing Programs Using Turbulent Flow Techniques," *J. Pet. Tech.* (May 1964) 503-508.
6. Slagle, K. A.: "Rheological Design of Cementing Operations," *J. Pet. Tech.* (March 1962) 323-328; *Trans.*, AIME, **225.**
7. Clark, C. R., and Carter, L. G.: "Mud Displacement With Cement Slurries," *J. Pet. Tech.* (July 1973) 775-783.
8. Parker, P. N., Ladd, B. J., Ross, W. N., and Wahl, W. W.: "An Evaluation of a Primary Cementing Technique Using Low Displacement Rates," paper SPE 1234 presented at SPE-AIME 40th Annual Fall Meeting, Denver, Colo., Oct. 3-6, 1965.
9. Teplitz, A. J., and Hassebroek, W. E.: "An Investigation of Oil Well Cementing," "*Drill. and Prod. Prac.*, API (1946) 76-101; *Pet. Eng. Annual* (1946) 444.
10. Jones, P. H., and Berdine, D.: "Oil-Well Cementing — Factors Influencing Bond Between Cement and Formation,"' *Oil and Gas J.* (March 21, 1940) 71; *Petroleum World* (June 1940) 26; *Drill. and Prod. Prac.*, API (1940) 45-63.
11. Marquaire, R. R., and Brisac, J.: "Primary Cementing by Reverse Circulation Solves Critical Problem in the North Hassi-Messaoud Field, Algeria," *J. Pet. Tech.* (Feb. 1966) 146-150.
12. Underwood, D., Broussard, P., and Walker, W.: "Long Life Cementing Slurries," paper presented at API Southwestern Dist. Div. of Production Spring Meeting, Dallas, March 10-12, 1965.
13. McLean, R. H., Manry, C. W., and Whitaker, W. W.: "Displacement Mechanics in Primary Cementing," *J. Pet. Tech.* (Feb. 1967) 251-260.
14. Graham, H. L.: "Rheology-Balanced Cementing Improves Primary Success," *Oil and Gas J.* (Dec. 18, 1972) 53.
15. Piercy, N. A. V., Hooper, M. S., and Winney, H. F.: "Viscous Flow Through Pipes With Cores," *Phil. Mag.* (1933) **15,** No. 99, 674.
16. Carter, L. G., Cook, C., and Snelson, L.: "Cementing Research in Directional Gas Well Completions," paper SPE 4313 presented at SPE-AIME European Spring Meeting, London, April 2-3, 1973.
17. Tausch, G. H., and McDonald, P.: "Permanent-Type Completions and Wireline Workovers," *Pet. Eng.* (Sept. 1956) B39.
18. Corley, C. B., Jr., and Rike, J. L.: "Tubingless Completions," *Drill. and Prod. Prac.*, API (1959) 7.
19. Huber, T. A., and Corley, C. B., Jr.: "Permanent-Type Multiple Tubingless Completions," *Pet. Eng.* (Feb. and March 1961).
20. Enloe, J. R.: "Amerada Finds Using Multiple Casing Strings Can Cut Costs," *Oil and Gas J.* (June 12, 1967) 76-78.
21. Rike, J. L., and McGlamery, R. G.: "Recent Innovations in Offshore Completion and Workover Systems," *J. Pet. Tech.* (Jan. 1970) 17-24.
22. Childers, M. A.: "Primary Cementing of Multiple Casing," *J. Pet. Tech.* (July 1968) 751-762; *Trans.*, AIME, **243.**
23. Frank, W. J., Jr.: "Improved Concentric Workover Techniques," *J. Pet. Tech.* (April 1969) 401-408.
24. Buster, J. L.: "Cementing Multiple Tubingless Completions," *Oil and Gas J.*, Part 1 (June 8, 1964) 121-125; Part 2 (June 15, 1964) 89-91.
25. *A Primer of Oil Well Drilling*, 3rd ed., Pet. Extension Service, U. of Texas, Austin (1957).
26. Hoch, R. S.: "Cementing Techniques Used for High Angle, S-Type Directional Wells," *Oil and Gas J.* (June 22, 1970) 88-93.
27. "Stresses on a Centralizer," Weatherford Oil Tool Co., Houston (1974).
28. Carter, L. G., and Slagle, K. A.: "A Study of Completion Practices To Minimize Gas Communication," *J. Pet. Tech.* (Sept. 1972) 1170-1174.
29. Christian, W. W., Chatterji, J., and Ostroot, G. W.: "Gas Leakage in Primary Cementing — A Field Study and Laboratory Investigation," paper SPE 5517, presented at SPE-AIME 50th Annual Fall Technical Conference and Exhibition, Dallas, Sept. 28-Oct. 1, 1975.
30. Gibbs, M. A.: "Delaware Basin Cementing — Problems and Solutions," *J. Pet. Tech.* (Oct. 1966) 1281-1285.
31. Slagle, K. A., and Smith, D. K.: "Salt Cement for Shale and Bentonitic Sands," *J. Pet. Tech.* (Feb. 1963) 187-194; *Trans.*, AIME, **228.**
32. Cheatham, J. B., Jr., and McEver, J. W.: "Behavior of Casing Subjected to Salt Loading," *J. Pet. Tech.* (Sept. 1964) 1069-1076.
33. Clegg, J. D.: "Casing Failure Study—Cedar Creek Anticline," *J. Pet. Tech.* (June 1971) 676-684.

Chapter 8

Deep-Well Cementing

8.1 Introduction

The technology for cementing deep wells has advanced greatly since 1965. Operational conditions once considered impossible or difficult are now dealt with as a matter of course. There are a great many wells deeper than 15,000 ft, and those with bottom-hole temperatures above 230°F should always be considered critical.[1-4] (Fig. 8.1 indicates the deep drilling conducted in the U. S. from 1971 through 1974.)

Most deep wells are started with 20- to 30-in. conductor or surface casing and are completed with 5-, 5½-, or 7-in. liners.[1,5,6] (For examples, see Figs. 8.2, 8.3A, and 8.3B.) In some deep holes, two liners (intermediate and production) are necessary to reach the ultimate drilling objective. Tie-back strings to surface are frequently used in West Texas to stabilize and reinforce the intermediate liner, which may be weakened from drilling. In a few areas a full casing string is used in preference to liners.

8.2 Cementing Considerations for Deep Wells

The basic procedures used in cementing deep wells (casing or liners) are generally the same as those used for shallower wells. However, the hole conditions and the working conditions encountered in deep wells are more critical and require greater emphasis on casing and slurry design. These conditions, which change with

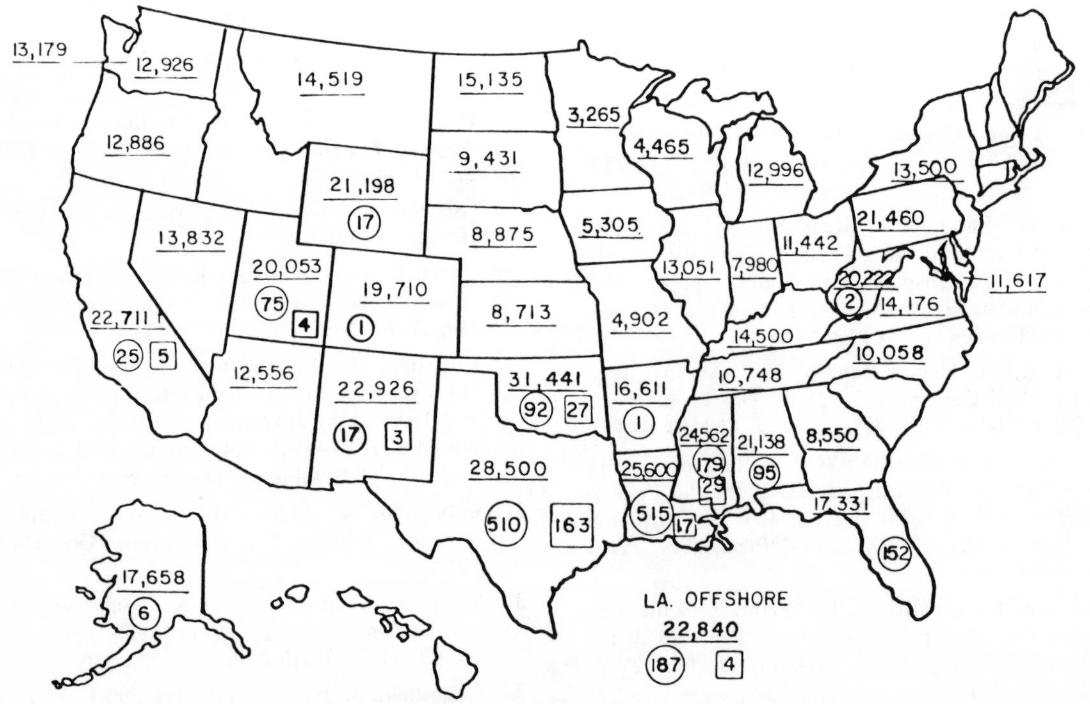

○ Wells deeper than 15,000 ft. Total (1971-1974), 1,879.
□ Wells deeper than 20,000 ft. Total (1971-1974), 248.
— Depth of deepest well in each state to Jan. 1, 1974.

Fig. 8.1 Deep wells drilled in U. S., 1971 through 1974.[4]

depth, include the following:

1. Higher temperatures, overpressured zones, and corrosive well fluids in many areas.
2. Increased casing length, reduced annular clearances, and difficulties in moving pipe during cementing.
3. Greater mechanical loads on the casing string and drilling rig.
4. Longer time intervals for tripping bits and running casing before cementing.
5. Heavier mud systems.
6. Increasing difficulty in effectively sealing the top of the liner to prevent gas leakage.

In planning a cementing job for a deep well, the following should be adhered to.

1. The cement slurry should be designed to allow adequate placement time. Most deep wells require a slurry with a 3- to 4-hour pumpability. Rig water, cements, and additives to be used on the job should always be pretested.

2. The best technique should be determined for displacing the mud with cement slurry. Where possible, the mud should have a low yield point and a low plastic viscosity. The drilling mud should be tested for its compatibility with whatever spacer is to be used between it and the cement during placement.

3. Optimum slurry properties (weight, viscosity, fluid-loss control, etc.) should be attained during the mixing process.

4. The setting properties of the cement should be designed to resist gas leakage, loss of strength, and corrosive environments encountered at high temperatures.

Restricted clearances (Table 8.1) and greater lengths of open hole in deep wells complicate mud removal. Cements tend to channel through the mud and follow the path of least resistance (see Fig. 7.18).[7] An increase in mud density can cause an increase in pressure differentials, tending to force the casing into the mud cake. This may result in differential sticking or frictional drag, should the casing remain stationary for a period of time.

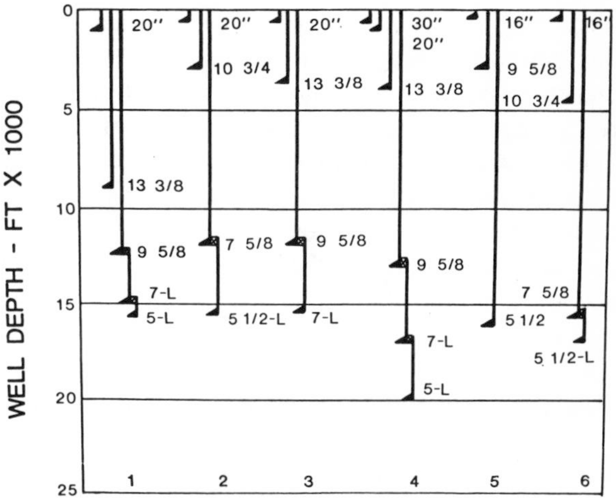

1. South Texas Wilcox — 15,000 ft.
2. Upper Texas Wilcox — 15,000 ft.
3. South Louisiana — 15,500 to 20,000 ft.
4. Offshore Louisiana — 20,000 ft.
5. Mississippi Smackover — 16,000+ ft.
6. Mississippi — 17,000+ ft.

Fig. 8.3A Casing programs for Texas, Louisiana, and Mississippi Gulf Coast wells.[1]

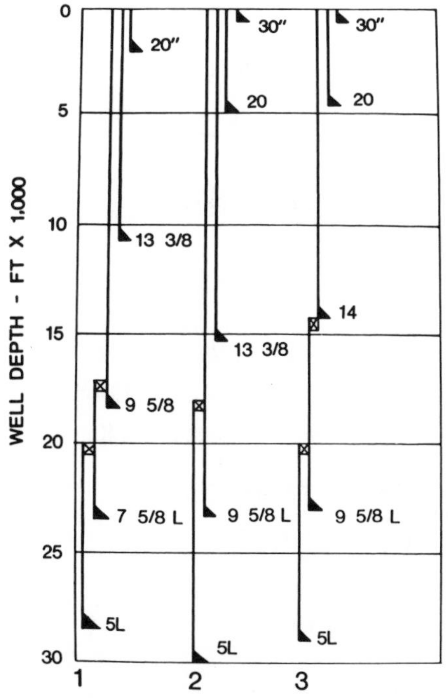

1. Ralph Lowe Estate Well (West Texas) — 28,500 ft.
2. Lone Star Baden Unit Well (Oklahoma) — 30,050 ft.
3. Lone Star Bertha Rogers No. 1 Well (Oklahoma) — 31,441 ft.

Fig. 8.2 Casing programs for some record wells (after Ref. 6).

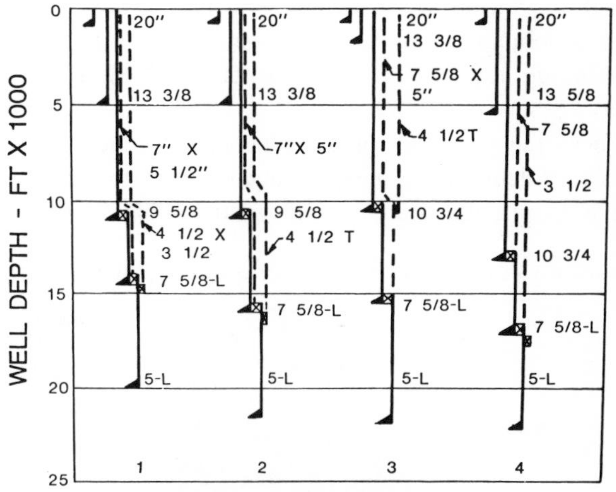

1. Ward County — 20,000 ft.
2. Reeves County — 21,000 ft.
3. Pecos County — 22,000 ft.
4. Winkler County — 22,500 ft.

Fig. 8.3B Casing programs for West Texas Delaware Basin wells.[1]

Where the hole is crooked, tool joints may wear a groove in the wall, and the casing, if forced into the groove, tends to stick.

Mechanical Aids

When casing is being run to the cementing depth, the use of centralizers and scratchers and high displacement rates provides the needed hole conditioning by allowing the maximum amount of annular mud to move into the fluid stream. Centralizers should be used on long casing strings if hole conditions are good and sticking is not a problem. In deep wells, scratchers should be the reciprocating type, although rotating scratchers have been used successfully in deviated holes at depths in excess of 15,000 ft. (See Section 6.6.)

Scratchers in deep wells are subjected to abuse and wear before reaching total depth. They should be designed to promote good flow characteristics to minimize plugging. Filter cake loosened from exposed permeable sections combines with gelled mud and increases the flow resistance of the mud. This can lead to formation breakdown and loss of returns. The decision to use centralizers or scratchers in deep wells is usually based on hole conditions and experience in the area.

Hole Conditioning

Any deep well should be circulated until the casing volume has been displaced.[8] Where practical, circulation should be continued until the return mud viscosity (plastic viscosity/yield point) has stabilized, indicating that most of the loosened filter cake and gelled mud has been displaced.

Prolonged circulation is particularly important before cementing gas wells. Gas trapped where holes are enlarged cuts the cement during setting and hydration.[7,9] (For more on gas leakage, see Section 7.8.)

Cement Displacement

High displacement rates and rapid mud removal are generally obtainable by using low-gel-strength mud and by thinning the cement with dispersants.[8] (See also Section 3.8.)

Where the geometry of the pipe and borehole prohibits turbulent rates, another effective way to remove mud is to put the cement in plug flow. Plug flow depends on the same criteria as turbulent flow, except that a different Reynolds number is used (see Chap. 11, Table 11.3). The one factor that most benefits any cementing operation is movement.[10] Any movement of the pipe during cementing will break up mud channels or pockets and promote a more homogeneous sheath of cement around the pipe.

TABLE 8.1 — TYPICAL HOLE/LINER RELATIONSHIPS

Size of Liner (in.)	Size of Hole (in.)	Size of Last Casing (in.)	Annular Area (sq in.)	Cement Sheath Thickness (in.)
9⅝	10⅝	11¾	15.9	7/16
7¾	9½	10¾	13.0	⅞
7⅝	9½	10¾	14.5	15/16
7¾	8½	9⅝	9.6	⅜
7⅝	8½	9⅝	10.4	7/16
7	8⅝	9⅝	19.9	13/16
5½	6⅝	7⅝	10.7	9/16
5	6⅝	7⅝	14.9	13/16
5	6⅛	7	9.9	9/16
4½	6⅛	7	13.6	13/16
3½	4¾	5½	8.1	⅝

8.3 Use of Liners in Deep Wells

Liners are widely used in deep wells to case off a section of open hole and thus eliminate a full string of casing.[11-13] In almost every deep-well completion in the U.S. today, liners are used to (1) case off the open hole to enable deeper drilling, (2) control water or gas production or hold back unconsolidated or sloughing formations, and (3) case off zones of lost circulation or zones of high pressure encountered during drilling.[12,14,15] (Fig. 8.4 shows a typical deep-well liner completion and Table 8.2 lists various types and uses of liners.)

Once the liner has been set, cement is circulated down the drillpipe, out the liner shoe, and up the outside of the liner. Plugs are used to prevent mixing of mud and cement in the liner just as they are in the full casing string during primary cementing. At the completion of the cementing operation, the excess

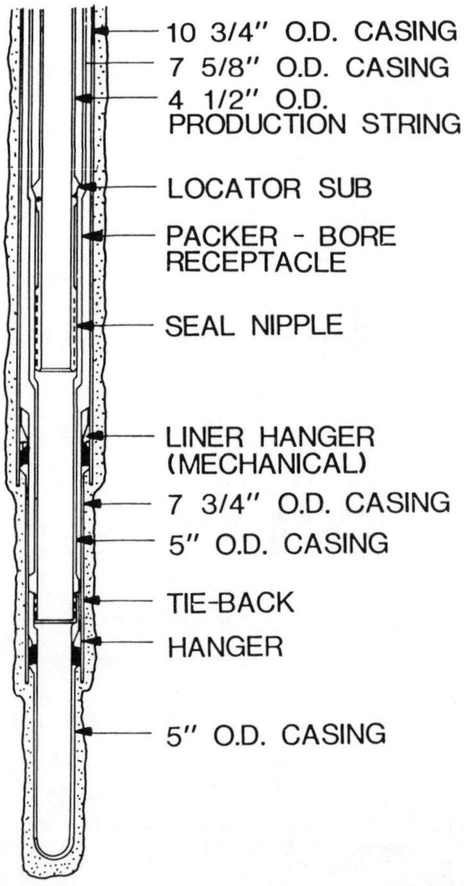

Fig. 8.4 Typical deep-well liner completion. (West Texas Gomez gas well with maximum bore from surface to total depth.)[18]

TABLE 8.2 — USES, CHARACTERISTICS, AND HANDLING OF VARIOUS TYPES OF LINERS

	Production Liner	Intermediate Liner	Scab Liner	Tie-Back Liner (Stub Liner)
Purpose	To serve as completion string.	To extend intermediate casing.	To repair damaged or parted casing.	To extend the lower liner into intermediate casing.
Advantages	Requires less casing. Permits larger tubing for greater flow capacity.	Cases hole to provide for change in mud density. Permits running tieback string later. Requires less casing.	Only a short section is needed.	Covers damaged casing.
Characteristics	May cover long area. Has small annular clearance. Is not moved during cementing. Restricts pumping rates. Requires careful control of cement thickening time to allow placement, yet set at liner top.	Covers long areas of hole. Is set to control gas. Usually requires heavy mud and cement (12 to 14 lb/gal).	Usually covers short section. Hung before being cemented. Not set in receptacle.	Liner hanger cannot be set before cementing. Is usually a short liner.
Cementing procedure	Liner hanger set before cementing. Cement circulated above liner. Liner tool removed from hole. Excess cement either reversed out through drillpipe or allowed to set and is drilled out later.	Same as for production liner.	Same as for production liner.	Hanger is not set before cementing; otherwise, same as for production liner.
Type of cement used	Depends upon hole conditions and mud densities; may be a combination slurry. Light-weight cement for density control and high fillup. Tailed out with API Class G or H, densified for high strength. Filtration control desirable in gas completions.	High-density, low water ratio, depending on mud. API Class G or H plus dispersant. (See production liner.)	High-density, low-water-ratio slurries for high strength. API Class G or H.	High-density, low-water-ratio slurries for high strength. API Class G or H plus dispersant.

cement is reversed out, or it is allowed to set above the liner and is later drilled out. (See Fig. 8.5.)

Tie-back liners, or shorter scab liners, or both, are run and cemented to protect the intermediate casing by (1) reinforcing the intermediate casing worn by drilling, (2) providing greater resistance to collapse stress from abnormal pressures, (3) providing protection from corrosion, and (4) sealing an existing liner that may be leaking gas.[16-18]

The tie-back liner may or may not extend to the surface, but does cover the top of an existing liner. When it extends from the top of a production liner to a point some distance up the hole, it is called a "stub" liner. This short liner may be set with its entire weight on the production liner, or it may be hung up the hole. When either a stub liner or a scab liner is set as a tie-back section, it connects to the liner in the well through a 3- to 6-ft-long polished or honed receptacle. It usually has a setting thread for holding and releasing the liner to permit insertion of a tie-back sealing nipple before, during, or after placement of the cement slurry around the tie-back liner. The diameter of the tie-back sealing nipple should not be less than that of the liner set below it.

8.4 Equipment Used in Hanging Liners

The liner hanger supports the suspended section of casing in tension to prevent pipe buckling until the cement sets. Most liner hangers, except those used to hang liners less than 200 ft long, are set with either a mechanical or a hydraulic device (Fig. 8.6). Both types of hangers contain slips that wedge against the casing wall to support the liner. When cementing is completed, the liner hanger tool is released from the liner and removed from the well. The selection of the

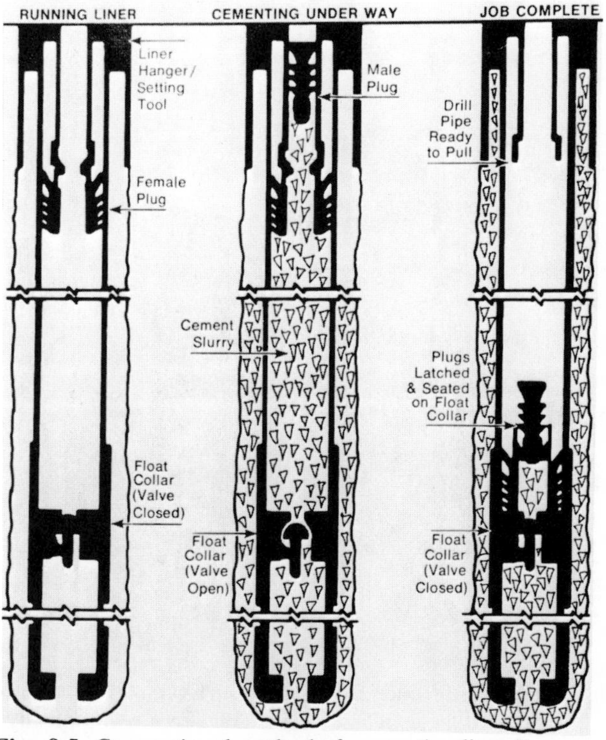

Fig. 8.5 Conventional method of cementing liner.[31]

proper liner hanger is governed by liner weight, annular dimensions, and bypass requirements, since each affects the hanger. For longer liners, there are heavy-duty hangers capable of carrying the liner load without severely restricting the flow past the hanger.[19]

The hydraulically set liner hanger, depending on the design, may be set by using a ball or a plug or by rotating the drillpipe. (See Examples 4 and 5 of Fig. 8.6.) The hydraulic pressure required to set the slips of the liner hanger is that pressure required to overcome the spring force or shear-pin load of the slip-actuating piston.

With the spring-tension type of hanger, after the liner is set, the hanger can be returned to the unset position by picking up the drillpipe. This allows the slips to snap back into the running-in position. Hydraulically set hangers eliminate problems presented by rotation or reciprocation but are relatively new and have not been so widely used as the mechanically set tools.

Mechanically set hangers are set by surface manipulation; i.e., rotating the pipe to disengage a J-slot and slacking off weight to engage the slips. A liner swivel below the hanger permits the setting string to be rotated counterclockwise without turning the liner.

Some mechanical liner hangers are set by vertical or reciprocal operation of their working parts. While the hanger is going in the hole, the slips are held away from the cone by a set of steel fingers latched under a shoulder in the tool. As the liner is run, care must be taken to limit the distance travelled while the rotary slips are removed, or the liner hanger will be set prematurely. Vertical drillpipe movement should be limited to less travel than is required to set the liner hanger. If the liner hanger is set while it is being run in the well, the drillpipe liner must be picked up and rotated to get the J-slot back in the running-in position.

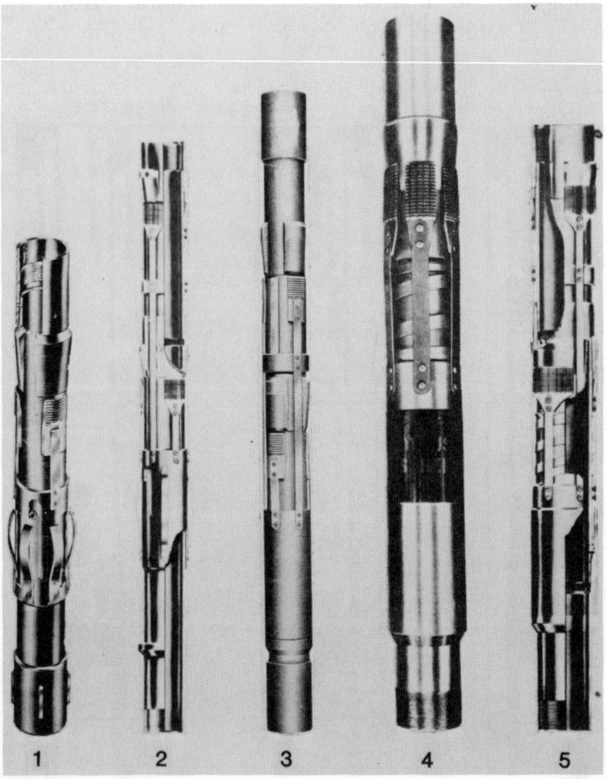

Fig. 8.6 Types of liner hanging tools. (Examples 1, 2, and 3 are mechanically set; 4 and 5 are hydraulically set.)[12,15,18,19]

There are mechanically set liner hangers that allow rotation during cementing. A "kelly" device permits a portion of the setting tool to be engaged during rotation, and a clutch prevents premature release from the liner. After the hanger is set and the clutch is engaged, further rotation to the right releases the setting tool from the liner.

Liner hanging and cementing practices may vary with specific well conditions and operator's requirements. There is a trend toward moving liners while cementing. Although it is common practice to hang the liner when it reaches bottom and allow it to remain stationary during the cementing operation, it is not the preferred method.

Of the two types of liner hangers, the hydraulically set type is preferred if a liner is already in the well. Hydraulically set hangers are designed to prevent breakage of the friction wiper springs and premature setting of the slips in entering the top of the liner.

8.5 Liner Cementing Through Fractured Formations

It is quite common to set and cement 6,000- to 8,000-ft liners in most deep-well drilling areas.[20-23] Any cementing composition that must cover this distance must be adequately retarded to be placed at the higher bottom-hole temperature yet develop reasonable strength at the lower temperatures at the liner top. It is not uncommon for this temperature difference to exceed 100°F. Overretardation of cement slurries to achieve excessive thickening times increases the possibility of gas cutting and may impair an otherwise successful cement job around the top of the liner.

Displacement studies have shown that cement slurries should possess water-loss control when annulur spaces are small; otherwise, filter cake deposited across permeable zones can create high friction pressure and resultant fracturing. To minimize such pressure on the weak zones, the properties and the placement of the slurry should be carefully controlled.[23] Stage-cementing tools may be used through fractured formations to help prevent the cementing slurry from exerting excessive pressure. If the fracturing pressure of a formation is exceeded before the cement has been placed, part or all of the slurry may be lost to the formation.

Placement limitations for liner cementing have been determined by the use of formation fracture gradients. Fracturing pressures (expressed as pressure gradients) can be determined quite accurately when fluid densities and fracture depths are known. Fracturing-pressure information can be obtained during well stimulation or squeeze cementing operations.

The fracture gradient (in psi/ft) can be obtained by dividing the bottom-hole fracturing pressure by the fracturing depth. It is usually constant for a given formation throughout a field. In some formations, fracture pressure gradients will remain the same through wide variations in depth.[23]

If the fracturing pressure or gradient of each formation penetrated in a well is known before cementing, placement design becomes a problem of hydraulics. The combination of mud, cement, and friction pressures may be regulated to avoid exceeding the fracturing pressure of any formation in the hole. In single-stage cementing, the slurry hydrostatic pressure plus the friction pressure cannot exceed the minimum formation fracture gradient. If the lowest formation fracture gradient is 0.70 psi/ft, the maximum slurry weight that can be placed is 13.5 lb/gal if friction pressure is negligible. Slurry weights and friction pressures can be varied, but the total effective gradient cannot safely exceed 0.70 psi/ft.

Pressure relations are more clearly expressed by plotting the formation fracture gradients as a wellbore profile (see Figs. 8.7 and 8.8).[23-25] Placement pressures are plotted as maximum, or plug-down, conditions. The wellbore profile in Fig. 8.8 illustrates that 16.0-lb/gal cement can be placed through the entire annulus in two stages, whereas only 10.5-lb/gal cement can be circulated in one stage. The only place the stage-cementing tool can be set is at the base of a weak section, which is where fracturing first occurs, since pressure buildup is from bottom to top.

8.6 Cementing Liners Through Abnormal-Pressure Formations

Some of the most difficult liner cementing problems are encountered opposite abnormal-pressure sections in the Delaware and Anadarko basins. In many instances, two liners are required through these high-pressure gas sections, which occur at depths from 10,000 to 18,000 ft.[22,23] The greater pore pressures dictate the use of heavy mud systems to control the gas until the wells can be cased off and cemented. In the lower sections of hole, the pore pressures drop rapidly and can be controlled with drilling muds of much lower densities (Fig. 8.9).[28]

These high-pressure gas sections usually have very low permeabilities and will support mud weights from 12 to 18 lb/gal. They are dense, hard zones that can produce a very low volume of gas. Underbalanced drilling is frequently used, with mud weights substantially lower than the pore pressures of the formation being penetrated. With such drilling, varying quantities of trip gas are produced, which can result in gas-cut cement and leakage at the top of the liner into the annulus. Hole caliper surveys are employed in planning and executing these liner cementing jobs. An excess of 20 to 30 percent of cement is used, flow is turbulent, and the plug is bumped on time. The result should be several hundred feet of good hard cement on top of

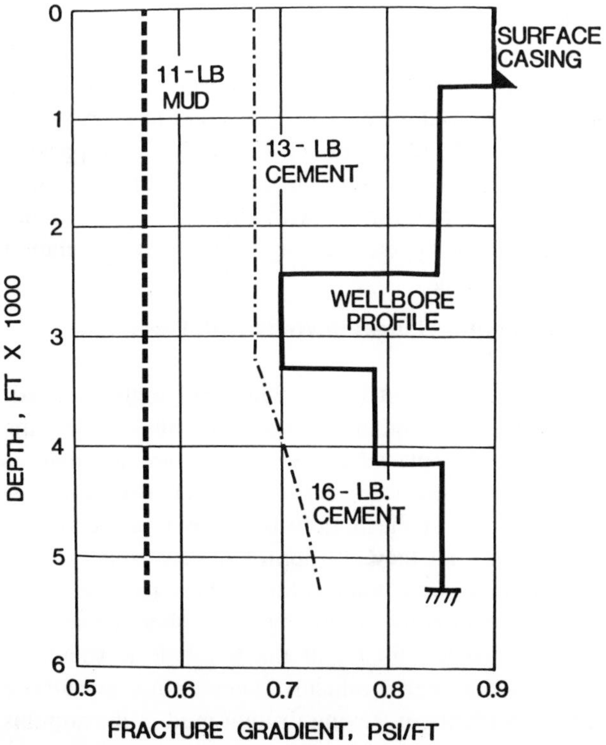

Fig. 8.7 Typical cement pressure profile, showing use of two different slurries.[23]

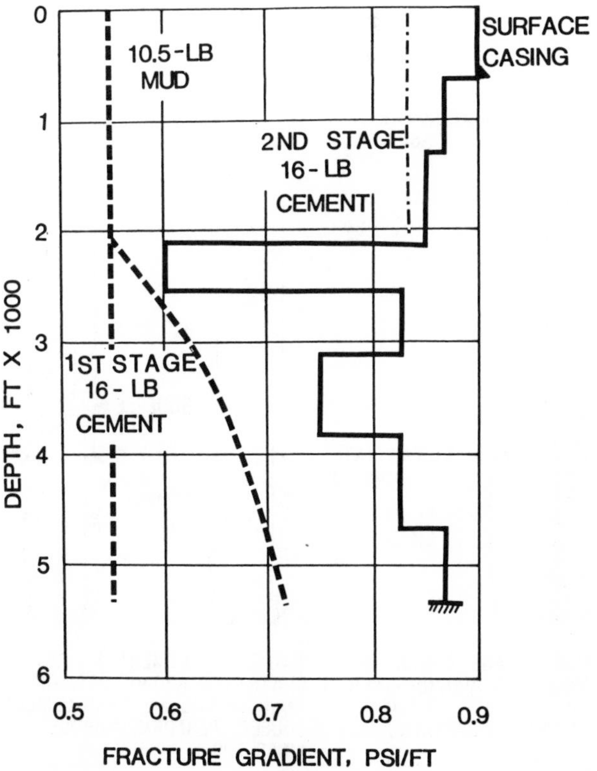

Fig. 8.8 Typical two-stage placement pressure profile, with stage collar at 2,100 ft.[23]

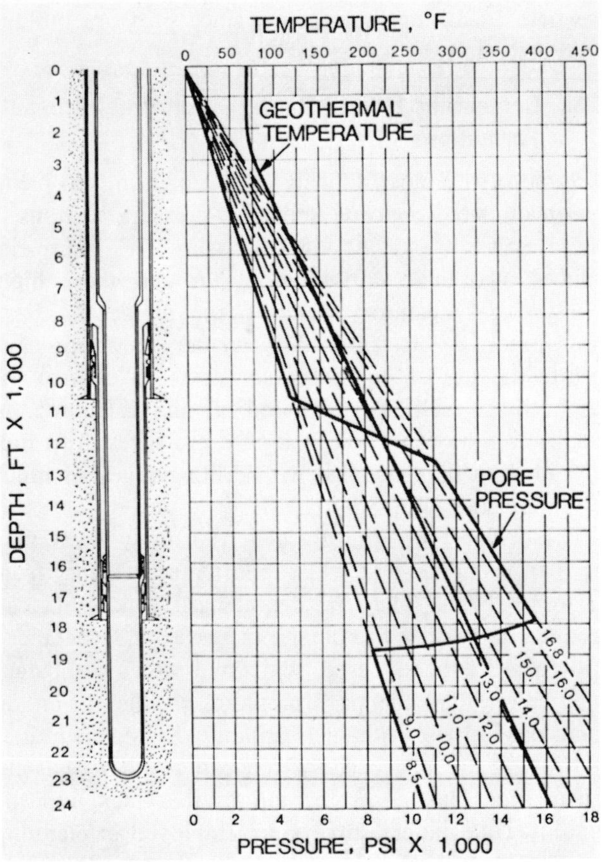

Fig. 8.9 Liner cementing through zones of high pore pressure in deep wells (dashed lines indicate cement weight in lb/gal). (After Ref. 28.)

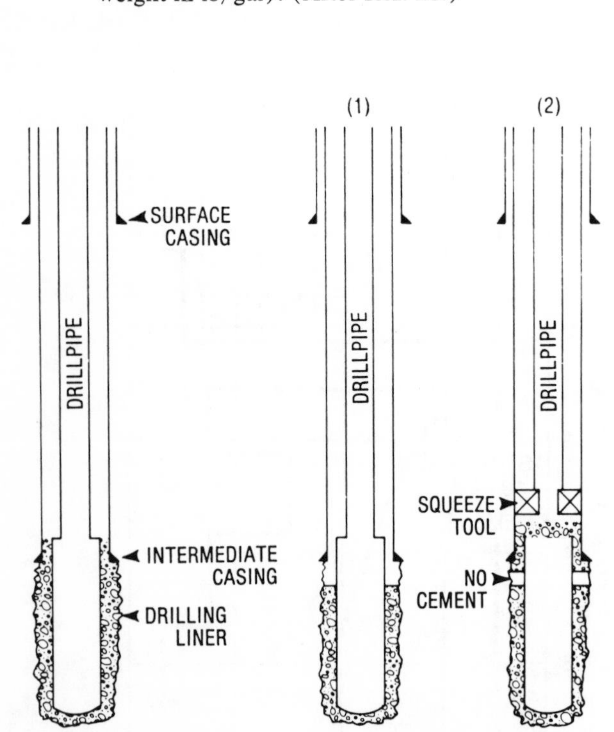

Fig. 8.10 Single-stage and two-stage methods of cementing liners in high-pressure zones.[28]

the liner. Frequently, however, although the job may have been executed according to plan, channeling or gas-cutting of cement will have caused irreparable leaks. Gas leaking from the top of the liner after the cement is drilled out cannot be squeezed off because cement slurry will not penetrate the microannulus through which the leakage occurs.

Methods for cementing liners in high-pressure formations include (1) placing cement around and over the top of the liner in a conventional single stage, and (2) placing cement over the lower half or two-thirds of the liner in a single stage, then pulling out of the hole, running a drillable or retrievable squeeze tool, and squeezing the top of the liner. (See Fig. 8.10.) The latter technique is preferable, but has the disadvantage of leaving a considerable gap between the two cemented intervals. High-pressure gas entrapped behind the liner subjects the unsupported section to pressure, environmental corrosion, and temperature fluctuation.

The squeeze variation in the single-stage liner cementing program, which enables the operator to hold pressure against the gas-producing section, is referred to as the "controlled reversing" method.[19,28] After the plugs are bumped, the setting string is raised a few stands above the liner and excess cement is reversed out of the hole by circulating down the casing and back through the drillpipe to exert pressure against the zone.

Because of inadequate hole clearance and the possibility of damaging the sealing elements with mud filter cake and other foreign matter carried by the circulated fluid, liner packers are not widely used to control gas in deep wells. Gas can remain under a liner packer for weeks with no indication of leakage, then migrate to the surface, where it can damage the rig or the well or injure personnel. Liner packers were primarily designed, not to hold high-pressure gas, but to keep hydrostatic heads off the formation.[28] Special liner packers have been designed to seal off leaks in liner tops; however, they are not run in with the liner and are not subjected to the type of damage caused by circulation during cementing.

8.7 Cementing Liners in Wells With Low Fluid Levels

When liners are set in low-fluid-level wells, delayed-set cementing is sometimes used.[29] In this method the cement slurry is placed in the well before the liner is run (Fig. 8.11 illustrates the procedure). During placement, pressures at the bottom of the hole are carefully controlled so that they are approximately equal to that of the normal static fluid column. This technique has advantages over conventional methods when the annular space between the liner and the borehole is small. In such cases, the liner is difficult to center in the wellbore and the possibility of leaving drilling mud in the annulus is increased.

The thickening time of cement may be controlled

TABLE 8.3 — EXAMPLE DELAYED-SET CEMENTING JOBS

	West Texas Job	South Louisiana Job
Specific conditions	Old wells	New well
Well depth, ft	3,470**	10,800
Casing depth, ft	2,900*	10,400
Casing size, in.	7*	5½
Liner depth, ft	600**	600
Liner size, in.	4½**	4½
Cement		
Type	API Class C	Pozzolan cement
Additives	4% gel, 1% dispersant, 1% fluid-loss-agent	6% gel, 1% fluid-loss-agent, 0.6% retarder
Thickening time, hours	24	18
Set time, hours	30	24
Fluid loss, cc	50 to 100	50 to 100
Strength	1,000 psi in 48 hours at 93°F	750 psi in 72 hours; 1,500 psi in 120 hours
Procedure	Spotted 500 gal 5% HCl, mixed and pumped 75 sk cement with drillpipe, lowered liner to bottom in slurry after 8 hours, completed wells by perforating and fracturing.	Mixed 35 bbl cement in batch mixer, spotted cement with drillpipe, reciprocated liner several times, drilled out hard cement after 48 hours.

*Open-hole completion.
**Recompletion.

to allow placement to depths of 10,000 to 12,000 ft where temperatures do not exceed 260°F. Movement when the liner is run allows any drilling mud that might remain in the annulus to commingle with the cement slurry and minimize or eliminate channels. The major disadvantage of this technique is the longer WOC time required. Examples of delayed-setting jobs designed for old and new wells are given in Table 8.3.

8.8 Factors To Consider in Designing Slurries for Deep Wells

Well Temperature

In any deep well, an accurate knowledge of the static temperature or the bottom-hole circulating temperature is necessary for selecting a cement composition.[20,29,31] Such data may be obtained from drillstem tests or logs, or with special temperature recording subs run on the bottom of the drillstring during hole conditioning. Many deep holes are not so hot as expected, and overestimating temperatures in the slurry design can result in gas cutting of the cement or in failure of the cement to set at the top of the liner. From 1971 through 1973 the API gathered down-hole circulating temperature data on deep wells (see Fig. 8.12). These data revealed that bottom-hole temperatures reach a minimum after mud has been circulated for 3 to 4 hours in the deepest wells.

In reporting temperature data to the laboratory designing the slurry, one should note the source of the well temperature (i.e., whether it has come from a log, a drillstem test, field experience, etc.). It is poor practice to overstate bottom-hole temperature merely to assure cement pumpability.

Pumpability

Temperature and pressure each influences the set of cement, whereas depth dictates the placement time (the greater the depth, the more time required). Pres-

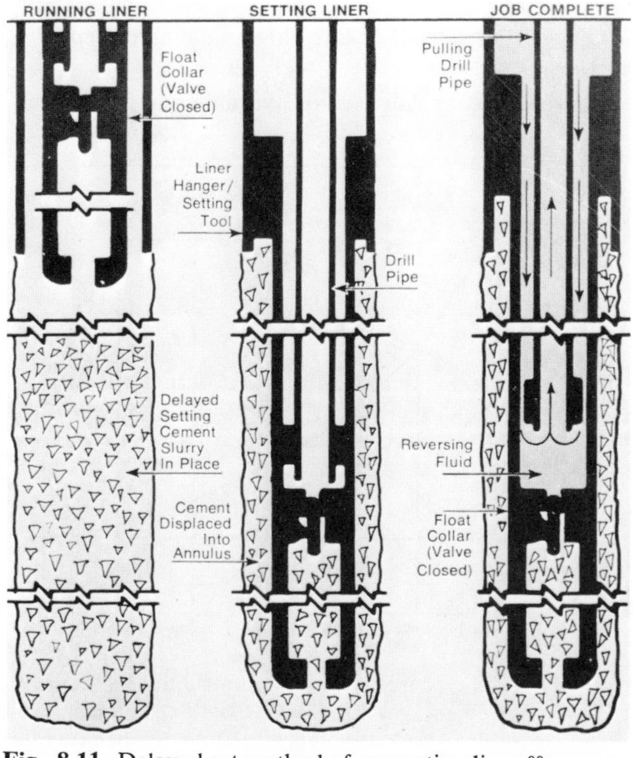

Fig. 8.11 Delayed-set method of cementing liner.[29]

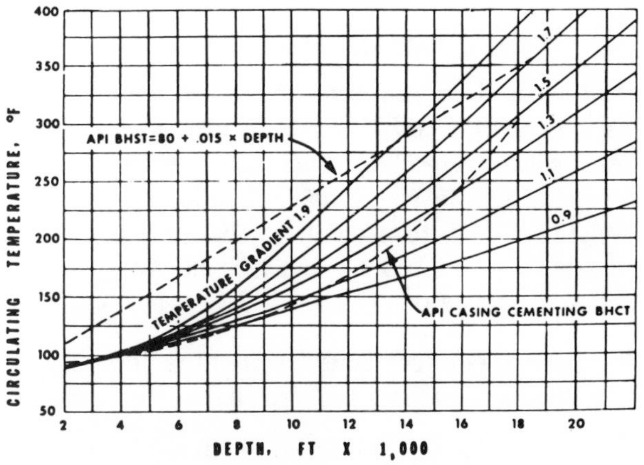

Fig. 8.12 Circulating temperature vs depth as a function of the temperature gradient.

sure alone accelerates the setting of cement more in deep wells than was previously thought (Fig. 8.13). This accelerating effect was not recognized until the development of the super pressure-temperature testing apparatus for cementing wells to depths of 40,000 ft and at static temperatures of 700°F.

With the development of uniform basic cements (API Classes G and H), retardation technology for deep-well conditions has been greatly simplified. (Table 8.4 shows typical cement thickening times attainable with various amounts of retarder at high temperatures.)

For reasons of safety the cementing materials and mixing water at the job site should be laboratory tested before they are used in deep wells. Where this is not possible, API Class D or E cement should be considered.

Strength Stability

All API Classes of cement lose strength and gain permeability when exposed to high temperature (see Section 3.10). A maximum strength is reached between 230°F and 260°F. Thereafter, the strength decreases as temperature increases. To inhibit that loss of strength where formation temperatures are above 230°F, cement should always contain 30 to 40 percent silica flour by weight of cement. The finely powdered silica reacts with cement at temperatures above 230°F to form a complex calcium silicate called tobermorite. This material is influenced by particle size, and the silica must be finely ground to impart high strength and low permeability to cement exposed to high temperatures. For heavy-weight cement slurries, coarse silica (60 to 140 mesh) can be substituted for the fine silica to reduce the additional water needed.

Field Mixing

The volume of cement used on most deep liners is usually rather small. Since slurry design is critical, batch mixing is sometimes preferred to promote uniformity. Since surface mixing time does not significantly influence the total thickening time of the cement, the slurry can be tested and its properties adjusted before it is pumped into the well.

Typical Composition

A typical deep-well cementing composition would consist of API Class G or H cement containing 35 percent silica flour, a dispersant, heavy-weight additives for density, KC1 or NaC1, and a retarder. It would have a weight of 17.5 to 19.5 lb/gal; its thickening time would be 3 to 4½ hours; and its fluid loss would be 60 cc or less at 1,000 psi.

8.9 Summary Check Lists for Running and Cementing Liners in Deep Wells

The following list comprises items that should be checked preparatory to running and cementing liners in deep wells. Table 8.5 constitutes an operational check sheet, to be used during the liner running and cementing procedure. It was drawn up by Lindsey and Bateman.[28,30]

TABLE 8.4 — TYPICAL CEMENT THICKENING TIMES AT HIGH TEMPERATURES

Cement: API Class H.
Additive: 35 percent silica flour.
Slurry weight: 18.5 lb/gal.

Simulated Conditions	Percent Retarder	Thickening Time (hours:min.)
At 500°F, 18,000 psi reached in 60 min.	4.0	2:10
	4.5	3:52
	5.0	6:03
	6.0	7:30
At 550°F, 18,000 psi reached in 60 min.	6.0	2:10
	7.0	4:22
	8.0	5:50
At 600°F, 25,000 psi reached in 90 min.	8.0	3:17

Hole Conditions

1. Depth of liner.
2. Mud weight.
3. Size of liner/hole annulus (Are there doglegs or obstructions?)
4. Temperature of hole — either at top and bottom of liner, or gradient.
5. Pore pressure of formations opposite liner.
6. Corrosiveness of well fluids and gas.
7. Maximum differential and circulation pressures on liner at time of landing plug.

Liner Design

1. Size, weight, and grade of steel.
2. Allowance for burst, collapse, tensile, or compressive forces.
3. Type of joint (Is it an integral or a collared pipe?)
4. Hanger
   a. Is it mechanically set, or hydraulically set?

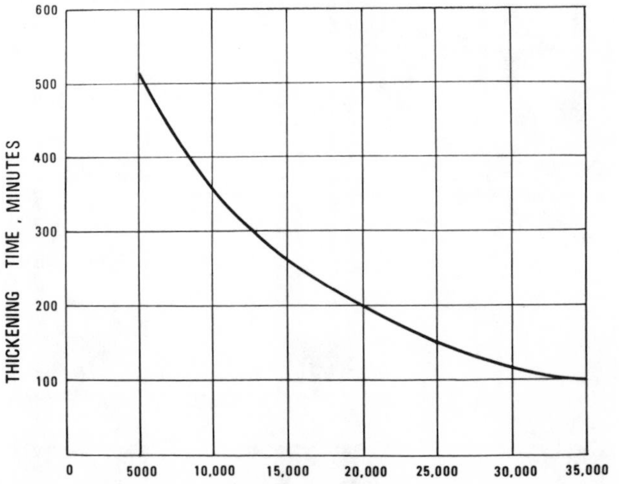

Fig. 8.13 Effect of pressure on pumpability of cement. (Cement — API Class H with 0.3 percent retarder; bottom-hole circulating temperature — 200°F.)

b. Is it rotating, or nonrotating?
c. How is it unlatched from the liner?

Cementing Equipment

1. Plugs.
2. Floating equipment.
3. Centralizers.
4. Scratchers.
5. Fillup equipment.

Cement Slurry Design

1. Density (Will it circulate, yet restrain well fluids or gas?)
2. Quality of mixing water.
3. Compatibility of flushes for mud/cement spacers.
4. Thickening time
   a. Is there a safety factor for placement?
   b. Has it been laboratory tested with the mixing water?
5. Viscosity (Is it low enough for the required displacement rate?)
6. Filtration control (Is it adequate to prevent dehydration?)
7. WOC time (Is it based on strength development at the top of the liner?)

Mixing

1. Method (Is is a small job, calling for batch mixing? a large job, requiring continuous mixing?)
2. Pumping and bulk equipment (Are they in good operating order?)
3. Water
   a. Is there enough for the expected rate of mixing and the volume of cement?
   b. Is there enough to displace a plug?
4. Preflush (Has a minimum of 20 bbl been prepared for displacement after the plug is released?)

Operational Considerations

1. Is the slurry to be reverse circulated, or is the drillpipe to be pulled without reversing?
2. Is displacement to be with rig pumps, service units, or both?
3. Have the pump strokes and the time to bump the plug been determined?
4. Is the pressure recording equipment ready for operation from the service unit and rig manifold?

References

1. "A Look at Deep Drilling," *World Oil* (May 1968) 57.
2. "How to Pre-Plan a Deep Cement Job," *Drilling-DCW* (April 20, 1972) 12.
3. "The Deep Ones," *Drilling-DCW* (July 1973) 26.
4. "U. S. Deep Wells Decline in 1974," *Pet. Eng.* (March 1975) 27.
5. Wheeler, R., Jr., and Moriarty, D. G.: "World's Longest/Strongest Casing Set," *Pet. Eng.* (May 1969) 105.

TABLE 8.5 — PROCEDURE FOR RUNNING AND CEMENTING LINERS[28]

Well Name and Location____________________.

1. Trip to condition hole for running liner. Temperature subs should be used where bottom-hole circulating temperatures are unknown. Drop hollow drift (rabbit) to check drillpipe ID for pump-down plug. Strap drillpipe to be used for running liner. Tie off other drillpipe on opposite side of board.

2. Run________feet of________liner with float shoe and float collar spaced________joints apart. Run plug landing collar________joints above float collar. Volume between float shoe and plug landing collar is________bbl. Run thread-locking compound on bottom five to eight joints. Sand blast lower 1,000 ft and upper 1,000 ft of liner. Pump through first few joints to make sure float equipment is working.

3. Fill each 1,000 ft while running, if fillup-type floats are not used.

4. Install liner hanger and setting assembly. Fill dead space (if packoff bushing is used instead of cups) between liner setting tool and the liner hanger assembly with inert gel to prevent foreign material from settling around setting tool.

5. Run liner on________(size, joint)________ (grade) drillpipe with________lb minimum overpull rating. Run 1 to 2 minutes per stand while in casing and 2 to 3 minutes per stand while in open hole. Circulate last joint to bottom with cement manifold installed. Shut pump down. Hang liner 5 ft off bottom. Release liner setting tool and leave 10,000 lb of drillpipe weight resting on setting tool and liner top.

6. Circulate bottoms-up with________bbl/min. rate to achieve________ft/min. annular velocity (approximately equal to previous drilling rate).

7. Cement liner as follows:

   A. If unable to continue circulation or cementing because of plugging or bridging in liner/open-hole annulus, pump on annulus between drillpipe and casing to maximum________psi and attempt to remove bridge. Do not overpressure and break down formation. If unable to start circulation, pull out of liner and reverse out any cement remaining in drillpipe.

   B. Slow down pumps just before pump-down plug reaches the liner wiper plug. Capacity of drillpipe is ________bbl. Watch for plug shear; recalculate or correct cement displacement and continue plug displacement plus________bbl maximum overdisplacement.

   C. If there is no indication of plug shearing, pump the calculated displacement plus________bbl (100 percent plus 1 to 3 percent).

   D. Pull out 8 to 10 stands or above cement, whichever is greater, and to prevent gas migration hold pressure on top of cement until cement hardens.

8. Trip out of hole.

9. Wait on cement________hours.

10. Run________in. OD bit, drill cement to top of liner. Test liner overlap with differential test, if possible. Trip out.

11. Run________in. OD bit or mill, drill out cement inside liner as necessary. Displace hole for further drilling, spot perforating fluid (if in production liner) or carry out other conditioning procedures as desired.

6. "Historic Oklahoma Test Bests 30,000 Feet," *Oil and Gas J.* (March 1972) 63.

7. Clark, C. R., and Carter, L. G.: "Mud Displacement With Cement Slurries," *J. Pet. Tech.* (July 1973) 775-783.

8. Lawrence, D. K., and Toland, T.: "Preplanning Deep Holes Pays Off for Sun," *Pet. Eng.* (March 1967) 63.

9. Carter, L. G., Cook, C., and Snelson, L.: "Cementing Research in Directional Gas Well Completions," paper SPE 4313 presented at SPE-AIME European Spring Meeting, London, April 2-3, 1973.

10. Leon, L., Hathorn, D. H., and Saunders, C. D.: "Completion Techniques in Very Deep Wells," *Proc.*, Eighth World Pet. Cong., Applied Science Publishers, London (1971) **3,** 159-166.

11. Dubrow, M. H.: "Deep-Well Cementing," *Oil-Well Cementing Practices in the United States,* API, New York (1959) 177.

12. West, E. R., and Lindsey, H. E., Jr.: "How to Run and Cement Liners in Ultra-Deep Wells," *World Oil* (June 1966) 101-106.

13. Davis, S. H.: "Cementing Liners," *Oil-Well Cementing Practices in the United States,* API, New York (1959) 187-188.

14. Scott, J.: "Deep Hole Upsurge Continues Despite Rise in Drilling Costs," *Pet. Eng.* (March 1967) 51.

15. Kastrop, J. E.: "Liner Hanging Cuts Deep Well Casing Costs," *Pet. Eng.* (Jan. 1962) 104.

16. Lindsey, H. E., Jr.: "Techniques for Liner Tie-Back Cementing," *Pet. Eng.* (July 1973) 40.

17. Kirk, W. L.: "Deep Drilling Practices in Mississippi," *J. Pet. Tech.* (June 1972) 633-642.

18. "Deep Gas Well Completion Practices Manual," Texas Iron Works, Inc., Houston (1972).

19. "Running and Cementing Liners in the Delaware Basin, Texas," *API Bull. D17,* 1st ed. (Dec. 1974).

20. Mahony, B. J.: "New Techniques Cut Drilling Costs in the Delaware Basin," *World Oil* (Nov. 1966) 117.

21. Beaupre, C. J.: "Drilling and Completion Programs and Problems, Rojo Caballos Penn Field, Pecos County, Texas," paper presented at API Div. of Production Spring Meeting, Fort Worth, Tex., March 1963.

22. Pugh, T. D.: "A Design for Cementing Deep Delaware Basin Wells," *Proc.,* Eleventh Annual Southwestern Pet. Short Course, Texas Technological College, Lubbock, April 23-24, 1964.

23. Gibbs, M. A.: "Delaware Basin Cementing — Problems and Solutions," *J. Pet. Tech.* (Oct. 1966) 1281-1285.

24. Gibbs, M. A.: "Primary and Remedial Cementing in Fractured Formations,'" paper presented at Southwestern Pet. Short Course, Texas Technological College, Lubbock, April 20-21, 1967.

25. Pugh, T. D.: "What To Consider When Cementing Deep Wells," *World Oil* (Sept. 1967) 52.

26. Mahoney, B. J., and Barrios, J. R.: "Cementing Liners Through Deep High Pressure Zones," *Pet. Eng.* (March 1974) 61.

27. "How 20,000-Foot Ellenburger Gas Wells Were Drilled," *World Oil* (May 1968) 58.

28. Lindsey, H. E., Jr., and Bateman, S. J.: "Improved Cementing of Drilling Liners in Deep Wells," *World Oil* (Oct. 1973) 65.

29. Glenn, E. N.: "Liner Cementing — Long Life Technique," paper presented at Southwestern Pet. Short Course, Texas Technological College, Lubbock, April 20-21, 1967.

30. Lindsey, H. E., Jr.: "Running and Cementing Deep Well Liners," *World Oil,* Part 1 (Nov. 1974); Part 2

31. *Liner Cementing Brochure,* Halliburton Services, Duncan, Okla. (March 1970).
(Dec. 1974); Part 3 (Jan. 1975).

Chapter 9

Squeeze Cementing

9.1 Introduction

Squeeze cementing — the process of applying hydraulic pressure to force or squeeze a cement slurry into a formation void or against a porous zone — is the most common type of down-hole remedial cementing. Its objective is to obtain a seal between the casing and the formation.

One of the earliest oilwell problems was to isolate down-hole water. The problem was partially solved by using cement slurry and squeeze pressure. It was observed that the higher the pressure, the greater the volume of cement that could be displaced and the more successful the isolation around the wellbore.[1] This high-pressure technique has been widely used for many years for remedial cementing.

The technical literature contains a wealth of material on squeezing wells[2-4]; still, many unanswered questions persist:

Where does the cement go on a squeeze job?

What is formation breakdown, and is it necessary?

Should water or mud be used for breakdown?

Will squeezed cement completely surround a wellbore?

Can perforations be plugged with cement?

Can the quantity of cement be controlled during placement?

Today, the most common use of squeeze cementing is to segregate a hydrocarbon-producing zone from those producing other fluids. The aim in squeezing, therefore, is to place the cement at the correct points to accomplish this. (See Fig. 9.1.)

9.2 Where Squeezing Is Required

The application of squeeze cementing technology has increased considerably with a better understanding of (1) the mechanics of fracturing rock and (2) the filtration properties of cement slurries pressured against a permeable medium.

Squeezing is widely used in wells for the following purposes.[4-6]

1. To control high GOR's. By isolating the oil zone from an adjacent gas zone, the GOR can usually be improved to help increase oil production.

2. To control excessive water or gas. Water or gas sands can be squeezed off below the oil sand to help decrease water/oil or gas/oil ratios. Independent water or gas zones can usually be squeezed to eliminate water or gas intrusion such as that illustrated in Fig. 9.2.

3. To repair casing leaks. Cement can be squeezed through corrosion holes in casing.

4. To seal off thief zones or lost-circulation zones.

5. To protect against fluid migration into a producing zone (block squeezing). (See Fig. 9.3.)

6. To isolate zones in permanent completions. It is common practice in many areas, after a well with multiple-producing-zone potential has been cased, to isolate the first zone by squeezing and produce the zone to depletion.[7]

7. To correct a defective primary cementing job. Problems resulting from channeling or insufficient fillup on the primary cementing job can often be overcome by squeeze cementing. Liners are commonly squeezed

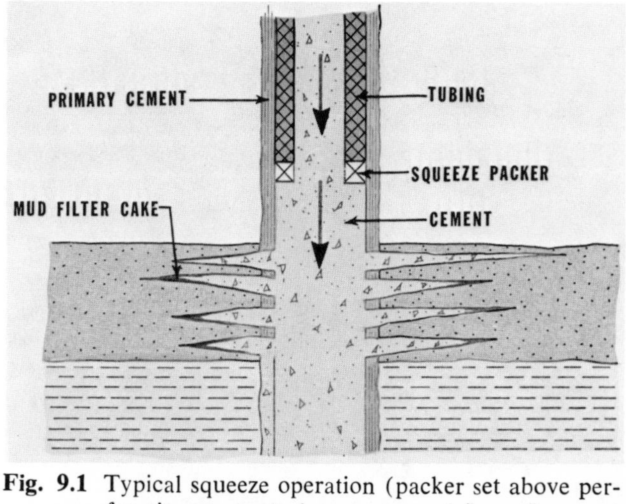

Fig. 9.1 Typical squeeze operation (packer set above perforations to control pressures and flow of cement slurry to formation).

from the top into the annulus or overlap.

8. To prevent fluid migration from abandoned zones or wells. Squeeze cementing is employed to seal old perforations or to plug depleted producing zones completed in open hole.

9.3 Squeeze Terminology

Squeeze cementing terms and their definitions vary from area to area. The meanings can be perplexing, so a discussion of terminology is appropriate.

Squeeze Pressure — Squeeze cementing objectives are usually defined by pressure requirements. The "high-pressure" technique (Fig. 9.4) involves breaking down the formation and pumping cement slurry or cement filtrate into the formation until a specific surface pressure can be maintained without bleedoff. The "low-pressure" technique (Fig. 9.5) involves placing cement over the interval to be squeezed and applying a pressure sufficient to form a filter cake of dehydrated cement in perforations, channels, or fractures that may be open.[8]

Block Squeezing — to block squeeze is to perforate above and below the pay section and then squeeze cement through the perforations. It is used to isolate the producing zone before completing a well. The technique normally involves two perforating steps, two squeeze steps, and drilling out.

Breakdown Pressure — Breakdown pressure is the pressure necessary to break down or fracture the formation so that it will accept fluid. In high-pressure squeezing, this is the pressure that must be achieved before putting cement slurry or cement filtrate into a formation. If the formation is permeable, filtrate will go into it at any pressure above the formation pore pressure. With the low-pressure technique, a satisfactory squeeze can be performed without breaking down the formation.

Fracture Gradient — Fracture gradient is usually defined as the pressure per foot of depth required to initiate a fracture. Less pressure is required to extend and prop a fracture than to create it.

Bottom-Hole Treating Pressure — Bottom-hole treating pressure is the pressure exerted on the formation during a squeeze operation. It is the surface pressure plus the hydrostatic pressure of well fluids minus the frictional pressure. To fracture a formation, this pressure must be exceeded (see Fig. 9.4). In Table 9.1 are

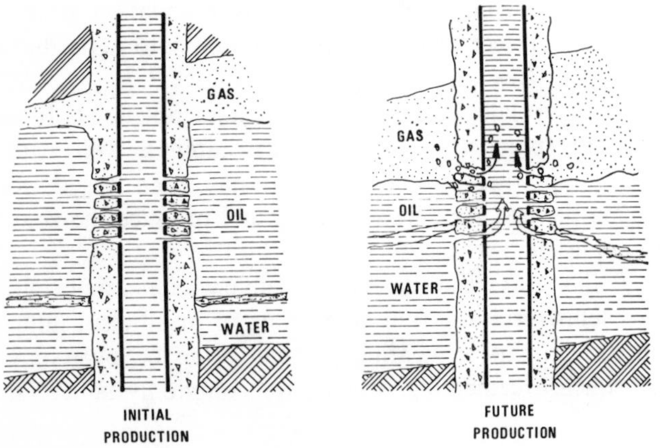

Fig. 9.2 Water or gas intrusion as oil zone is depleted.

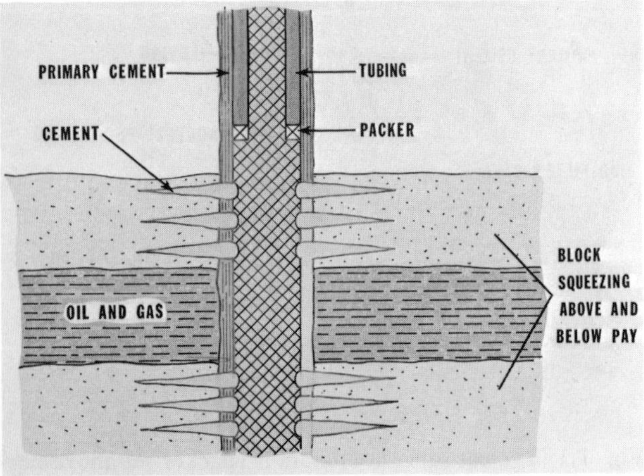

Fig. 9.3 Block squeezing.

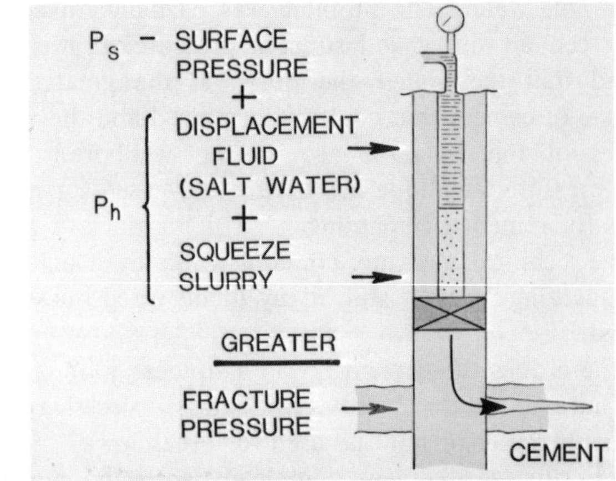

Fig. 9.4 High-pressure squeeze (bottom-hole treating pressures greater than fracture pressure of formation).

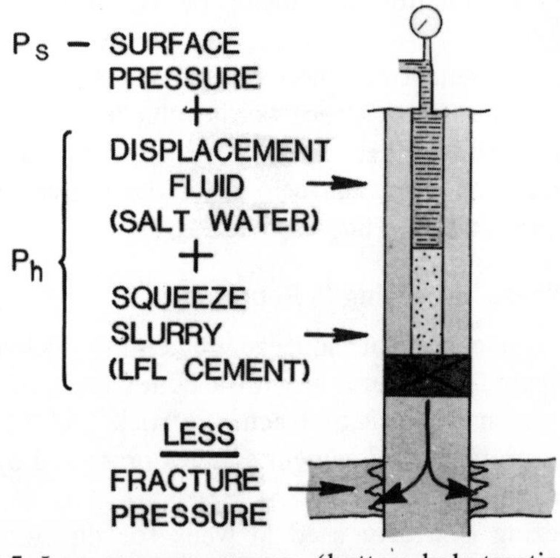

Fig. 9.5 Low-pressure squeeze (bottom-hole treating pressure less than fracture pressure of formation).

some representative bottom-hole treating pressures and fracture gradients in the U. S.

Cement Dehydration — In dehydration the water is squeezed from the cement slurry and a filter cake of solid particles forms on the face of the formation. If excessive pressure is exerted, the formation will fracture and some slurry will be forced into the fractures during the squeeze.

9.4 Squeeze Techniques

Bradenhead Squeeze Method

The original method of squeezing was the bradenhead method, which is accomplished through tubing or drillpipe without the use of a packer. Pressure is built up by closing the blowout preventers or wellhead control valves after the cement has been pumped to near the bottom of the cementing string. A predetermined amount of slurry is mixed and pumped to a specific height outside the tubing or drillpipe. The tubing or drillpipe is then pulled out of the slurry and the wellhead is packed off at the surface. Displacing fluid is pumped down the tubing until the desired squeeze pressure is reached or until a specific amount of the fluid has been pumped. The wellhead is closed at the surface. As pumping continues, the cement slurry is forced to move into or against zones of weakness since it can no longer circulate up the annulus. This method is used

TABLE 9.1 — BOTTOM-HOLE TREATING PRESSURES AND FRACTURE GRADIENTS IN VARIOUS AREAS OF THE U. S.[5]

Field	Formation	Depth (ft)	Bottom-Hole Treating Pressure (psi)	Fracture Gradient (psi/ft)
	Southwest Texas			
East Mathis	Frio	5,785-92	3,860	0.668
Cosden	Slick Wilcox	7,050	3,260	0.463
Seven Sisters	Argo	2,479-89	1,855	0.75
Stratton	Frio	6,226-30	4,225	0.68
Stratton	Frio	7,262-69	5,910	0.81
N. W. Freer	Wilcox Sand	6,814-25	4,980	0.73
Magnolia City	Frio Sand	5,626-33	3,630	0.645
	North Texas			
Grayback	Strawn	3,700	—	0.62
Electra	Cisco	1,600	—	1.07
Kamay	KMA	4,000	—	0.58
Graham	Strawn	3,121	—	0.69
North St. Jo	Strawn	2,400	—	0.58
Sherman	Davis	8,800	—	1.03
Corsicana	Wolf City	1,100	—	1.23
East Texas	Woodbine	3,600	—	0.725
Walnut Bend	Hudspeth	4,451	3,028	0.78
Dye Mound	Conglomerate	6,596	4,080	0.672
Burkburnett	Gunsight	1,550	1,824	1.14
Burkburnett	Canyon	2,140	1,874	0.88
Electra	Cisco	1,040	996	0.96
Kamay	KMA	3,900	2,660	0.68
	Rocky Mountain			
Elk Basin	Tensleep	6,100	4,395	0.72
Lost Cabin	Windriver	3,800	—	0.625
Garland	Basal Amsden	4,145	3,410	0.818
Red Wash (Uintah, Utah)	Green River	—	—	0.625
Red Desert (Sweetheart)	Almond (Mesa Verde)	—	—	0.80
Farmington-San Juan	Pictured Cliffs	2,340	1,565	0.67
Farmington-San Juan	Dakota	7,030	4,960	0.71
	West Texas			
Midland County	Lower Spraberry	8,180	4,330	0.529
Runnels County	Upper Gardner	3,870	1,570	0.41
	Oklahoma			
N. W. Stroud	Red Fork	3,485	2,200	0.631
Yale	Bartlesville	3,100	1,350	0.435
N. Bristow	Peru	2,350	1,100	0.468
Merrick	Wilcox	5,085	3,110	0.613
SE St. Louis	Earlsboro	3,206	2,165	0.675
Kiowa	Dolomite	775	890	1.15
E. Marlow	Helms	4,900	2,750	0.76
Sholem Alechem	Springer	5,400	3,800	0.70
Cement	Fortuna	2,352	1,930	0.82
Sholem Alechem	Springer	5,360	3,770	0.70
Lindsay	Hart	8,330	4,900	0.47
	Kansas			
Ellsworth County	Pennsylvanian Conglomerate	3,215	—	0.45
Rice County	Quartzite	3,125	—	1.31
McPherson County	Viola	3,383	—	0.64
Reno County	Simpson Sand	3,872	—	0.42

extensively in squeezing shallow wells, in plugging, and sometimes in squeezing off zones of partially lost circulation during drilling.

When shallow wells are squeezed by this method, fluids in the tubing are displaced into the formation ahead of the cement. In deeper wells, the cement may be spotted halfway down the tubing before the casing valve at the surface is closed. The applicability of bradenhead squeezing is restricted because the casing must be pressure tight above the point of squeezing and because maximum pressures are limited by the burst strength of the casing. Also, it is difficult to spot the cement very accurately across the target interval without using a packer.

Squeeze-Packer Method

The squeeze-packer method uses a retrievable or a nonretrievable tool run on tubing to a position near the top of the zone to be squeezed. This technique is generally considered superior to the bradenhead method since it confines pressures to a specific point in the hole. Before the cement is placed, a pressure test is conducted to determine the formation breakdown pressure. In certain instances, the section below the perforations to be squeezed must be isolated with a bridge plug. When the desired squeeze pressure is obtained, the remaining slurry is reversed out. Squeezing objectives and zonal conditions will govern whether high pressures or low pressures are used.

9.5 Squeeze Pressure Requirements

Most squeeze jobs are defined by the pressure required to accomplish a down-hole seal or shutoff.

The high-pressure technique (Fig. 9.4) uses a quantity of salt water (or chemical wash) to determine the breakdown pressure of the formation to be squeezed. Mud should not be used as a breakdown fluid as it can plug or damage the formation. After breakdown, a slurry of cement and water is spotted near the formation and pumped at a low rate. As pumping continues, injection pressures begin to build up until surface pressure indicates that either cement dehydration or a squeeze has occurred. Pressure is held momentarily on the formation to verify static conditions and then released to determine if the cement will stay in place. The excess slurry above the perforations is then reversed out.

If the desired squeeze pressure is not obtained, a hesitation, or staging period, is often employed. This method involves mixing one batch of cement (30 to 100 sk), placing it against the formation, waiting at least until the initial set, and repeating the operation as many times as required. (Fig. 9.6 illustrates the steps and shows the surface pressures in a hesitation squeeze.)

The low-pressure technique has become the more efficient method of squeezing with the development of controlled-fluid-loss cements and retrievable packers. With this technique, formation breakdown is avoided. Pressure is achieved by shutting down or hesitating during the squeeze process. In this hesitation method, the cement is placed in a single stage, but in alternate pumping/waiting periods. The controlled-fluid-loss properties of the slurry cause filter cake to collect against the formation or inside the perforations while the parent slurry remains in a fluid state inside the casing (Fig. 9.7).

Fluid loss of neat cement slurries (cement and water) is usually very rapid and cement may build up in the casing before the slurry can completely cover a given area of formation. The result can be a cement plug across open perforations at the top of a zone and no coverage of cement across the lower perforations.

Beach *et al.*[8] studied the influence of filter cake on pressure buildup during high-pressure squeeze operations. High surface pressures on a particular well indicated the successful down-hole coverage of a 40-ft depleted interval at approximately 9,800 ft. Coring of the set cement inside the casing revealed that the upper 34 ft of perforations was sealed, but the lower 6 ft required an additional squeeze because the coverage

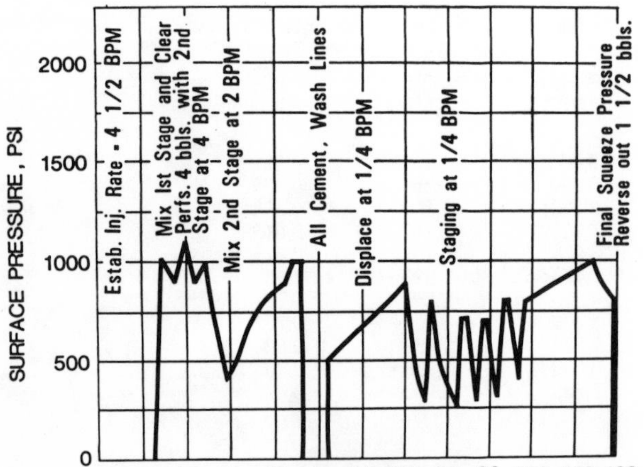

Fig. 9.6 Steps in a hesitation squeeze.[8]

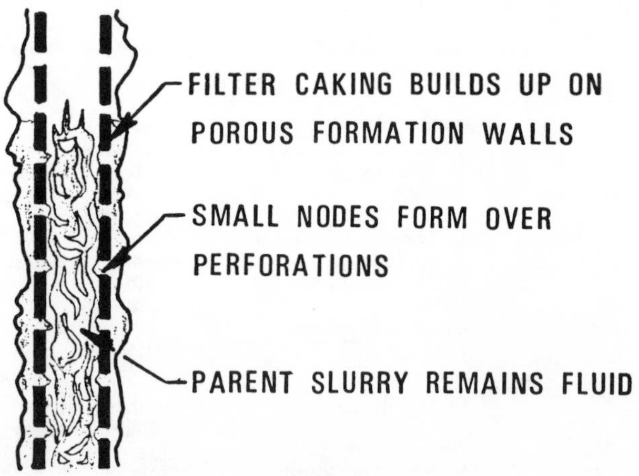

Fig. 9.7 Low-pressure squeezing.[8]

was incomplete. The cement had dehydrated in the upper, more open section, causing a resistance to flow. (See Fig. 9.8.)

Filtration control helps avoid both premature loss of fluid from the slurry and rapid buildup of cement solids in the casing. Cement containing a fluid-loss-control additive loses filtrate to the formation much more slowly than does neat cement (Fig. 9.9), so the filter cake is denser and more pressure resistant. As fluid loss occurs in the formation, little or none is taking place in the casing; therefore, it is often possible to obtain cement plugging or dehydration in the formation and across perforations and still have sufficient time to reverse excess slurry from the casing, avoiding the time and expense of drilling out.

9.6 Squeezing Fractured Zones

In squeezing fractured limestone and dolomite formations, greater emphasis must be placed on effectively sealing the fracture network or channel system behind the casing.[9,10] It is necessary to modify the slurry design from that used to squeeze permeable sandstones, where the prime interest is slurry behavior within the perforations or nominal penetration of the perforations. In squeezing a fractured carbonate formation it is more important that the cement fill the fracture or channels than that it build up a filter cake. Larger volumes of slurry are required than for squeezing permeable sandstone reservoirs.

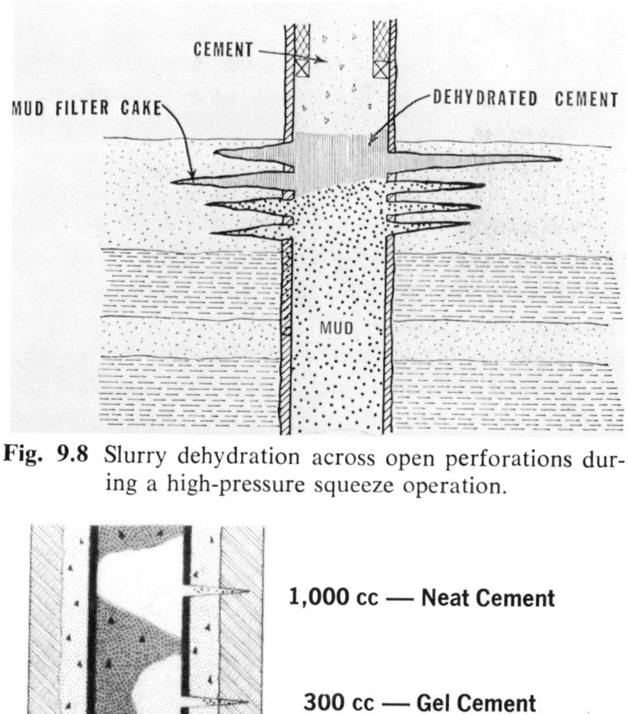

Fig. 9.8 Slurry dehydration across open perforations during a high-pressure squeeze operation.

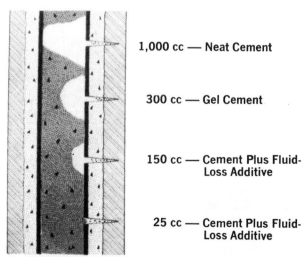

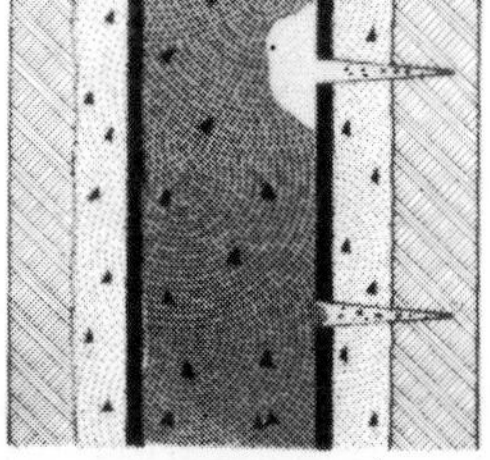

Fig. 9.9 Cement node buildup effect with filtration control (showing API fluid loss in cc/30 minutes at 1,000 psi).

In the most successful squeeze technique, two cement slurries are used: (1) a highly accelerated slurry and (2) a moderate-fluid-loss slurry.

Accelerated slurries designed to set up shortly after reaching the formation are pumped into areas of least resistance and allowed to take an initial set. Once this has occurred, moderate-fluid-loss slurries can be forced into less accessible fractures.[9] The prime objectives in designing the slurries are that they be compatible with bottom-hole conditions and that they take an initial set within 10 to 15 minutes of placement. Pumping times for these slurries vary with bottom-hole conditions, and volumes of accelerated cement range from 35 to 100 sk. Because of the low permeability of most carbonate formations, cement slurries with moderate fluid-loss characteristics give satisfactory results.

In some instances a moderate-fluid-loss cement can be used as a lead slurry to fill the primary existing fractures and channel extremities. This slurry is followed by a high-strength slurry incorporating bridging agents.

9.7 Erroneous Squeeze-Cementing Theories

There are three predominant theories about squeeze cementing that contribute most to misapplication and improper field procedures.[11]

1. Whole cement enters the formation. This misconception leads to emphasizing the quantity of cement pumped behind the pipe and the amount of pressure applied, when actually these factors affect cementing results very little. The truth is that in low-pressure squeezing, cement filtrate — not whole cement — enters the formation (see Fig. 9.10). When the formation is fractured by exceeding the fracturing pressure, then cement slurry can be squeezed into the fractures.

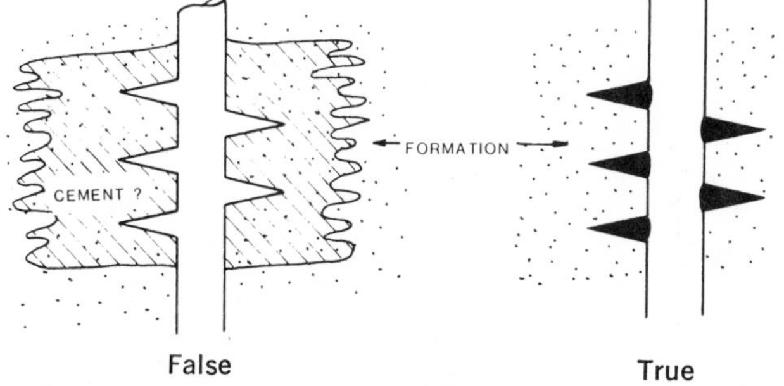

Fig. 9.10 Misconception 1 — Whole cement is squeezed into formation.[11]

2. Breakdown from injecting whole mud automatically opens all the perforations. In reality, it is rare to find all perforations open and receptive to fluid. To achieve this requires considerable effort. (See Fig. 9.11.)

3. A single, horizontal pancake or wedge of cement is formed around the wellbore. Indications are, rather, that because whole cement cannot enter the formation, the filtrate emanates from the perforations. When the formation is fractured, the cement slurry may enter in a series of irregular wedges. (See Fig. 9.12.) The orientation of these fractures depends upon the compressive forces on the zone being squeezed; in many instances it is northeast-southwest.

9.8 Job Planning

Planning is the most important single step in any squeeze operation.[12,13] Well conditions should be studied and objectives should be carefully established, as squeeze cementing can be complicated and expensive. In planning, the following questions should be posed:

1. Why are we squeezing? (Are we isolating a zone? repairing casing? filling up to seal the well?)
2. If we are not performing a bradenhead squeeze, what tools will we use?
3. Should the packer be drillable, or retrievable?
4. How far should we set the packer from the zone of interest?
5. Should we use high pressure, or low pressure?
6. How should we pump? (In hesitation stages? slowly? fast?)
7. What kind of fluid is in the well? (Acid? water? drilling mud?)
8. What kind of slurry should we use? (How much? with what characteristics?)
9. What mechanical equipment and other restrictions must we contend with?

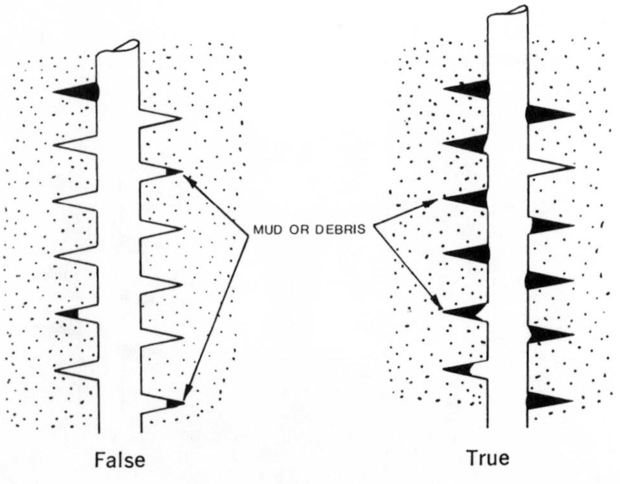

Fig. 9.11 Misconception 2 — All perforations are opened.[11]

10. What are the well conditions? (The fluids in the hole? the bottom-hole pressure and temperature?)
11. Is the target formation fractured? What is the fracture gradient?
12. What is the WOC time?
13. How will we test the squeeze job?

Every effort should be made to enhance hole conditions before and during the squeezing operation. The casing and tubing string should be as clean internally as possible — free of rust, paraffin scale, and perforation burrs. Wellhead packoff equipment should be used, or blowout preventers should be tested to the pressure expected to be exerted on them.

If squeeze work is to be performed through casing it is necessary to calculate the internal yield pressure and joint strength unless the casing is cemented to the surface. If the casing is not cemented to the surface, the critical stresses at the squeeze point can be calculated. If squeezing is to be performed through tubing set inside the casing, calculations must be made for the tubing and the casing, allowing for the collapse resistance of the tubing.

9.9 Slurry Design

The following factors should be considered in designing the cement slurry for any squeeze operation.

Temperature and Pressure

In squeezing, as in primary cementing, both temperature and pressure influence the placement and thickening time of a cement slurry. Squeeze pressure also affects the dehydration of the slurry.

Temperatures encountered in squeezing can be higher than those on primary jobs because the well usually has not been circulated by enough fluid to decrease the bottom-hole temperature. Table 7.5 illustrates the time at which the first sack of cement reaches bottom-hole conditions on a squeeze job and the static vs circulating temperatures at various well depths according to normal API testing schedules. Table 9.2 compares the thickening times of a given cement slurry for casing cementing with those for squeezing.

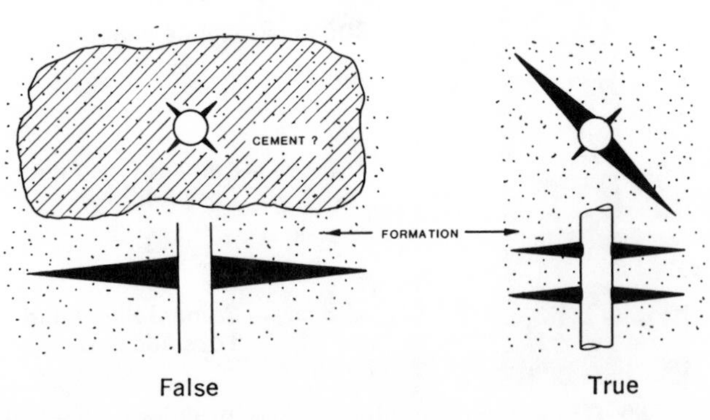

Fig. 9.12 Misconception 3 — Squeezing produces a horizontal pancake of cement.[11]

If a shallow cavity is to be filled or if perforations are to be abandoned to move back up the hole, the slurry can be designed for a fairly short pumping time. But a hesitation low-pressure squeeze may require a pumping time of 4 to 6 hours. A cement slurry must obviously remain fluid long enough not only to be placed properly but also to achieve the squeeze pressure and to be reversed out.

Type of Cement

For most squeeze operations, API Class A, G, or H cement can be used since these cements are manufactured for squeeze conditions to 6,000 ft where bottom-hole static temperatures do not exceed 170°F.[14] (See Table 2.5.) For deeper wells, Class G or H cement can be adequately retarded on the basis of the estimated time required to perform the squeeze. A slurry can be designed for any squeeze conditions, but unless it provides a good seal, it can be worthless, despite a good pressure buildup.

Filtration Control

Filtration is important in designing cement for a squeeze job.[8,15] When cement is squeezed against a permeable medium, differential pressure forces water from the cement solids, forming a filter cake.[16,17] The cake is rather soft and can be removed by jetting; but it is not pumpable and considerable pressure is required to force it through a small aperture. The thickness of the filter cake depends upon the permeability of the cake or of the formation (whichever is lower), the fluid-loss characteristics of the slurry, the differential pressure, and the length of time the differential pressure is maintained.[15]

The API filter loss of neat cement ranges from 600 to 2,500 cc in 30 minutes; in fact dehydration occurs so rapidly it is difficult to measure. Filter loss can be reduced to low values — 25 to 100 cc in 30 minutes — by adding bentonite and dispersing agents, or polymers.[17-19] (See Fig. 9.9 and Table 9.3.)

Fig. 9.13 shows perforated samples against which cement was squeezed to study cement hydration.[20] The filtration effects of a high-fluid-loss system can be seen in Specimen 1, where rapid dehydration occurred against a sandstone medium. In Specimens 2, 3, 4, and 5, which were squeezed with the same cement slurry containing a filtration-control additive, the rate at which solids were deposited was much slower, allowing them to build up after 1 to 4 hours.

Quantity of Cement

The quantity of cement to be used in a given squeeze operation can vary from a few sacks on a wireline job to several hundred sacks on a difficult stage job. The average volume ranges between 100 and 200 sk; however, the specific amount will depend on whether perforations or channels are being squeezed. Besides being wasteful, excess cement can be detrimental to the productivity of the formation being squeezed.

The volume of cement slurry to be squeezed cannot be controlled precisely, and experience in the vicinity of the job is the best guide. However, there are some useful indexes and rules of thumb:

1. The volume should not exceed the capacity of the run-in string.
2. Two sacks of cement should be used per foot of perforated interval.
3. The minimum volume should be 100 sk if an injection rate of 2 bbl/min. can be achieved after breakdown; otherwise it should be 50 sk.

TABLE 9.2 — CEMENT THICKENING TIMES — CASING CEMENTING VS SQUEEZING

Depth: 8,000 ft.
Temperatures
Casing cementing: 125°F.
Squeezing: 159°F.
Cement: API Class H.
Water Ratio: 4.3 gal/sk.

Fluid-Loss Agent (percent)	Thickening Times (hours:min.) Casing Cementing	Squeezing
0.0	2:16	1:15
0.4	4:00	2:16
0.6	5:32	4:15
0.8	6:15	4:58

TABLE 9.3 — COMPARISON OF CEMENT SLURRY FILTRATION LOSS, FILTER CAKE PERMEABILITY, AND FILTER CAKE DEVELOPMENT TIME

API Filtration Loss at 1,000 psi (cc/30 min.)	Permeability of Filter Cake at 1,000 psi (md)	Time To Form 2-in. Cake (min.)
1,200	5.00	0.2
300	0.54	3.4
100	0.09	30.0
50	0.009	100.0

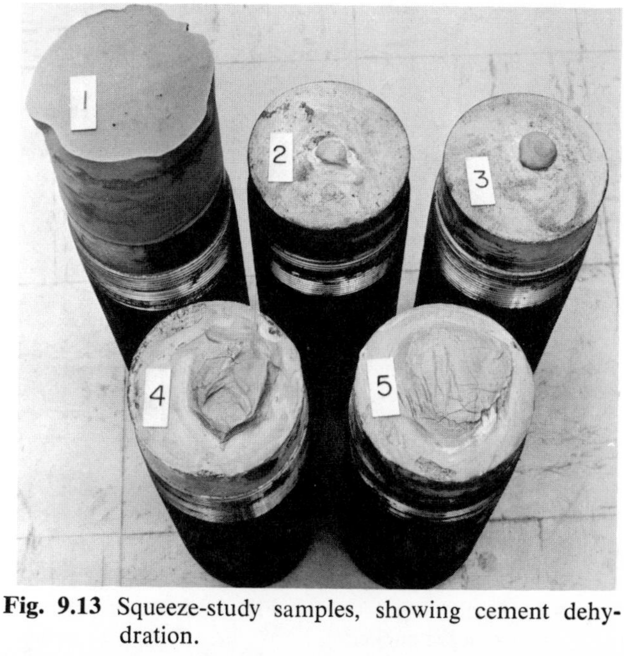

Fig. 9.13 Squeeze-study samples, showing cement dehydration.

4. The volume should not be so great as to form a column that cannot be reversed out.

Workover Fluids

Where well conditions permit and where it is obtainable, salt water or fresh water is the preferred workover fluid for both low-pressure and high-pressure squeeze jobs. However, even if clean fluids are used and pressures are great enough to fracture the formation, if a well has been perforated in mud, or if perforations have ever been contacted by mud, some mud plugs may remain and more than one squeeze job may be needed.[7,8] Should squeezing be necessary in mud-filled or partially plugged perforations, one of the best ways to insure a uniform deposit of cement is to run a weak hydrochloric or acetic acid solution ahead of the cement. The acid shrinks the clay particles and allows the cement slurry to penetrate farther.

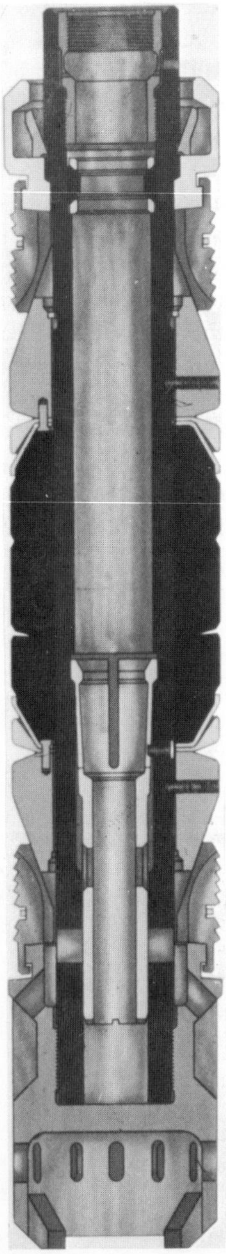

Fig. 9.14 Drillable squeeze packer.

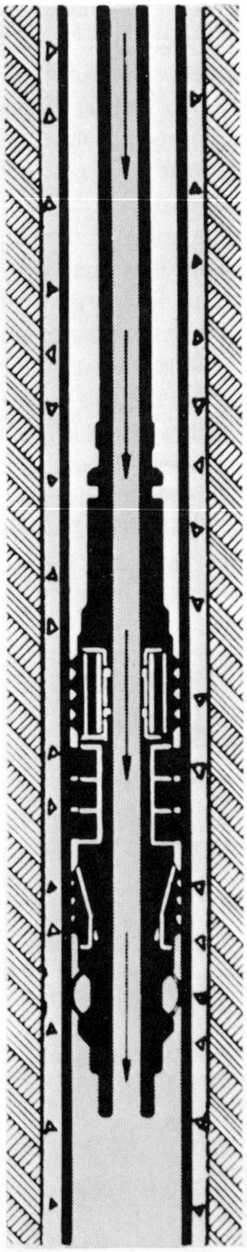

Fig. 9.15 Retrievable squeeze packer.

9.10 Squeeze Packers

The use of squeeze packers makes it possible to apply higher pressure to specific down-hole points than can be applied with the bradenhead method.[3,13] The two commonly used packers are the drillable and the retrievable. Drillable packers, which are expendable, are left in the well and can be drilled out after the squeeze operation. The retrievable packer is rented on a job basis and, after the squeeze job, is removed from the well. While each packer has its specific application, the ratio of usage is approximately 40/60, drillable/retrievable.

Drillable Packers

The drillable packer (Fig. 9.14) contains a poppet-type backpressure valve to prevent backflow at the completion of displacement and a sliding valve for when it is desirable to hold pressure in either or both directions. The sliding valve makes it possible to support the weight of the hydrostatic fluid column and relieve the cement of this weight while it is setting. Excess cement can be reversed out of the tubing or drillpipe without applying circulating pressure to the squeezed area below the packer. The tubing or drillpipe can also be withdrawn from the well without endangering the squeeze job. Drillable packers are easily drilled. Another advantage is that they can be set close to the perforations or between sets of perforations.

Retrievable Squeeze Packer

Unlike drillable packers, retrievable packers (Fig. 9.15) can be set and released as many times as necessary. The major objections to the use of retrievable packers are that (1) flowback cannot be prevented when the displacement pressure is released unless a backpressure valve is employed, and (2) reversing of excess cement can disturb the slurry that has already been squeezed.

The design of retrievable packers varies; however, most of them are of a nondrillable material and are available in most API sizes and weights.

Whether to use a drillable or a retrievable packer depends largely on well conditions and squeezing pressure requirements. If there is doubt about the choice, it is best to consult a service company that specializes in squeeze jobs.

Straddle Packers

To perform a straddle operation, packers can be used in the following combinations:

1. A retrievable packer with a retrievable bridge plug (Fig. 9.16).
2. A retrievable packer with a drillable bridge plug.
3. A drillable packer with a drillable bridge plug.

Conventional straddle packers are not recommended for squeeze cementing operations.

The positioning of the packer, or packers, depends on the type of squeeze job and on experience in a given area. When the packer is positioned at a point too far above the squeeze interval, the large volume of fluid that precedes the cement slurry must be pumped into the squeeze zone. It is best not to have more than 75 to 100 ft of blank pipe above the perforations, and less than that if the volume of fluid to displace the cement slurry can be calculated reliably.

9.11 Squeeze Pressure Calculations

Sample calculations for a low-pressure squeeze with filtration control and a high-pressure squeeze without filtration control are given in Appendix B. Selection of the final pressure to be reached in squeezing is very important because this will define when the job is completed. There are many ways to estimate final pressures, but experience in a given field is probably the best, especially where zones of extremely high or low pressure are encountered for a particular well depth. If cement inside the casing dehydrates, applied pressure is exerted only against the casing. If the squeeze is successful and the applied pressure is high for the depth, there is a tendency to set that pressure as the minimum final pressure required for a good job in the particular area. However, a successful squeeze job can often be attained with a considerably lower final pressure.[5,8,9]

For safety, it should be assumed that any pressure exerted below the packer is applied to the outside of the casing because a channel may exist that will allow this pressure to be transmitted against the casing above the packer. One should always consider the maximum collapse pressure that the casing will safely withstand. The difference between this pressure and the maximum final squeeze pressure is the amount of backpressure required on the tubing/casing annulus to protect the casing. This "backup" pressure should be calculated on the basis of down-hole treating pressure.

Fig. 9.17 illustrates the casing-collapse mechanism involved in a squeeze job.

9.12 WOC Time

The time spent waiting for cement to set after a squeeze job should be governed by the strength required of the set cement. The cement must be strong enough to withstand the shocks of drilling, to resist the flow of down-hole fluids, and to isolate a producing interval during a fracturing operation.

In field practice, a waiting period of 4 to 12 hours

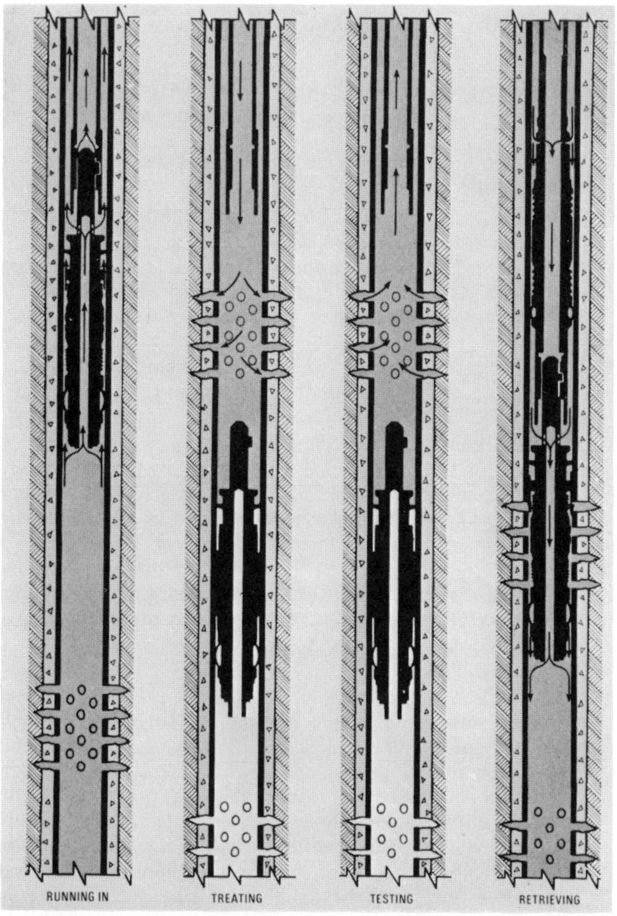

Fig. 9.16 Retrievable packer and retrievable bridge plug under various operating conditions.

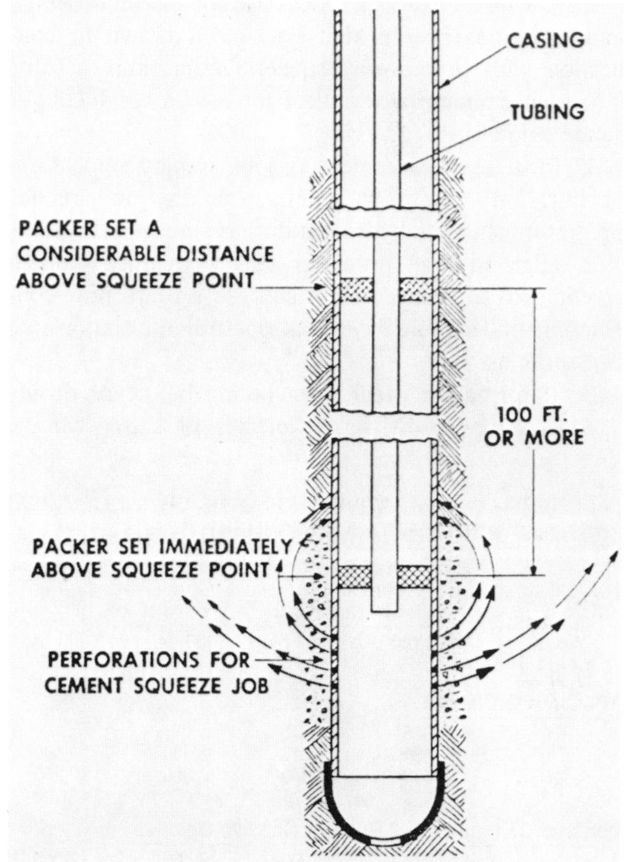

Fig. 9.17 Mechanism of casing collapse during a squeeze operation.[13]

is usually allowed between squeeze treatments or after the final squeeze pressure has been reached.

Once in place, dehydrated cement filter cake will develop strength faster than a slurry that has not lost fluid under pressure. Dehydrated filter cake (nodes in perforations) will develop a strength of several thousand pounds in the first 8 hours. (See Table 9.4.) Washing or flushing between stages can damage squeezed zones if they are agitated or disturbed within 4 hours after squeezing.

9.13 Testing Squeeze Jobs

The apparent success or failure of a squeeze job should be confirmed by applying pressure to the set cement. Although squeeze jobs are most commonly tested by pump pressure, a better way is to create a pressure differential in the wellbore. This can be accomplished by swabbing, by artificially lifting fluid from the well, or by circulating oil or a lighter fluid down the tubing and closing the circulating ports above the packer. This differential pressure should not exceed the drawdown expected in the well when it is put on production.

9.14 Summary

Squeeze cementing technology has improved considerably through the years, and although there still are a number of problems, the success ratio becomes increasingly higher.

Following are some practices that have been observed and some conclusions that have been drawn in connection with present-day squeezing oprations. (Table 9.5 is an example work sheet for use in conducting a squeeze job.)

1. Most squeeze cementing jobs employ some form of filtration control to maintain fluidity and reduce premature slurry dehydration during placement.

2. High squeeze pressures that were once thought essential to the success of a squeeze job are now considered undesirable when a controlled-filtration-rate cement is used.

3. If formation breakdown occurs before or during a squeeze operation, large volumes of slurry can be pumped before a shutoff occurs.

4. The low-filtration technique provides better control in directing the flow of the cement into narrow channels or voids behind the pipe. It also reduces the quantity of cement required for a squeeze operation.

5. Slurries can be designed for fractured limestone or dolomite zones for maximum sealing efficiency (i.e., a fast-setting cement can be followed by a low-water-loss slurry).

6. Testing and design work for squeeze jobs are generally performed under API conditions using a 325-mesh screen at 1,000-psi pressure differential, since formation cores are not readily available for such tests.

7. The benefit of hesitating during the pumping operation is that the deposition of cement solids against the formation can be more readily controlled. As a general rule, the faster the deposition, the sooner the squeeze job can be successfully completed.

8. In questionable situations a drillable packer offers better control than a retrievable packer because it contains a backpressure valve.

9. An effective way to break up mud particles in plugged perforations is to spot an acid solution ahead of a squeeze job.

10. WOC time after squeeze pressure is achieved need not necessarily exceed 8 to 24 hours, since dehydrated filter cake, with its low water/cement ratio, builds up strength rapidly.

TABLE 9.4 — COMPARATIVE STRENGTHS OF SET CEMENTS AND DEHYDRATED CEMENT CORES

Fluid Loss Agent (percent)	Strength of Set Cement After Curing 24 Hours at 800 psi 95°F	Strength of Set Cement After Curing 24 Hours at 3,000 psi 140°F	Strength of Dehydrated Cores After Curing 8 Hours at 800 psi 95°F	Strength of Dehydrated Cores After Curing 8 Hours at 3,000 psi 140°F
API Class G Cement				
0.0	2,085	4,545		
0.8	980	3,515	2,400	12,400
1.0	800	3,440	2,080	12,200
1.2	580	3,525	400	12,100
Portland Cement With 2 Percent Calcium Chloride				
0.8	2,075	4,000+	3,160	12,000+
1.0	1,975	4,000+	3,400	12,000+
1.2	1,920	4,000+	3,280	12,000+

TABLE 9.5 — WORK SHEET — RECOMMENDED CEMENT SQUEEZE JOB PROCEDURE

1. Kill well by circulating hole with salt water.
2. Pull tubing.
3. Run________ on tubing, with approximately ________ ft of tailpipe.
4. Set________ at approximately________ ft and establish feed rate (2 to 3 bbl/min. is desirable. If necessary, acid flush may be spotted to increase feed rate.)
5. Batch mix________ sacks of cement containing ________ filtration-control agent and________ at________ lb/gal (________ bbl of slurry).
6. Circulate cement to bottom (less 1 bbl).
7. Reset________ and squeeze approximately ________ sacks (________ bbl) of cement into perforations, gradually reducing rate to approximately ⅓ bbl/min. and attempt to obtain squeeze.
8. If necessary, stage in 5- to 15-minute intervals until a 10-minute static pressure of approximately________ psi is obtained. (This slurry has a thickening time of approximately________ hours for these well conditions.)
9. Check backflow.
10. While holding approximately________ psi backpressure, reverse circulate excess cement, lowering tailpipe to approximately________ ft, then pull packer to approximately________ ft and reset. Repressure squeeze to original standing pressure. If still holding after 10 minutes, bleed off to approximately ________ psi wellhead pressure; WOC________ hours.
11. If desired, the success of the squeeze may be evaluated by a pressure test (not to exceed________ psi) or by swab testing.
12. Pull________, rerun tubing.

9.15 Helpful Formulas for Squeeze Cementing

1. *Maximum surface squeeze pressure that can safely be applied to the annulus between the squeeze string and the casing.*

$$p_{b\max} = p_{\mathrm{WBS}} - (D_{\max})\ (g_m - g_{m2}), \quad . \quad . \quad . \quad (9.1)$$

where

$p_{b\max}$ = maximum annular backup pressure, psi,

p_{WBS} = "working burst strength" of the casing, psi (book value ÷ safety factor of 1.33),

$D_{\max}$ = maximum depth of casing,* ft,

g_m = pressure gradient of mud in annulus, psi/ft,

g_{m2} = pressure gradient of mud in annulus outside casing, psi/ft.

2. *Hydrostatic pressure of fluids inside the squeeze string.*

$$p_{hs} = (g_m)(h_m) + (g_c)(h_c) + (g_w)(h_w), \quad . \quad . \quad . \quad (9.2)$$

where

p_{hs} = hydrostatic pressure of squeeze column, psi (this value will usually decrease as the cement is squeezed away and replaced with lighter-weight mud),

g_m = pressure gradient of mud (or other fluid) used to squeeze, psi/ft,

h_m = height of mud column, ft,

g_c = pressure gradient of cement slurry, psi/ft,

h_c = height of cement column, ft,

g_w = pressure gradient of water, psi/ft,

h_w = height of water column, ft.

3. *Pressure to resist collapse of the casing above the packer.*

$$p_s = p_{b\max} + (g_m)(D_p), \quad . \quad . \quad . \quad . \quad . \quad (9.3)$$

where

p_s = support pressure to resist collapse above packer, psi,

$p_{b\max}$ = maximum annular backup pressure, psi,

g_m = pressure gradient of mud in annulus, psi/ft,

D_p = depth to packer, ft.

4. *Maximum allowable squeeze pressure (high-pressure squeezing).*

$$p_{s\max} = p_s = p_{\mathrm{WCS}} - p_{hs}, \quad . \quad . \quad . \quad . \quad . \quad . \quad (9.4)$$

where

$p_{s\max}$ = maximum allowable surface squeeze pressure, psi,

p_s = support pressure to resist collapse, psi,

p_{WCS} = "working collapse strength" of weakest casing within 1,000 ft above packer, psi,

p_{hs} = hydrostatic pressure of squeeze column, psi.

5. *Hydrostatic pressure.*

$$p_h = 0.052 \times w_m \times D, \quad . \quad . \quad . \quad . \quad . \quad . \quad (9.5)$$

where

p_h = hydrostatic pressure, psi,

w_m = mud weight, lb/gal, and

D = depth, ft.

*Where more than one grade of casing in involved, the maximum allowable should be calculated for each. The lowest value would be the safe one to use.

References

1. Millikan, C. V.: "Cementing," *History of Petroleum Engineering,* API Div. of Production, Dallas (1961) Chap. 7.
2. Howard, G. C., and Clark, J. B.: "Factors To Be Considered in Obtaining Proper Cementing of Casing," *Drill. and Prod. Prac.,* API (1948) 257-272; *Oil and Gas J.* (Nov. 11, 1948) 243.
3. Howard, G. C., and Fast, C. R.: "Squeeze Cementing Operations," *Trans.,* AIME (1950) **189,** 53-64.
4. Montgomery, P. C., and Smith, D. K.: "Oil Well Cementing Practices and Materials," *Pet. Eng.* (May and June 1961).
5. Shryock, S. H., and Slagle, K. A.: "Problems Related to Squeeze Cementing," *J. Pet. Tech.* (Aug. 1968) 801-807.
6. Abadie, H. G.: "Oil Well Repair by Scabbing Methods," *Oil Weekly* (Dec. 2, 1940) **99,** No. 13, 18-28.
7. Tausch, G. H.: "Squeeze Cementing With Permanent-Type Completions," *Oil-Well Cementing Practices in the United States,* API, New York (1959) 161-175.
8. Beach, H. J., O'Brien, T. B., and Goins, W. C., Jr.: "Formation Cement Squeezed by Using Low-Water-Loss Cements," *Oil and Gas J.* (May 29 and June 12, 1961).
9. Goolsby, J. L.: "A Proven Squeeze Cementing Technique in a Dolomite Reservoir," *J. Pet. Tech.* (Oct. 1969) 1341-1346.
10. Boice, D., and Diller, J.: "A Better Way To Squeeze Fractured Carbonates," *Pet. Eng.* (May 1970) 79-82.
11. Rike, J. L.: "Obtaining Successful Squeeze Cementing Results," paper SPE 4608 presented at SPE-AIME 48th Annual Fall Meeting, Las Vegas, Nev., Sept. 30-Oct. 3, 1973.
12. Young, V. R.: "Well Workover With Remedial Rig," Petroleum Engineer Refresher Course No. 4 — Individual Well Analysis, Los Angeles Basin Section of SPE-AIME (1967).

13. Hodges, J. W.: "Squeeze Cementing Methods and Materials," *Oil-Well Cementing Practices in the United States,* API, New York (1959) 149-159.

14. "Recommended Practice for Testing Oil-Well Cements and Cement Additives," *API RP 10B,* 19th ed., API Div. of Production, Dallas (1973).

15. Binkley, G. W., Dumbauld, G. K., and Collins, R. E.: "Factors Affecting the Rate of Deposition of Cement in Unfractured Perforations During Squeeze-Cementing Operations," *Trans.,* AIME (1958) **213,** 51-58.

16. Shell, F. J., and Wynne, R. A.: "Application of Low-Water-Loss Cement Slurries," paper 875-12-I presented at API Rocky Mountain Dist. Div. of Production Spring Meeting, Denver, Colo., April, 1958.

17. Dumbauld, G. K., Perry, D., Brinkley, G. W., and Brooks, F. A., Jr.: "An Accelerated Squeeze-Cementing Technique," *Trans.,* AIME (1956) **207,** 25-29.

18. Beach, H. J., O'Brien, T. B., and Goins, W. C., Jr.: "Controlled Filtration Rate Improves Cement Squeezing," *World Oil* (May 1961) 87-93.

19. Stout, C. M., and Wahl, W. W.: "A New Organic Fluid-Loss-Control Additive for Oilwell Cements," *J. Pet. Tech.* (Sept. 1960) 20-24.

20. Carter, L. G., Harris, F. N., and Smith, D. K.: "Remedial Cementing of Plugged Perforations," paper SPE 759 presented at SPE-AIME 34th Annual California Regional Fall Meeting, Santa Barbara, Oct. 23-25, 1963.

Chapter 10

Open-Hole Cement Plugs

10.1 Introduction

Almost always, at some time in the life of an oil, gas, or water well, down-hole plugging is required. In most areas, regulatory bodies list rules governing the method of plugging a drilled hole for abandonment. Where no rules exist, the method is left to the discretion of the operator.

Each plugging operation presents a problem because a relatively small volume of cement slurry is placed in a large volume of wellbore fluid. Mud can contaminate the cement and the result, even after a reasonable WOC time, is a weak, diluted, or unset plug. For these reasons, both mechanical and chemical technology are necessary to successful plugging.

Most of the techniques for placing plugs in open holes have been developed in the process of whipstocking, of plugging back open holes before setting casing, and of testing formations.[1-3] As many as five or six attempts have been made to place open-hole plugs in highly treated mud systems before a satisfactory plug has been obtained. Numerous wells are plugged for abandonment with as many as 15 plugs between the bottom of the hole and the surface.[4]

Normally, an operator does not check the position of the plugs, but assumes they are where they should be. And in fact, wells that have been plugged for abandonment and re-entered years later have shown every plug to be located in the hole exactly as recorded. The best way to ascertain the position and condition of a plug is to feel it. This can be done by running tubing string or drillpipe down the hole and touching or "tagging" the plug with it. Such tagging is common practice in offshore operations.

10.2 Uses of Cement Plugs

There are several reasons for performing plugging operations; some of them are described in Fig. 10.1 and discussed in the following paragraphs.

Abandonment

To seal off selected intervals of a dry hole or a de-

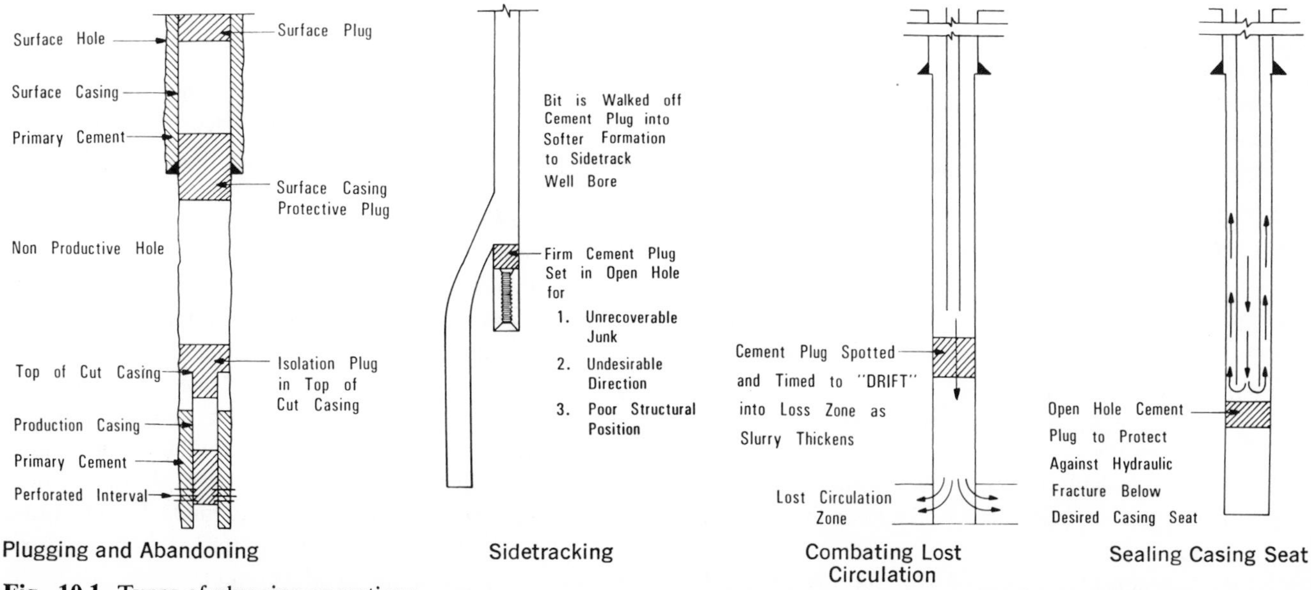

Fig. 10.1 Types of plugging operations.

pleted well, a cement plug placed at the required depth helps prevent zonal communication and any migration of fluids that might infiltrate underground fresh-water sources. Local regulations governing plugging operations for well abandonment should be consulted.

Zone Isolation

One of the most common reasons for plugging is to isolate a specific zone. The purpose may be to shut off water, to recomplete a zone at a shallower depth, or to protect a low-pressure zone in an open hole before squeezing. In a well that has two or more producing intervals, it is sometimes beneficial to abandon a depleted zone or unprofitable producing zone by placing a permanent cement plug to isolate this zone, thus helping to prevent possible production losses into or fluid migration from another interval.

Directional Drilling

In sidetracking a hole around a nonretrievable fish — such as a stuck, broken drillstring — it is necessary to place a cement plug at the required depth to help support the whipstock so the bit can be guided in the desired direction. Such a plug is used to change directions in shoreline drilling, relief-well drilling, drilling under salt domes, and drilling toward any other inaccessible target.

Lost-Circulation Control

When mud circulation is lost during drilling it is sometimes possible to restore lost returns by spotting a cement plug across the thief (lost-circulation) zone and then drilling back through the plug. Generally, this is less expensive than a squeeze-cementing job. Reinforcing fibers and lost-circulation-control additives incorporated in the cement plug minimize shattering and disintegration of the residual cement as the cement plug is being drilled out, and assure a more successful job.

Formation Testing

Cement plugs are frequently placed in the open hole below a zone that is to be tested and that is a considerable distance off bottom where it is not possible or practical to place a sidewall anchor or a bridge plug. Cement plugs should be long enough to keep from sliding down the hole when abnormal weights are applied to them.

10.3 Placement Precautions

In placing a cement plug in open hole, one must consider carefully the type of formation through which the plug is to be placed. Plug failures can be prevented by taking the following precautions[5]:

1. Selecting, with the help of a caliper log, a gauge section of the hole.
2. Carefully calculating cement, water, and displacement volumes, and always planning to use more than enough cement.
3. Using a densified cement (API Class A, G, or H) that will tolerate considerable mud contamination.
4. Preceding the cement with sufficient flush.
5. Rotating the tubing using tail pipe with centralizers and scratchers while placing the cement.
6. Using drillpipe wiper plugs and plug catchers.
7. Placing the plug with care and moving the pipe slowly out of the cement to minimize mud contamination.

For maximum bonding, a clean, hard formation should be selected, particularly for zone isolation or abandonment. For directional drilling, an easily drilled formation should be selected; placing the plug in a hard zone can mean having to redrill the cement plug in the open hole. In drillstem testing or in settting bottom plugs to seal off water or other well fluids, the cement should be placed across the fluid interval and extend up through a very hard, impermeable, gauge hole. In plugging wells for abandonment, a plug should be set in or below the lowest fresh-water zone or at the base of the surface pipe, or both places (Fig. 10.1).[4] Caliper logs should always be consulted in selecting plugging locations.

10.4 The Mud System

Before a cement plug is placed, the mud and its properties should be studied. After the well is cored and tested, the mud system may be contaminated and yet, particularly if it is to be abandoned, the hole may not warrant reconditioning. It has been estimated that more than 70 percent of the mud systems in use contain some ferrochrome lignosulfonate, which can interfere with the proper setting of the cement.[6-8] Some muds do not gel or thicken when in contact with cement and they therefore allow the heavier cementing slurry to lubricate or slide down the hole.

With native or simple water-base mud systems, a thick gel may form where the cement contacts the mud, and the cement will channel as it is being placed in the open hole. To obtain a good cement bond between the plug and the formation, thick, soft filter cake should be removed.

Open-hole plug-back studies show that the preferred mud system has a Marsh funnel viscosity of 45 to 80 seconds, a plastic viscosity of 12 to 20 cp, a yield point of less than 5 lb/100 sq ft, and a water loss of less than 15 cc.[5] Operators in some areas have eliminated problems of mud/cement contamination by spotting a freshly mixed nonchemically treated mud consisting only of bentonite, water, and weighting material (such as barite) below and throughout the zone in which the cement plug is to be placed. Since most plugs are placed by the balance method, it is important that the mud be circulated long enough to ensure that the entire mud system is uniformly weighted. Otherwise, the cement slurry may become overbalanced or underbalanced and the plug might not be placed at the desired depth or might become contaminated by mud.

10.5 Cement Volume and Slurry Design

The amount of cement for a particular job is controlled by the length of the plug and the diameter of the hole. (State and federal regulations also have a bearing: a specific number of feet of cement may be legally required to be left in an abandoned hole.) In some instances, 10 to 20 sk may be sufficient to shut off bottom-hole water. For whipstocking, on the other hand, upwards of 200 ft may be needed to minimize the contamination that can occur at the top of the plug. Plugs for abandoning hole or for whipstocking probably average between 125 and 150 sk. Larger quantities of cement improve the chances of success when a hole is to be deviated. When plugs are placed through drillpipe or tubing, fluid spacers should be used both ahead of and behind the slurry to minimize the mixing of cement and drilling mud. Water is by far the most common spacer fluid, although crude oil, diesel oil, and mixtures of water and gel are also used.

In selecting the volume of cement for a given plugging operation, allowances are usually made for dressing off the mud-contaminated cement at the top of the plug. For small jobs, batch mixing of the slurry may be preferred to continuous mixing to assure homogeneity. Where facilities for dry mixing are not readily available, the additives — dispersants, accelerators, retarders — can be put in the mixing water. Table 10.1 lists volumes of cement slurry required for open-hole plugs of various heights. Calculations used in connection with cement plugging are set out in Appendix C.

The selection of a cement composition for an open-hole plug will depend on well depth, temperature, and mud properties. API test schedules developed from field studies on a large number of plugging jobs show that the first sack of cement on a plug reaches bottom rather quickly. For a 100-sk cement plug, the placement time in a 16,000-ft hole is approximately 40 min.[9]

Mud contamination is always a possibility during open-hole plug placement. In the cementing system it can cause retardation and dilution of the cement plug; therefore, densified or reduced-water-ratio cements (API Classes A, G, and H) generally produce more successful results.[10,11] (Table 10.3 shows typical compressive strengths of densified Classes G and H cements used for open-hole plugging.)

TABLE 10.1 — VOLUME OF SLURRY REQUIRED FOR OPEN-HOLE CEMENT PLUGS

Hole Size (in.)	Volume (cu ft) Required for Plug of Following Height			
	50 ft	100 ft	200 ft	500 ft
4	4.36	8.72	17.44	43.65
6	9.81	19.63	39.26	98.15
8	17.45	34.91	79.82	174.50
10	27.27	54.54	109.08	272.70
12	39.27	78.54	157.08	392.70
14	53.45	106.90	213.80	534.50

TABLE 10.2 — EFFECT OF MUD CONTAMINATION ON STRENGTH OF CEMENT

Curing time: 12 hours.
Curing temperature: 230°F.
Slurry weight
Cement A: 15.6 lb/gal.
Cement B: 17.4 lb/gal.

Mud Contamination (percent)	Compressive Strength (psi)	
	Cement A	Cement B*
0	2,910	7,010
10	2,530	5,005
30	1,400	2,910
60	340	2,315

*Contains dispersant.

TABLE 10.3 — TYPICAL COMPRESSIVE STRENGTHS OF API CLASSES G AND H CEMENTS

Slurry Weight (lb/gal)	Compressive Strength (psi) at API Curing Conditions of			
	110°F 1,600 psi	140°F 3,000 psi	170°F 3,000 psi	200°F 3,000 psi
		After 12 Hours		
16.5	2,075	4,000	7,800	9,035
17.0	2,850	6,535	8,375	10,025
17.5	3,975	6,585	8,550	10,675
		After 24 Hours		
16.5	5,475	8,985	9,750	10,460
17.0	6,035	9,060	11,075	12,660
17.5	7,025	10,125	11,860	12,875

10.6 Placement Techniques

The Balanced Method

The balanced method (Fig. 10.2) involves pumping a desired quantity of cement slurry through drillpipe or tubing until the level of cement outside is equal to that inside the string. The pipe or tubing is then pulled slowly from the slurry, leaving the plug in place. The method is simple and requires no special equipment other than a cementing service unit. The characteristics of the mud are very important in the balancing of a cement plug in a well, particularly the ability to circulate freely during placement. When the purpose of a plug is to control lost circulation, the plug is often spotted

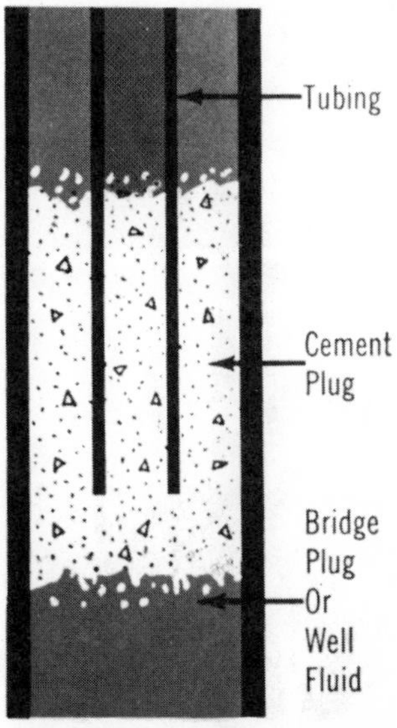

Fig. 10.2 Balance method of placing cement plug.

with a "drift plug" technique; that is, the plug is "timed" so that when it reaches the lost-circulation zone down the hole it sets and seals off the zone (see Fig. 10.1).

Movement of well fluids while the cement plug is setting may affect the quality of a plug. Even a small amount of gas migrating slowly through a cement plug can disturb it enough to prevent it from setting. In some areas, there are artesian flows that tend to move the cement plug up the hole or wash it out of the well. It is necessary, therefore, to check the cement system very carefully to see that the well is in a static state — neither gaining nor losing returns. The amount of mud, wash, and cement slurry must be carefully calculated to ensure equal volumes of fluid ahead of and behind the cement plug as it is being balanced in the hole.

When it is difficult to establish the top of a cement plug, it may be necessary to run an excess of cement, then pull the running-in string to the desired plug top and reverse out the cement above that point. A loss of fluid to the formation below this point may cause movement of the plug.

The Dump Bailer Method

The dump bailer method (Fig. 10.3) is usually employed at shallow depths; but with the formulation of retarded cementing compositions, it has been used to depths exceeding 12,000 ft. The dump bailer, containing a measured quantity of cement, is lowered on a wire line. A limit plug, cement basket, permanent bridge plug, or gravel pack is usually placed below the desired plugging location. The bailer is opened by touching the bridge plug and is raised to release the cement slurry at this location. The method has certain advantages in that the tool is run on wire line and the depth of the cement plug is easily controlled. The cost of a dump bailer job is usually low compared with one using conventional pumping equipment.

Some disadvantages of the dump bailer method are that (1) it is not readily adaptable to setting deep plugs; (2) mud can contaminate the cement unless the hole is circulated before dumping (this is also true of the balanced method); and (3) there is a limit to the quantity of slurry that can be placed per run, and an initial set may be required before the next run can be made.

The Two-Plug Method

In the two-plug method (Fig. 10.4), top and bottom tubing plugs are run to isolate the cement slurry from the well fluids and displacement fluids (as in standard primary cementing practice). A bridge plug is usually run at the cement plugging depth. A special baffle tool is run on the bottom of the string and placed at the depth desired for the bottom of the cement plug. This tool permits the bottom tubing plug to pass through and out of the tubing or drillpipe. Cement is then pumped out of the string at the plugging depth and begins to fill the annulus. The top tubing plug, following the cement, is caught in the plug-catcher tool, causing a sharp rise in the surface pressure, which indicates that the plug has landed. The latching device holds the top tubing plug to help prevent cement from backing up into the string, but permits reverse circulation. (This design allows the string to be pulled up after cement placement to "cut off" the cement plug at the desired

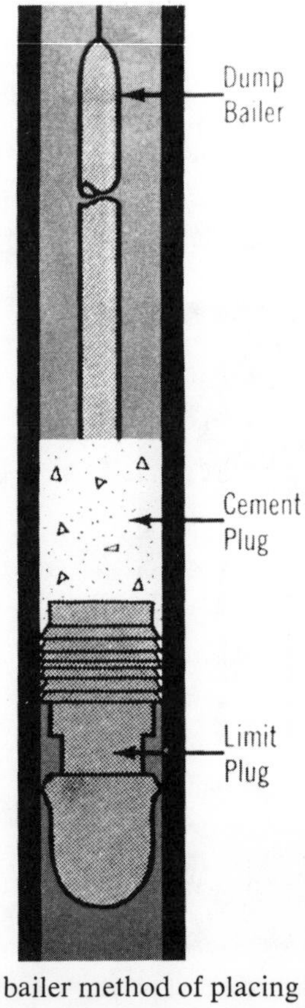

Fig. 10.3 Dump bailer method of placing cement plug.

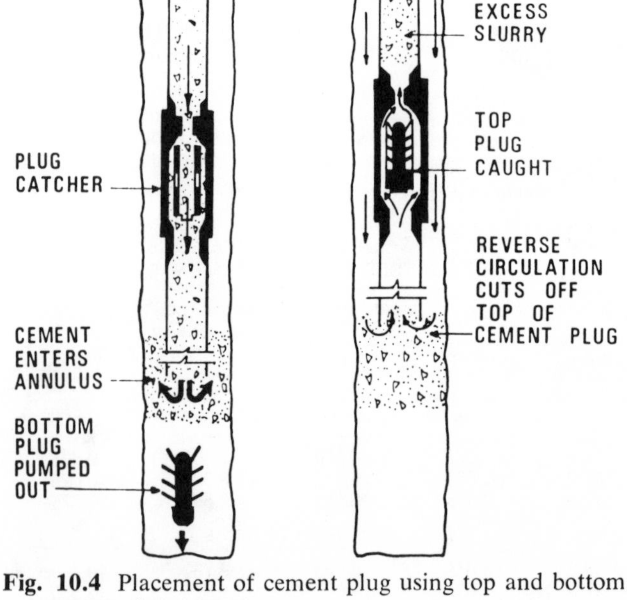

Fig. 10.4 Placement of cement plug using top and bottom wiper plugs.

depth by establishing reverse circulation through the plug catcher, thus allowing excess cement to be reversed up and out of the tubing.) The string is then pulled, leaving a cement plug that should last indefinitely and provide good, hard support for any subsequent operation.

To minimize contamination, centralizers and rotating scratchers can be put at the lower end of the bottom drillpipe or tail pipe.[13] The rotation of the scratchers cleans the wellbore — thus promoting better bonding — and allows bypassed mud to mix uniformly with the cement, eliminating mud channels in the unset cement.

Advantages of the two-plug method are that (1) it minimizes the likelihood of overdisplacing the cement; (2) it forms a tight, hard cement structure; and (3) it permits establishing the top of the plug. All in all, the two-plug method of plugging is preferred to the balanced method.

10.7 Testing Cement Plugs

There is no simple method of testing down-hole plugs. In most cases, plugs for abandonment or for sealing off bottom water are never tested. Plugs set to control lost circulation or for whipstocking are tested by determining the hardness of the plug. The most common approach is to run drillpipe — either open ended or with a bit — back into the hole to locate the plug by applying weight. This method is commonly used after the plug has been allowed to set some 12 to 24 hours.[14] Although it is not always satisfactory, at least it gives some indication of whether some degree of plugging has been achieved in the desired location. A plug might be hard on top, but soft farther down, so that in time fluids can migrate past it.

Normal WOC time after placement of plugs is from 12 to 36 hours; however, with the use of densified cement and accelerators a very hard plug can be achieved in 8 to 18 hours. Where temperatures are above 230°F, silica flour functions as a stabilizing agent and as a catalyst for producing high-strength cement plugs in minimum time after placement.[15]

10.8 Barite Plugs

Plugs composed of barite, water, and a thinner are commonly used for pressure control.[12] Formulations of barite and water (18 to 24 lb/gal) can be mixed by adding 0.2 to 0.7 lb/bbl of complex phosphate and adjusting the pH to 8-10 with caustic soda. (See Table 10.4 for various data on barite slurries.)

Barite plugs are placed through the drillpipe and spotted near the active zone. They are most successful when placed immediately after an active zone has been opened. The drill bit need not be removed.

A barite plug will seal the wellbore because (1) it has a high density, which increases the hydrostatic head and restrains the active zone; and (2) the barite slurry has a high filtrate loss, which causes it to dehydrate rapidly and form a plug. (This high filtrate loss can also cause sloughing and bridging in the hole, as well as dehydration and settling of the plug.)

The viscosity and yield point of the barite slurry must be kept low if the barite is to settle and form a plug. Mixing the slurry with mud in the hole should be avoided, since the barite settles rapidly and can plug and stick the drillstring. Once the barite plug is placed, the drillpipe should be removed quickly to avoid sticking or plugging.[12]

The only source of mud contamination is from inside the drillpipe; fluid and mud from the annulus will remain on top of the barite plug. Mud can be kept out of the barite slurry by underdisplacing the slurry in the drillpipe. Because the barite plug weight is greater in the drillpipe than in the annulus, there will be a flow from the drillpipe to the annulus that will help prevent premature setting. Overdisplacing the barite plug from the drillpipe is certain to get mud into the barite plug and should be avoided.

Fig. 10.5 shows how barite plugs are placed in open hole.

10.9 Summary

Plug-back operations should be carefully planned, with emphasis given to placement, volume of cement, borehole irregularities, and errors in pipe measurement, as well as to the calculated volumes of cement to be placed in the hole. For a successful plug-back operation, the following measures should be taken (see Fig. 10.6 and the Check List, Table 10.5).

1. Place the plug in a competent formation (i.e., place a strong cement against a hard formation).
2. Use ample cement.
3. Use tail pipe through plug-back intervals.
4. Use scratchers or wipers and centralizers on tail pipe where the hole is not excessively washed out.

TABLE 10.4 — WEIGHT/VOLUME RELATIONSHIPS OF BARITE SLURRIES*

Slurry Density (lb/gal)	Water (gal/sk)	Sacks Barite per Barrel Slurry	Slurry Yields: Barrels per Sack	Barrels per 200 sk	Barrels per 300 sk	Barrels per 400 sk	Cubic Feet per Sack
18.0	5.10	5.30	0.189	37.8	56.2	75.5	1.060
20.0	3.70	6.43	0.156	31.1	46.6	62.1	0.873
21.0	3.20	6.95	0.144	27.8	43.2	57.5	0.807
22.0	2.75	7.50	0.133	26.6	40.0	53.3	0.748

*Based on water density of 8.33 lb/gal, a barite absolute volume of 0.0284 gal/lb (corresponding to a density of 35.2 lb/gal), and a sack weight of 100 lb.

5. Use a drillpipe plug and a plug catcher.
6. Circulate the hole sufficiently before running the job. Use a mud of low yield point and low plastic viscosity, of sufficient weight to control the well.
7. Ahead of the cement, run a flush that is compatible with the mud.
8. Use densified cements with dispersant to combat mud contamination.
9. Where hard plugs are desired, use sand or similar materials in the cement.
10. Allow ample time for the cement to set.

References

1. Parsons, C. P.: "Plug-Back Cementing Methods," *Trans.*, AIME (1936) **118,** 187-194.
2. Howard, G. C., and Scott, P. P., Jr.: "Plugging Off Water in Fractured Formations," *Trans.*, AIME (1954) **201,** 132-137.
3. Goins, W. C., Jr.: "Open Hole Plugback Operations," *Oil-Well Cementing Practices in the United States,* API New York (1959) 193.
4. Banister, J. A.: "Methods and Materials for Placing Cement Plugs in Open Holes," paper presented at meeting of Interstate Oil Compact Commission, Yellowstone, Wyo., June 10-12, 1957.
5. Anderson, F. M.: "A Study of Surface Casing and Open-Hole Plug-Back Cementing Practices in the Mid-Continent District," *Drill. and Prod. Prac.,* API (1955) 312-325.
6. Morgan, B. E., and Dumbauld, G. K.: "Use of Activated Charcoal in Cement to Combat Effects of Contamination by Drilling Muds," *Trans.,* AIME (1952) **195,** 225-232.
7. Beach, H. J., and Goins, W. C., Jr.: "A Method of Protecting Cements Against the Harmful Effects of Mud Contamination," *Trans.,* AIME (1957) **210,** 148-152.
8. Anderson, F. M.: "Effects of Mud-Treating Chemicals on Oil Well Cements," *Oil and Gas J.* (Sept. 29, 1952) 283-284.
9. "Recommended Practice for Testing Oil-Well Cements and Cement Additives," *API RP 10B,* 19th ed., API Div. of Production, Dallas (1973).
10. Horton, H. L., Morris, E. F., and Wahl, W. W.: "Improved Cement Slurries by Reduction of Water Content," paper 909-9-D presented at API Southwestern Dist. Div. of Production Spring Meeting, Midland, Tex. (March 1964) preprint.

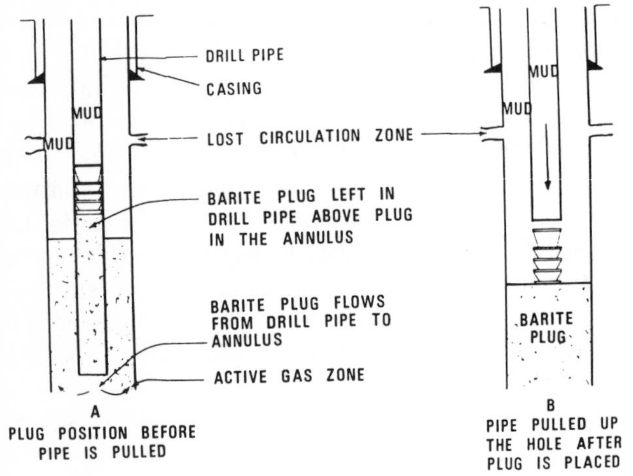

Fig. 10.5 Placing barite plugs in open hole (after Ref. 12).

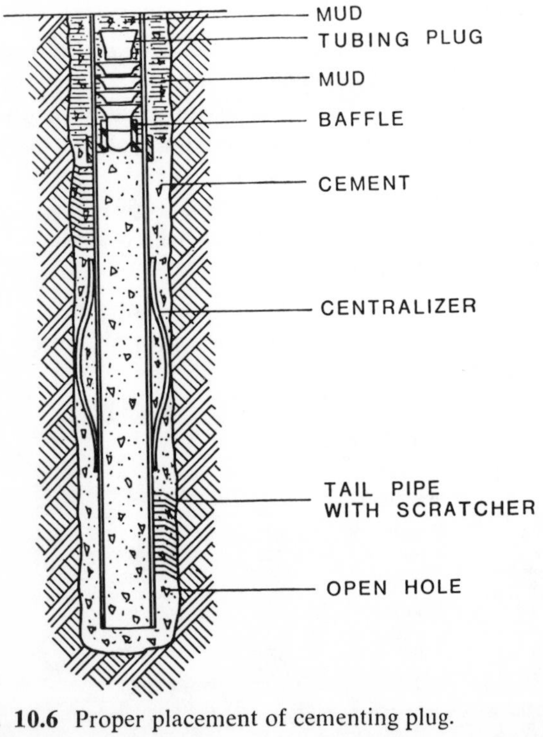

Fig. 10.6 Proper placement of cementing plug.

TABLE 10.5 — PLUG-BACK CHECK LIST

Hole size (in. or mm)__________. Casing size (in. or mm)__________. Drillpipe or tubing size (in. or mm)__________. Drillpipe or tubing thread:__________. Top plug from__________(ft or m) to__________(ft or m).

Type of cementing material and additives:__________.

Slurry weight (lb/gal or kg/liter)__________.

Mixing water (gal/sk or liters/sk)__________.

Total water (bbl or cu m)__________.

Yield (cu ft/sk or liters/sk)__________.

Slurry volume (bbl or cu m)__________.

Pumping time for cement__________hours__________minutes.

Bottom-hole temperature (°F or °C)__________.

Compressive strength (psi or kPa)__________ after__________hours.

Fluid to displace cement to equalization (bbl or cu m)__________.

Water to be pumped ahead of cement (bbl or cu m)__________.

Water to be pumped behind cement (bbl or cu m)__________.

11. Waggoner, H. F.: "Additives Yield Heavy High-Strength Cements With Low Water Ratios," *Oil and Gas J.* (April 3, 1964) 109-111.

12. Messenger, J. U.: "Barite Plugs Simplify Well Control," *World Oil* (June 1969) 83.

13. Teplitz, A. J., and Hassebroek, W. E.: "An Investigation of Oil-Well Cementing," *Drill. and Prod. Prac.*, API (1946) 76-101; *Pet. Eng. Annual* (1946) 444.

14. Pugh, T. D.: "What to Consider When Cementing Deep Wells," *World Oil* (Sept. 1967) 52.

15. Carter, L. G., and Smith, D.K.: "Properties of Cementing Compositions at Elevated Temperatures and Pressures," *J. Pet. Tech.* (Feb. 1958) 20-28.

Chapter 11

Flow Calculations

11.1 Introduction

An understanding of flow mechanics in a wellbore can be helpful in selecting pumping equipment, cementing compositions, and placement techniques. With the development of the multispeed rotational viscometer for measuring flow properties, the rheological properties of well fluids have become more widely understood and analysis has advanced beyond the empirical methods used in the past.[1-4]

The use of rheological parameters of cement slurry and drilling fluid allows the following factors to be computed during the cementing operation (Fig. 11.1):

1. Annular velocity and pumping rate required to establish plug, laminar, or turbulent flow.
2. Cement slurry velocity inside casing.
3. Frictional pressure in pipe and annulus from flow of both drilling mud and cement slurry.
4. Expected wellhead pressure as the top plug moves down the pipe.
5. Hydraulic horsepower needed at the wellhead.
6. Slurry volume for a given contact time.
7. Time required to complete the cementing operation.

11.2 The Flow Properties of Wellbore Fluids

The flow properties of wellbore fluids (water, mud, cement slurries, mud spacers, and displacement fluids) are conventionally classified as Newtonian or non-Newtonian. Newtonian fluids are fluids such as oil, syrup, or water. They exhibit a direct and constant proportionality between shear rate (which is related to flow velocity or rate) and shear stress (which is related to flowing pressure drop) as long as the flow regime is laminar. In a fluid of this type, viscosity is independent of the shear rate at constant temperature and pressure. A Newtonian fluid will begin to flow immediately when pressure (force) is applied. When the pressure is released, the fluid returns immediately to its previous state before pressure or force was applied. (See Fig. 11.2.)

The term non-Newtonian describes all fluids whose behavior is different from that of a Newtonian fluid — for example, drilling fluids, cement slurries, and heavy asphaltic oils. These are rheologically complex, frequently particle-bearing fluids that are usually described as Bingham plastics or power-law fluids. Non-Newtonian fluids (Figs. 11.2 and 11.3) do not exhibit a direct proportionality between pressure loss and flow rate at constant temperature and pressure. Some types of non-Newtonian fluids, such as drilling mud, do not start to move immediately when a force is applied, but will go through stages of flow: plug, laminar, turbulent.[1,5,6]

Some non-Newtonian fluids in a static state possess thixotropy — a property of a fluid that causes it to build up a rigid or semirigid structure that breaks down with applied shear. Once the gel structure is broken, the fluid will flow as long as pressure or shear stress is applied. This structure can rebuild if the fluid is allowed to rest.

The two mathematical models commonly used in describing the behavior of drilling fluids and cement slurries are the Bingham-plastic model and the power-law model.[3]

The Bingham-Plastic Model

This is the model more widely used in the oil industry because of its early identification with drilling fluids.[5] It assumes that cement slurries and drilling fluids behave like an ideal Bingham plastic and that all rheological calculations can be made from a linear relationship between shear stress and shear rate. This relationship, called "apparent viscosity" (rather than simply "viscosity") can be obtained with a rotational viscometer, or Fann VG meter (Fig. 11.4). The Fann instrument is available in a variety of models for field and laboratory use and is designed to operate at six rotation speeds (600, 300, 200, 100, 6, and 3 rpm). For this instrument, the shear stress (in pounds per square foot) may be expressed as follows:

$$S_s = \text{shear stress} = \frac{\text{dial reading} \times \text{instrument spring factor} \times 1.066}{100} \quad . . . . (11.1)$$

(Note that the 1.066 may vary, depending on the combination of rotor and bob.)

The shear rate is a function of the speed of rotation and the dimensions of the rotor and bob. For the standard instrument,

$$S_r = \text{Shear Rate} = 1.703 \times \text{rpm} \quad . . \quad (11.2)$$

See Table 11.1.

The two terms used in describing the fluid in a Bingham-plastic model are plastic viscosity and yield point (Fig. 11.3). The plastic viscosity is expressed as the slope of the straight-line extrapolation; the yield point is the intercept of this straight line on the shear-stress axis. The Fann VG meter is designed to easily calculate

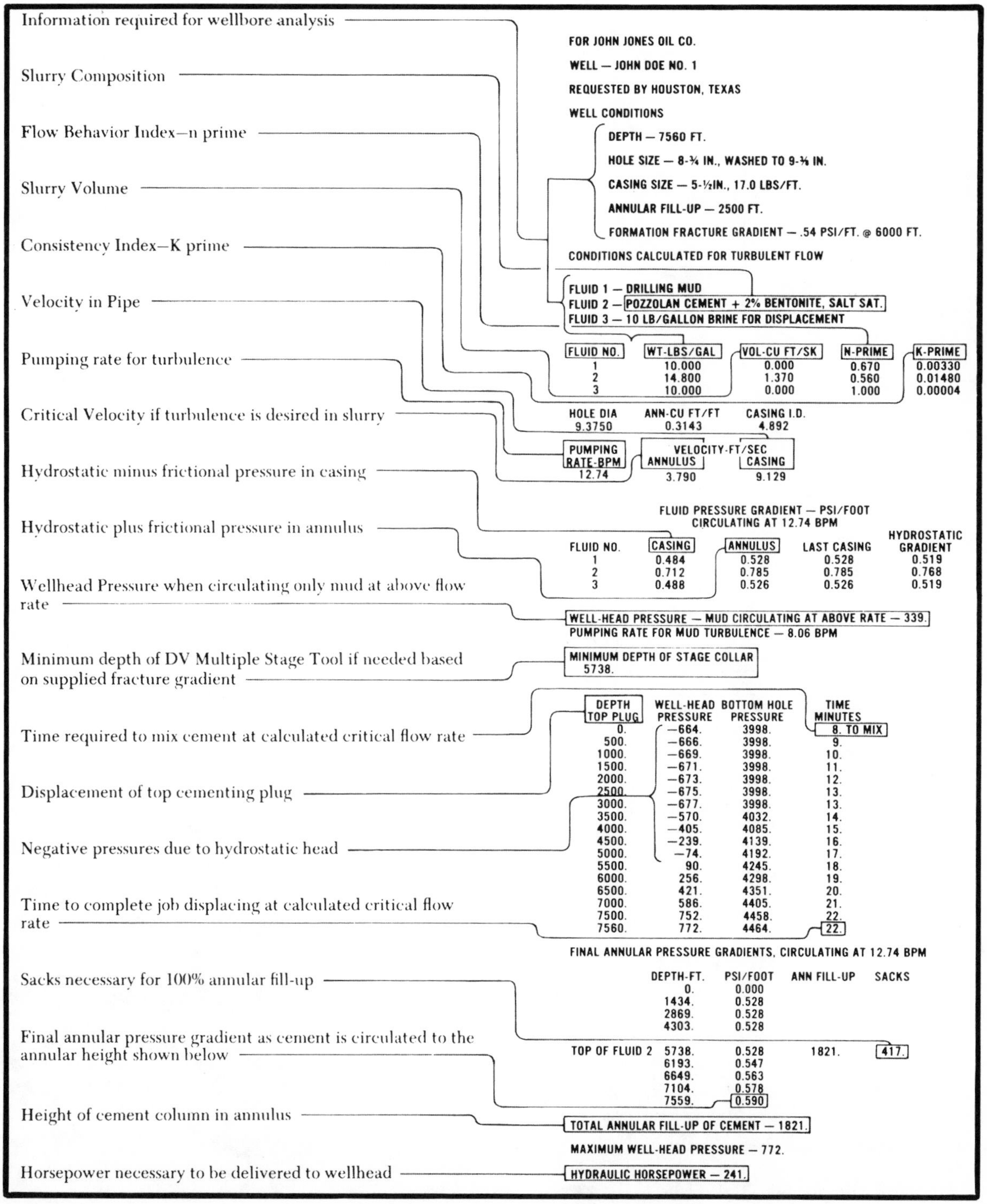

Fig. 11.1 Computer calculation of wellbore hydraulics during a cementing operation (adapted from printout).

plastic viscosity and yield point. (The VG meter exerts the required shear stresses at two different shear rates: 600 and 300 rpm.) The basic equation describing the Bingham-plastic model is, therefore,

$$S_s = Y + 2.088555 \times 10^{-5} (\mu_p)(S_r), \quad \ldots\ldots\ldots \quad (11.3)$$

where

S_s = shear stress, lb_f/ft^2,

Y = yield point, lb_f/ft^2,

μ_p = plastic viscosity, cp,

S_r = shear rate, sec^{-1}.

This equation yields a straight line and from it, on the basis of the instrument calibration, plastic viscosity and yield point can be determined. Plastic viscosity is the 300 reading subtracted from the 600 reading, and yield point is the plastic viscosity subtracted from the 300 reading.

It is important to understand the difference between plastic viscosity and apparent viscosity. Plastic viscosity is the actual slope of the linear part of the shear-rate/shear-stress curve as shown in Fig. 11.3 and is constant. Apparent viscosity, which is dependent on shear rate, is the slope of a straight line from the origin to a point on the shear-rate/shear-stress curve (that is, it is the quotient of shear stress divided by shear rate).

The Power-Law Model

The power-law model, another non-Newtonian mathematical approach, was popularized by Metzner and Reed.[7] It is based on the assumption that the fluid (cement slurry) exhibits a proportionality between the logarithm of pressure loss and the logarithm of flow rate, with a starting stress in the region of streamline or laminar flow.

The equations of the power-law model are more complex but also more accurate than those of the Bingham-plastic model; therefore, the results are closer to the exact behavior of the cement slurries in the well.[3] By knowing the characteristics of the shear-stress vs

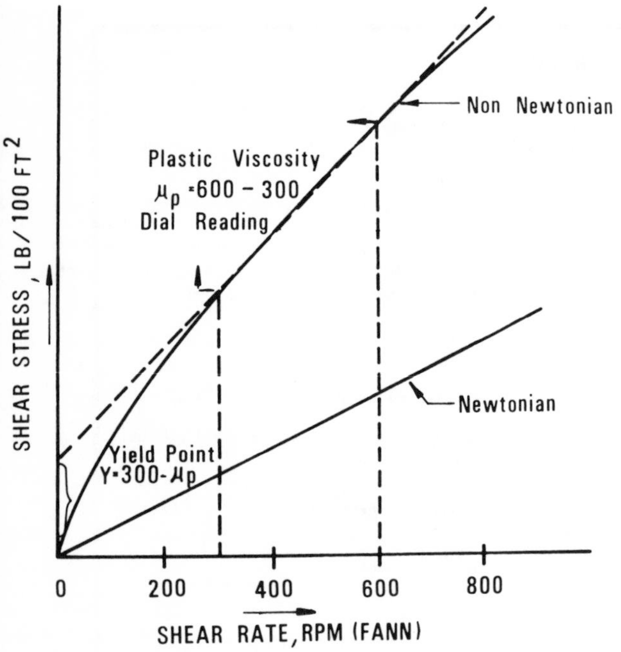

Fig. 11.2 Flow-rate/shear-stress curves of Newtonian and non-Newtonian fluids.

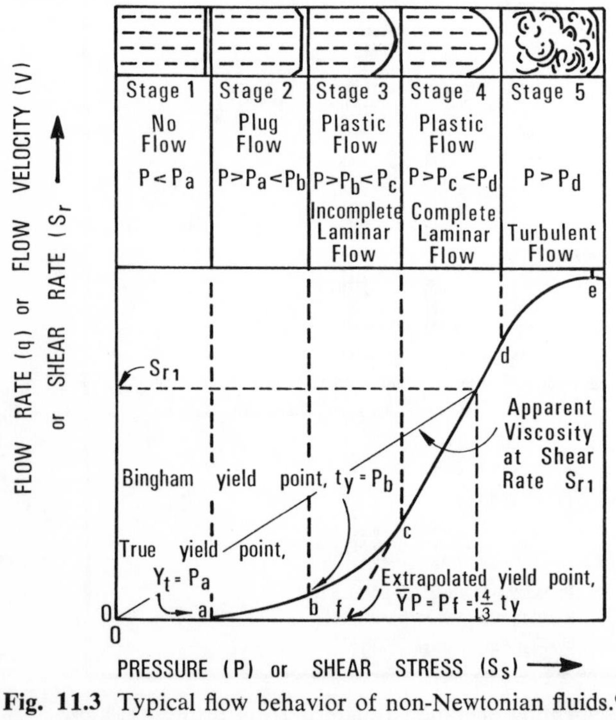

Fig. 11.3 Typical flow behavior of non-Newtonian fluids.[6]

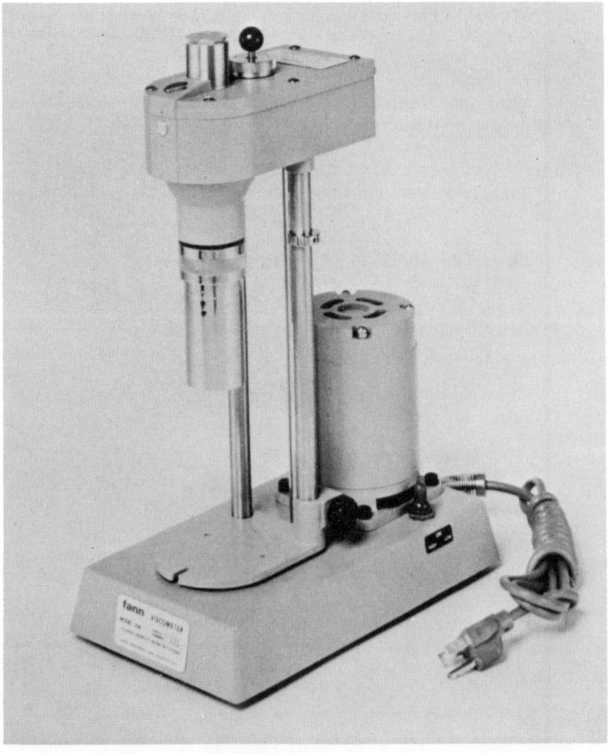

Fig. 11.4 Fann VG meter.

shear-rate curve, it is possible to calculate the apparent viscosity of the cement slurry at the observed shear rate. It is this viscosity that is used in the hydraulic equations, rather than the plastic viscosity in the Bingham-plastic model.[8,9]

The Fann VG meter is still required to obtain data for the power-law model. The equation for the power-law model is

$$S_s = K' (S_r)^{n'}, \quad . \quad . \quad . \quad . \quad . \quad . \quad . \quad . \quad . \quad (11.4)$$

where

K' = intercept of lines (fluid consistency index), $lb_f sec^{n'}/ft^2$

n' = slope of the shear-stress/shear-rate curve (flow behavior index), dimensionless.

Whereas the Bingham model uses only the 600 and the 300 readings to describe the shear-stress/shear-rate curve, the power-law model can use either the 600 and 300 dial readings or, for greater accuracy, the 600, 300, 200, and 100 readings. The data derived with the power-law model yield a straight line on log/log paper:

$$\log_{10}(S_s) = \log_{10}K' + n' \log_{10}(S_r). \quad . \quad . \quad . \quad . \quad (11.5)$$

The two parameters required to define the power-law model fluid are usually denoted by the symbols n' and K'. (See Table 11.2.) For the purpose of this discussion, n' is the slope of the log(shear stress) vs log (shear rate) curve, or "flow behavior index," and K' becomes the intercept of this line at unit shear rate and is referred to as the "consistency index." If these two indexes are known, it is possible to calculate the Reynolds number and the critical velocity at which departure from laminar flow begins.

For a power law fluid, apparent viscosity at any specific shear rate is given by

$$\mu = 47880\ K'\ S_r^{n'-1}, \quad . \quad . \quad . \quad . \quad . \quad . \quad . \quad (11.6)$$

where units are centipoises, lb $sec^{n'}/ft^2$, and sec^{-1}, respectively. Clearly, if $n' = 1$, then $\mu = 47880\ K'$, a constant, and the fluid is Newtonian.

11.3 Instruments Used To Predict Fluid Flow Properties

For flow calculations the accurate measurement of the viscosity of a fluid is essential. The two principal instruments used for rheological measurements are the pipeline (capillary-tube) viscometer and the rotational Fann VG viscometer.[2,3,8] The capillary-tube viscometer (which is used to measure the relation between the pressure drop and the flow rate of a fluid) is the preferred method for determining the flow behavior index and consistency index for non-Newtonian fluids.[8] Having pressure-drop data at various flow rates, one can prepare a log/log plot of shear rate vs shear stress. For fluids that do not exhibit time dependency (that is, fluids that are not thixotropic), these data will usually produce a straight line. The flow behavior index, n', is the slope of this line, and the consistency index, K', is the intercept (the value) at unity shear rate (Fig. 11.5). The n' will be the same as determined by each instrument. The K' values, however, will differ and will require a correction from the Fann VG meter to either pipe or annular geometry.[2]

It is difficult to keep the cement slurry uniform and pumpable long enough to obtain measurements from the pipe viscometer. Therefore the n' and K' data are normally obtained with the readily available direct-indicating rotational viscometer (Fann VG meter), using Eqs. 11.1 and 11.2 to convert instrument rpm and dial reading to shear rate and shear stress. As with the pipe viscometer, a logarithmic plot of shear rate vs shear stress yields essentially a straight line, where the slope

TABLE 11.1 — COMPARISON OF FANN READINGS, SHEAR STRESS, AND APPARENT VISCOSITY USING A SPECIFIC MUD

Fann Data			Shear Stress		Apparent Viscosity
Rotation (rpm)	Shear Rate (sec^{-1})	Dial Reading ($lb/100\ ft^2$)	$dynes/cm^2$	lb_f/ft^2	(cp)
600	1,022	17	86.7	0.181	8.49
300	511	12	61.2	0.128	11.99
200	340	10	51.0	0.107	15.0
100	170	8	40.8	0.085	24.0
6	10.2	3.5	17.9	0.037	175.1
3	5.11	3	15.3	0.032	299.6

TABLE 11.2 — POWER-LAW VS BINGHAM-PLASTIC MODEL INFORMATION OBTAINED WITH ROTATIONAL VISCOMETER

API Cement Slurry	Weight (lb/gal)	Power Law		Bingham Plastic	
		n'	K'	Plastic Viscosity (cp)	Yield Point ($lb_f/100\ ft^2$)
Class H	15.6	0.30	0.1950	29	97
Class H	16.5	0.36	0.2185	58	147
Class C	14.8	0.25	0.1441	12	55
Class C	14.1	0.43	0.0300	15	28
Class E	16.2	0.50	0.0472	44	62
Class G					
+ 4% gel	14.1	0.10	0.9500	12	164
+ 8% gel	13.1	0.10	0.9000	12	155
+ 12% gel	12.8	0.10	0.7600	10	131
+ 12% gel and dispersant	13.2	0.07	0.9987	7	146

TABLE 11.3 — FLOW RATES FOR TURBULENT AND PLUG FLOW* WITH AND WITHOUT CEMENT DISPERSANTS

API Cement Composition	Dispersant (percent)	Flow Rate, bbl/min			
Hole Size, in.:		6¾	6¾	8¾	9⅞
Casing Size, in. OD:		2⅞	4½	5½	7
Class H	0.00	18.18	13.58	23.29	24.93
15.4 lb/gal		(3.03)	(2.27)	(3.88)	(4.16)
	0.50	14.32	11.28	18.66	20.21
	0.75	6.57	5.86	8.91	9.93
Class A with 4% gel	0.00	25.17	17.58	31.54	33.26
14.1 lb/gal		(5.07)	(3.54)	(6.35)	(6.70)
	0.50	15.50	11.21	19.65	20.88
	0.75	6.58	5.30	8.65	9.41
Class A with 12% gel	0.00	23.55	16.45	29.51	31.12
12.8 lb/gal		(4.74)	(3.31)	(5.94)	(6.51)
	0.75	14.08	10.26	17.89	19.05
	1.00	2.93	2.88	4.10	4.67

*The plug flow is in parentheses below the corresponding turbulent flow.

of the straight-line portion and the intercept at unity shear rate again are the two descriptive parameters.[3,8]

The complexity of the chemical reactions occurring when cement and water are first mixed makes it extremely important that slurry preparation methods be standardized so that reasonably reproducible data can be obtained with the rotational viscometer.[3,9] Standard testing procedures have been published in *API RP 10B*.[10]

11.4 Displacement Theories — Plug Flow vs Turbulent Flow

In placing cement down hole the preferred method, where hole conditions permit, is to thin the mud and cement slurry so that turbulent flow can be induced at moderate pumping rates.[3,11,12] The injection rate is adjusted to a level that will overcome the resistance of the mud. This is accomplished by adding friction reducers or dispersants to the cement slurry or by increasing the volume of mixing water.[13,14] (See Table 11.3 and Fig. 11.6.) The displacement rate will be moderately high and high displacement efficiency should result. Turbulent flow offers other advantages over plug flow in that radial components of velocity are present and these exert resisting forces as well as driving forces and therefore promote mixing at the interfaces (Fig. 11.7).[12,13] This increases the probability of removing mud from hole restrictions and reduces the contamination of the cement sheath. The leading and following interfaces may be highly contaminated with mud, but by using large volumes of slurry, competent cement can be placed across the zone of interest.

The literature abundantly illustrates the efficiency of turbulent flow in removing circulatable mud from a wellbore[1,3,11-16]; however, there are situations in which

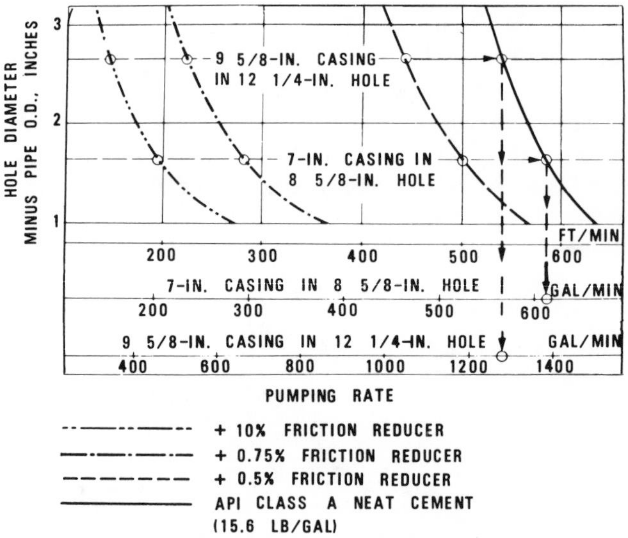

Fig. 11.6 Pumping rate to achieve turbulent flow in annulus.

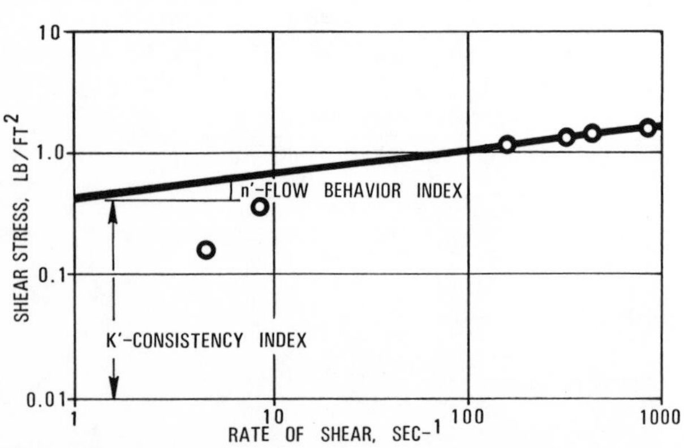

Fig. 11.5 Power-law plot of non-Newtonian slurry.[3]

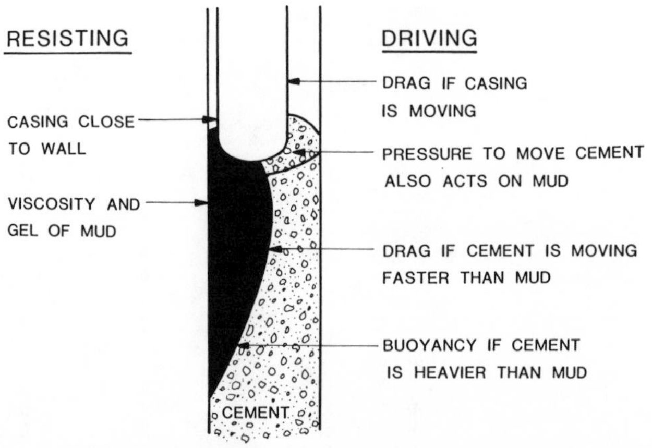

Fig. 11.7 Forces on mud channel during cementing.[9,12]

such high displacement rates are not feasible. The geometry of the hole, the size of the annulus, the flow properties of both the cement and the drilling mud, and pressure restrictions may dictate the use of plug flow.[15,16] The benefits achieved from centralization and pipe movement to "puddle" static mud pockets that are not effectively removed by displacement should not be overlooked.[12,17,18]

11.5 Equations Used in Flow Calculations

There is very little difference between results of calculations performed by the Bingham-plastic method and results of calculations performed by general non-Newtonian methods, except for the pressure drop in laminar flow. A much greater difference can result from different laboratory methods of preparing the slurry or from variations in brands of cement than from the two methods of calculation.

Turbulent-flow calculations are usually based on some form of the Reynolds Number correlation (modified for non-Newtonian fluids) or other generalized correlations. The rotational viscometer may be used, with variations in mathematical approach, to develop data for calculating either turbulent or plug flow behavior of non-Newtonian fluids. Following are example equations for a fluid that may be considered a power-law fluid:

From Fann Viscometer

$$n' = 3.32^* \times \left(\log_{10} \frac{600 \text{ rpm dial reading}}{300 \text{ rpm dial reading}}\right), \quad \ldots \quad (11.7)$$

$$K' = \frac{N \times (300 \text{ rpm dial reading}) \times 1.066^*}{100 \times (511)^{n'}}, \quad \ldots \quad (11.8)$$

where

N = range extension factor of the Fann torque spring.

For calculations where plastic viscosity and yield point are known:

$$n' = 3.32 \times \log_{10}\left(\frac{2\mu_p + Y}{\mu_p + Y}\right), \quad \ldots \quad (11.9)$$

$$K' = \frac{N \times (\mu_p + Y) \times 1.066^*}{100 \times (511)^{n'}}, \quad \ldots \quad (11.10)$$

where

μ_p = plastic viscosity, cp,

Y = yield point, $lb_f/100\ ft^2$.

For Newtonian fluids,

$$n' = 1.0, \quad \ldots \quad (11.11)$$

$$K' = \frac{\text{viscosity (cp)}}{47880} \quad \ldots \quad (11.12)$$

*These factors change for different combinations of rotor and bob.

Following are equations used in calculating flow.

1. Displacement Velocity

$$\overline{v_d} = \frac{17.15\, q_b}{d_i^2} = \frac{3.057\, q_{cf}}{d_i^2}, \quad \ldots \quad (11.13)$$

where

$\overline{v_d}$ = average displacement velocity, ft/sec,

q_b = pumping rate, bbl/min.,

q_{cf} = pumping rate, cu ft/min.,

d_i = inside diameter of pipe, in.

For the annulus, $d_i^2 = d_{iop}^2 - d_{oip}^2$,

where

d_{iop} = inside diameter of outer pipe, or hole size, in.,

d_{oip} = outside diameter of inner pipe, in.

2. Reynolds Number

$$N_{Re} = \frac{1.86\, v^{(2-n')}\, \rho}{K'\, (96/d_i)^{n'}}, \quad \ldots \quad (11.14)$$

where

N_{Re} = Reynolds number, dimensionless,

v = velocity, ft/sec,

ρ = slurry density, lb/gal.

For the annulus, $d_i = d_{iop} - d_{oip}$; or

$$d_i = \frac{4 \times \text{area of flow}}{\text{wetted perimeter}}.$$

This applies also for Items 4, 6, and 7.

3. Casing/Open-Hole Annular Area

$$A = 0.7854\, (d_h^2 - d_o^2), \quad \ldots \quad (11.15)$$

where

A = area, sq in.,

d_h = diameter of hole, in.,

d_o = outside diameter of casing, in.

4. Velocity at Which Turbulence Begins (N_{Re} = 2,100)

$$v_c^{2-n'} = \frac{1{,}129K'\, (96/d_i)^{n'}}{\rho}$$

$$v_c = \left[\frac{1{,}129K'\, (96/d_i)^{n'}}{\rho}\right]^{\frac{1}{2-n'}}, \quad \ldots \quad (11.16)$$

where

v_c = critical velocity for turbulence, ft/sec.

5. Hydrostatic Pressure

$$p_h = 0.052\, \rho h, \quad \ldots \quad (11.17)$$

where

p_h = hydrostatic pressure, psi,

h = height of column, ft,

ρ = fluid density, lb/gal.

6. Frictional Pressure Drop

$$\Delta p_f = \frac{0.039 L \rho v^2 f}{d_i}, \quad \ldots\ldots \quad (11.18)$$

where

Δp_f = frictional pressure drop, psi,

L = length of pipe, ft,

f = Fanning friction factor, dimensionless.

7. Velocity at Some Specific Reynolds Number
For generalized calculations,
N_{Re} for plug flow = 100 (value commonly used),
N_{Re} for beginning of turbulence = 2,100 (value commonly used).

$$v^{2-n'} = \frac{N_{Re} K' (96/d_i)^{n'}}{1.86 \rho},$$

$$v = \left[\frac{N_{Re} K' (96/d_i)^{n'}}{1.86 \rho}\right]^{\frac{1}{2-n'}}, \quad \ldots \quad (11.19)$$

where terms are as defined in the preceding equations.

In calculating pressure drop in turbulence by the Bingham-plastic equation, the Fanning friction factor is obtained from the usual Reynolds-number/friction-factor correlation curve[3] for Newtonian fluids in commercial pipe, shown in Fig. 11.8.

For both the Bingham-plastic and the power-law methods of calculation, annulus calculations are based on an equivalent diameter (d_e). For a simple annulus (single string of pipe), d_e is the hole diameter minus the outside diameter of the casing.

The data shown in Table 11.4 indicate the small variations that exist between the two methods of calculation. In general, calculations of critcal velocity for the two methods are within 10 percent of each other. In the turbulent-flow region, calculations of frictional pressure drop by the two methods deviate an average of ±10 percent. The largest discrepancy between the two systems occurs in calculations of frictional pressure drop for the laminar flow region.

11.6 Summary

Results of flow calculations performed by the Bingham-plastic method and by the power-law method are in reasonably good agreement except for frictional pressure drop in the laminar flow region. Much larger discrepancies can result from variations in brand of cement and method of slurry preparation.[3]

The technical literature indicates that the flow behavior of cement slurries is non-Newtonian and that it is more fundamentally described by the n' and K' method of analysis. Although the Bingham-plastic method is sometimes used in determining the flow calculations of cementing slurries, it is not so accurate as the

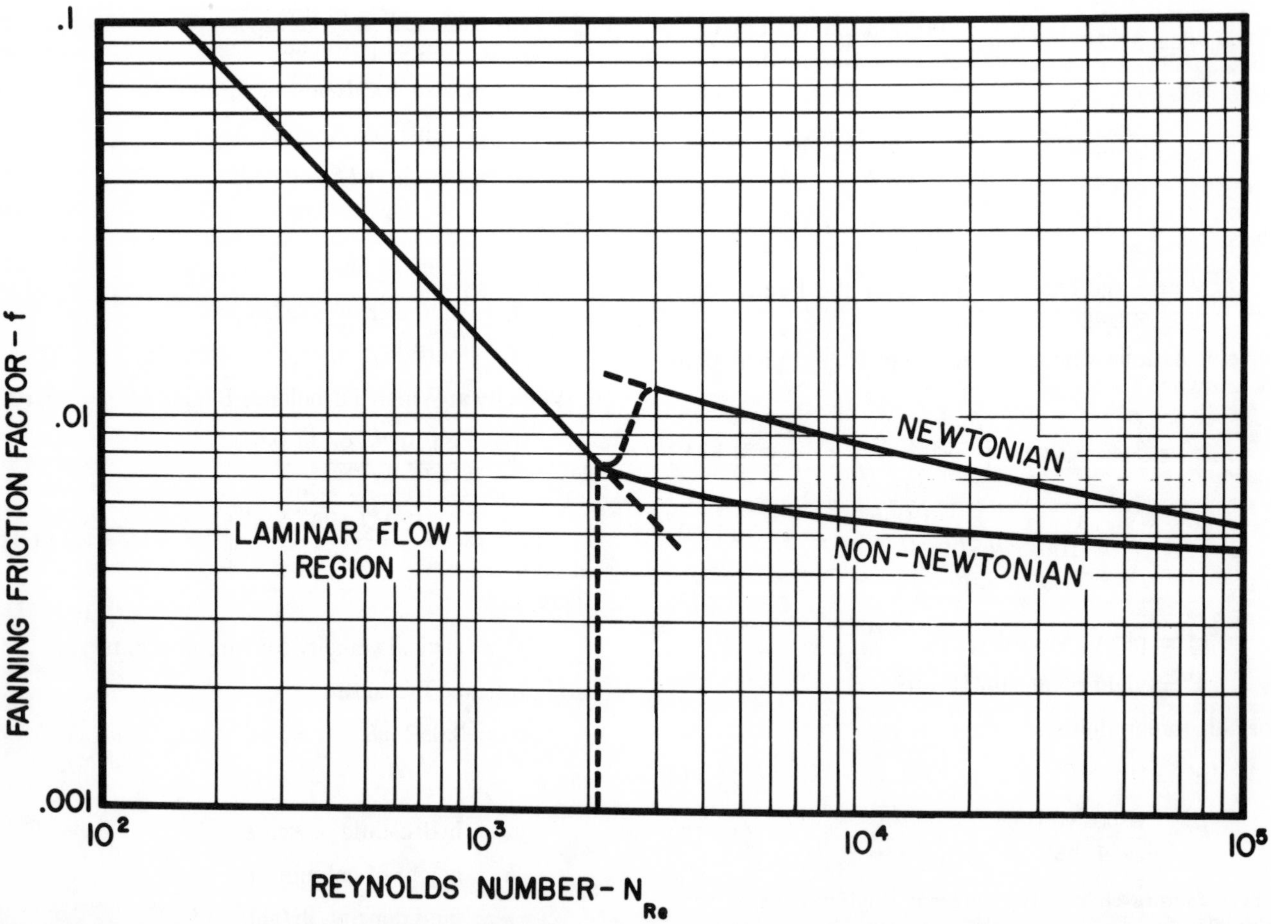

Fig. 11.8 Reynolds-number/friction-factor correlation.[3]

TABLE 11.4 — COMPARISON OF FLOW CALCULATIONS USING BINGHAM-PLASTIC AND POWER-LAW MODELS

API Cement Slurry	Weight (lb/gal)	Fann Viscometer Reading at Speed (rpm) of				Rheological Parameters*			
						Bingham Plastic		Power Law	
		600	300	200	100	μ_p	Y	n′	K′
Class A	15.6	211	157	134	111	54	103	0.41	0.134
Class A + 0.3% retarder	14.8	99	65	54	40	34	31	0.54	0.025
Class A + 10% diatomaceous earth	13.2	138	114	101	86	24	90	0.27	0.228

	Critical Velocity (ft/sec in 2-in. tubing)		Frictional Pressure Drop (psi/1,000 ft through 2-in. tubing) at Flow Rates** of					
			50 gal/min (5.13 ft/sec)		150 gal/min (15.38 ft/sec)		300 gal/min (30.77 ft/sec)	
	Bingham Plastic	Power Law	Bingham Plastic	Power Law	Bingham Plastic	Power Law	Bingham Plastic	Power Law
Class A	11.76	11.32	276	204	498	491	1,706	1,677
Class A + 3% retarder	6.86	6.51	98	77	425	411	1,480	1,452
Class A + 10% diatomaceous earth	10.93	10.20	222	163	366	403	1,246	1,343

* Dimensions
μ_p (plastic viscosity): centipoises
Y (yield point): lb/100 ft^2
n′ (flow behavior index): dimensionless
K′ (consistency index): lb_f sec^{-1}/ft^2
** All slurries were in laminar flow at 50 gal/min and in turbulent flow at 140 and 300 gal/min.

power-law method.

Regardless of which concept is used for hydraulic calculations, the basic flow properties are measured by the same equipment: either the capillary viscometer or the rotational viscometer (Fann VG meter). The power-law model *n′* and *K′* values are determined by laboratory procedures using Fann data at 600, 300, 200, and 100 rpm. The Bingham-plastic method uses slurry parameters of plastic viscosity (μ_p) and yield point (*Y*).

References

1. Howard, G. C., and Clark, J. B.: "Factors To Be Considered in Obtaining Proper Cementing of Casing," *Drill. and Prod. Prac.*, API (1948) 257-272; *Oil and Gas J.* (Nov. 11, 1948) 243.
2. Savins, J. G., and Roper, W. F.: "A Direct-Indicating Viscometer for Drilling Fluids," *Drill. and Prod. Prac.*, API (1954) 7.
3. Slagle, K. A.: "Rheological Design of Cementing Operation," *J. Pet. Tech.* (March 1962) 323-328; *Trans.*, AIME, **225.**
4. Hedstrom, B. O. A.: "Flow of Plastic Materials in Pipes," *Ind. and Eng. Chem.* (1952) **44,** 651.
5. Rogers, W. F.: *Composition and Properties of Oil Well Drilling Fluids,* 1st ed., Gulf Publishing Co., Houston, Tex. (1948).
6. "Flow of Drilling Fluids," technical bulletin, Imco Drilling Mud, Inc. (Dec. 19, 1970).
7. Metzner, A. B., and Reed, J. C.: "Flow of Non-Newtonian Fluids — Correlation of the Laminar, Transition, and Turbulent Flow Regions," *AIChE Jour.* (1955) **1,** No. 4.
8. Garvin, T. R., and Moore, P. L.: "A Rheometer for Evaluating Drilling Fluids at Elevated Temperatures," paper SPE 3062 presented at SPE-AIME 45th Annual Fall Meeting, Houston, Oct. 4-7, 1970.
9. Lord, D. L., Hulsey, B. W., and Melton, L. L.: "General Turbulent Pipe Flow Scale-Up Correlation for Rheologically Complex Fluids," *Soc. Pet. Eng. J.* (Sept. 1967) 252-258; *Trans.*, AIME, **240.**
10. *Testing Oil-Well Cements and Cement Additives,* 17th ed., *API RP 10B,* API (April 1971).
11. Brice, J. W., Jr., and Holmes, B. C.: "Engineered Casing Cementing Programs Using Turbulent Flow Techniques," *J. Pet. Tech.* (May 1964) 503-508.
12. Clark, C. R., and Carter, L. G.: "Mud Displacement With Cement Slurries," *J. Pet. Tech.* (July 1973) 775-783.
13. Graham, H. L.: "Rheology — Balanced Cementing Improves Primary Success," *Oil and Gas J.* (Dec. 18, 1972) 53.
14. Buster, J. L.: "Plan Turbulence Into Your Cement Job," *Pet. Eng.* (May 1962).
15. McLean, R. H., Manry, C. W., and Whitaker, W. W.: "Displacement Mechanics in Primary Cementing," *J. Pet. Tech.* (Feb. 1967) 251-260.
16. Parker, P. N., Ladd, B. J., Ross, W. H., and Wahl, W. W.: "An Evaluation of a Primary Cementing Technique Using Low Displacement Rates," paper SPE 1234 presented at SPE-AIME 40th Annual Fall Meeting, Denver, Colo., Oct. 3-6, 1965.
17. Garvin, T. R., and Slagle, K. A.: "Scale-Model Displacement Studies To Predict Flow Behavior During Cementing," *J. Pet. Tech.* (Sept. 1971) 1081-1088.
18. Childers, M. A.: "Primary Cementing of Multiple Casing," *J. Pet. Tech.* (July 1968) 751-762; *Trans.*, AIME, **243.**

Chapter 12

Bonding, Logging, and Perforating

12.1 Introduction

Once the primary cementing operation has been completed, bonding, logging, and perforating must be considered. It is the purpose of this chapter to discuss factors influencing the bonding of cement, methods of locating the cement behind the casing, and methods of perforating and its effect on the cement sheath.

12.2 Bonding Considerations

Cementing recommendations are often based on the compressive or tensile strength of set cement,[1-3] on the assumption that a material satisfying strength requirements will also provide an adequate bond. Field experience, however, has shown that this assumption is not always valid.

In a wellbore, shear bond and hydraulic bond are the two forces to be considered for effective zonal isolation along the cement/casing and cement/formation interfaces.[4,5] (See Figs. 12.1 and 12.2.) Shear bond mechanically supports pipe in the hole, and is determined by measuring the force required to initiate pipe movement in a cement sheath. This force divided by the cement/casing contact surface area yields the shear bond in pounds per square inch. Fig. 12.3 compares compressive strength and shear bonding strength.

Hydraulic bonding blocks the migration of fluids or gas in a cemented annulus and is usually measured by applying pressure at the pipe/cement interface until leakage occurs (Fig. 12.2).

For zonal isolation, hydraulic bonding is of greater significance than shear bonding, since the cement composition for most jobs will provide adequate mechanical support to hold the pipe in place. If an optimum cement bond is to be achieved between pipe and formation, the effect of various hole conditions and completion techniques must be considered.

12.3 Bonding of Cement to Pipe

Shear, hydraulic, and gas bond strengths are directly affected by the surface finish of the pipe against which the cement is placed.[6] (See Table 12.1.)

To determine the strength of cement or height of cement column required to support any given casing string one may use the following equation:[2]

$$S_t = \frac{(C_L)\ (C_W)}{9.69\ (d_o)\ (h_c)},$$

where S_t = tensile strength of cement, psi; C_L = casing string length, ft; C_W = casing weight, lb/ft; d_o = casing outside diameter, in.; h_c = cement column height, ft.

Where pipe/cement bonding is critical, a resin-sand coating applied to the outside of the pipe will improve the bond as well as the resistance to gas migration. On equivalent pipe finishes, oil-wet surfaces provide the poorest bond. By and large, the rougher and drier the surface of the pipe, the better the bond. As Fig. 12.4

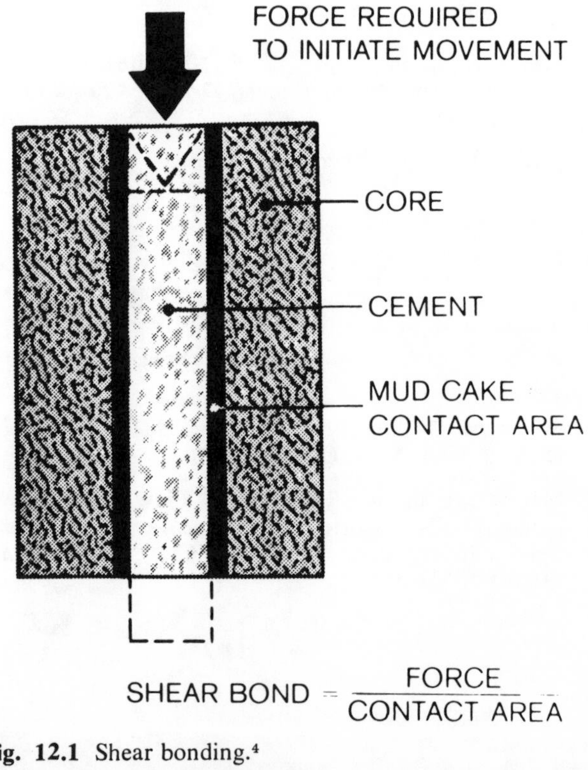

Fig. 12.1 Shear bonding.[4]

TABLE 12.1 — BONDING PROPERTIES OF VARIOUS PIPE FINISHES

Cement: API Class A.
Water content: 5.2 gal/sk.
Curing temperature: 80°F.
Curing time: 24 hours.
Casing size: 2-in. inside 4-in.

	Bond Strength		
Type of Finish	Shear (psi)	Hydraulic (psig)	Gas (psig)
Steel Pipe			
New (mill varnish)	74	200 to 250	15
New (varnish chemically removed)	104	300 to 400	70
New (sandblasted)	123	500 to 700	150
Used (rusty)	141	500 to 700	150
New (sandblasted, resin-sand coated)	2,400	1,100 to 1,200	400+
Plastic Pipe			
Filament wound (smooth)	79	210	—
(rough)	99	270	—
Centrifugally cast (smooth)	81	220	—
(rough)	101	310	—

shows, the bond of cement to an oil-wet surface is approximately half that to a water-wet or dry surface.[6]

Other factors influencing casing/cement bonding are the direction in which pressure is applied and the length of time pressure is held on the bonded interface. Closed-in pressure after completion of a primary cement job can be detrimental to both hydraulic and shear bonding at the cement/pipe interface.[2,4]

During the setting of a cement, the heat of hydration can produce an effect similar to internal pressuring of the casing and cause expansion of the pipe. Normally, heat begins to build up inside the casing as the cement hydrates and takes its initial set. After the cement sets, its temperature decreases, causing the casing to contract. This expansion and contraction places additional stress on the casing and cement sheath, which can decrease the shear and hydraulic bond strength. The greatest damage to the bond can occur if the casing is closed in while the temperature is rising inside the casing. (Fig. 12.5 shows how the shear-bond strength increases as the closed-in pressure declines.)

Hydraulic bond failure is a function of time, cement properties, applied pressure, and viscosity of the pressuring medium. Investigations[4] have shown that the rate of bond failure with water ranges from 1.125 to 1.250 ft/min. Normally, pressure from gas — which has a lower viscosity — causes bond failure to progress up the pipe faster than pressure from water, oil, or mud.

Vertical bond failure will normally occur 30° each side of the pressure-application point when there is a uniform cement sheath around the pipe. Unequal distribution of cement can cause bond failure at the weakest plane (which could account for communication in multiple-string tubingless completions).

The intrusion of casing attachments such as collars, centralizers, and scratchers has little influence on hydraulic or gas bond failure pressure.

In considering pipe/cement bonding, the following should be borne in mind:

1. A change in internal pressure on the casing will cause a corresponding change in hydraulic and shear bond strength. If the casing string is closed in while the cement is setting, the heat of hydration causes a pressure buildup that lowers the strength of the bond

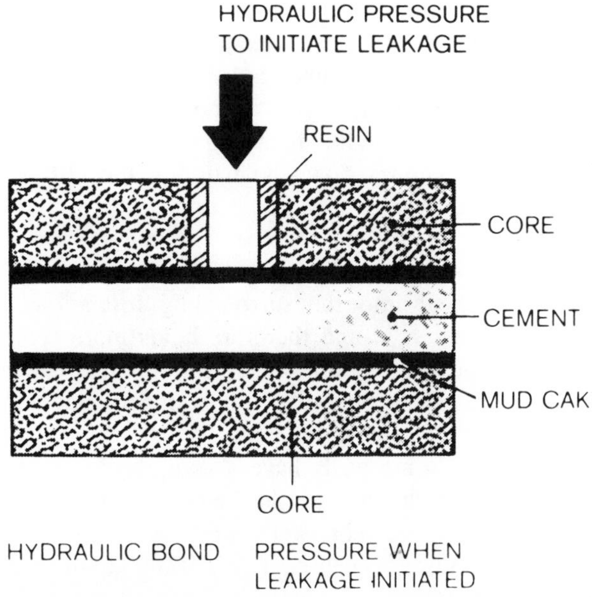

Fig. 12.2 Hydraulic bonding.[5]

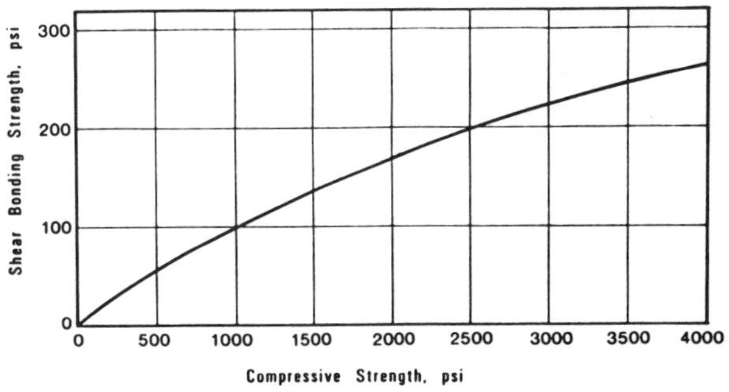

Fig. 12.3 Shear bonding strength vs compressive strength.

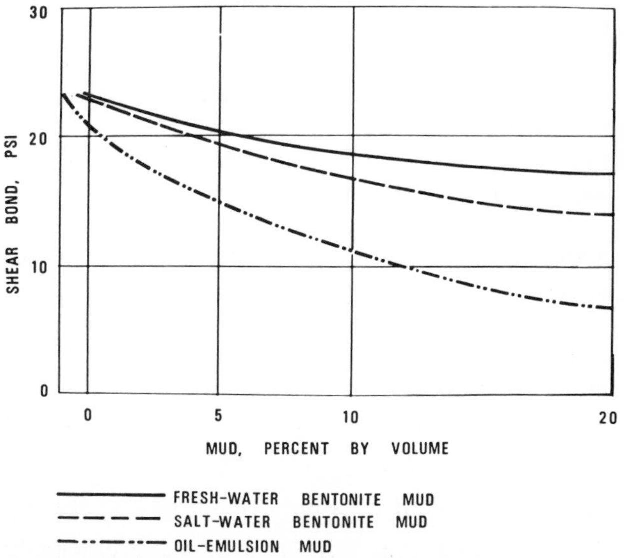

Fig. 12.4 Bond strength after contamination with mud.[5]

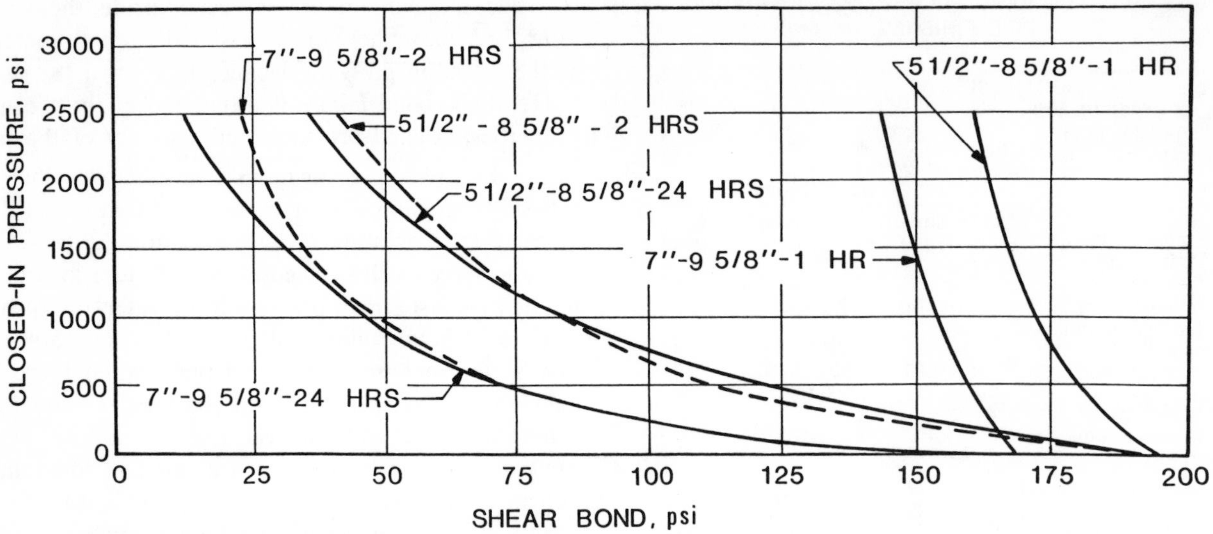

Fig. 12.5 Shear bond vs closed-in pressure (API Class A cement cured 24 hours at 150°F; closed-in pressure held for 1, 2, and 24 hours; 5½-in. pipe cemented inside 8⅝-in. pipe, and 7-in. pipe inside 9⅝-in. pipe).[6]

and can readily create a microannulus through which gas can easily migrate.

2. Hydraulic and shear bond strengths increase with surface roughness.

3. As the viscosity of the pressuring fluid increases, the pressure increases, hastening failure or communication of fluid where the pipe and cement come in contact.

4. Oil-wet pipe surfaces reduce the hydraulic shear strength of the cement/pipe bond.

5. Hydraulic bond failure is a function primarily of pipe expansion or contraction.

12.4 Bonding of Cement to Formation

The bond between the cement and the formation is what normally determines whether or not there will be gas or fluid communication in the annulus. Cement sets better against a clean formation than against one coated with mud cake.

The following general statements apply in cement/formation bonding:

1. A good hydraulic bond to the formation depends upon intimate contact between the cement and the formation.

2. A thick mud layer at the cement/formation interface greatly reduces hydraulic bonding.

3. Higher bond strengths can be expected on more permeable formations if the mud cake has a uniform thickness.[4,5]

4. The bond strength ultimately attained on a dry formation or a formation free of filter cake will approach or exceed the formation strength.

5. Failure to remove mud can be more detrimental to formation bond than to pipe bond.

12.5 Methods of Locating Cement Behind the Pipe

Three means commonly used to locate cement behind pipe are temperature surveys, radioactive-tracer surveys, and acoustic or bond logs. Although temperature surveys and radioactive tracers have been in use longer, they do not provide the quantitative data that bond logs do, so they are not used as widely.

Temperature Surveys

A temperature survey (or log) measures the heat generated during the setting of cement behind pipe (Fig. 12.6). This heat of hydration raises the temperature in the wellbore enough that a device set opposite the cemented zone can detect and record that increase in temperature. Such a survey can usually locate the top of the cement with reasonable accuracy, provided the recorded temperature anomaly is clear enough. Some causes of poorly registered temperature increases are (1) cement with low heat of hydration, (2) a temperature survey run either too early or too late, and (3) excessive contamination and dilution of the cement slurry. For best results, a temperature survey should be run within the first 12 to 24 hours[7] (see Table 12.2).

The amount of heat liberated during the setting of cement depends upon well conditions, cementing systems, and surface conditions during the mixing process. Typical temperatures down hole during the setting of cement are shown in Figs. 12.7 and 12.8.[3]

The poor quality of a temperature survey run too long after the cement is placed can be due to the dissipation of heat into the surrounding formations. A hole that is enlarged and therefore holds more cement can cause the temperature survey line to slope steeply above or below the enlarged interval and may erroneously imply a cement top.

Surveys on some wells have shown the increase in temperature at the top of the cement to be as high as 35° to 40°F; on most wells, however, the temperature will be lower because of interfacial dilution and mud contamination.

Once the top of the cement has been located, the

TABLE 12.2 — OPTIMUM NUMBER OF HOURS AFTER CEMENTING TO RUN TEMPERATURE SURVEYS AT DIFFERENT WELLBORE TEMPERATURES

Cement	Additive (percent)	Time (hours) 100°F	120°F	140°F	160°F
API Classes A, B, G, H, with gel	0	8 to 12	8 to 12	6 to 9	4 to 8
	4	8 to 12	8 to 12	6 to 9	4 to 8
	8	9 to 12	9 to 12	6 to 9	6 to 9
	12	9 to 12	9 to 12	9 to 12	9 to 12
Pozzolan with gel	0	8 to 12	8 to 12	6 to 9	4 to 8
	2	8 to 12	8 to 12	8 to 12	6 to 9
	4	8 to 12	8 to 12	8 to 12	6 to 9

High-Temperature Cements

		Time (hours) 140°F	160°F	180°F	220°F	260°F
API Class G or H with retarder	0.3	15 to 18	12 to 15	9 to 12	8 to 12	6 to 9
	0.5	16 to 24	16 to 24	12 to 18	9 to 12	8 to 12
Retarded cements		16 to 24	16 to 24	12 to 18	9 to 12	9 to 12

fillup efficiency can be calculated. The fillup efficiency, which is derived by dividing the volume of the annulus by the volume of cement slurry, is useful as an indicator of the amount of channeling present.

Radioactive-Tracer Surveys

The radioactive-tracer method of detecting cement behind casing uses a radioactive material having a short half-life. The two most commonly used tracers are Iodine 131 and Scandium 46, which have half-lives of 8 and 84 days, respectively.

With this method, the top of the cement can be accurately determined if the uppermost portion (i.e., the first portion encountered on the way down the hole) is rendered radioactive. The tracer is added as a soluble salt to the cement mixing water and, for best results, is thoroughly mixed throughout the slurry. Adding a radioactive tracer to the uppermost portion of the cement is good insurance in case too much time elapses after cement is placed for a temperature survey to be

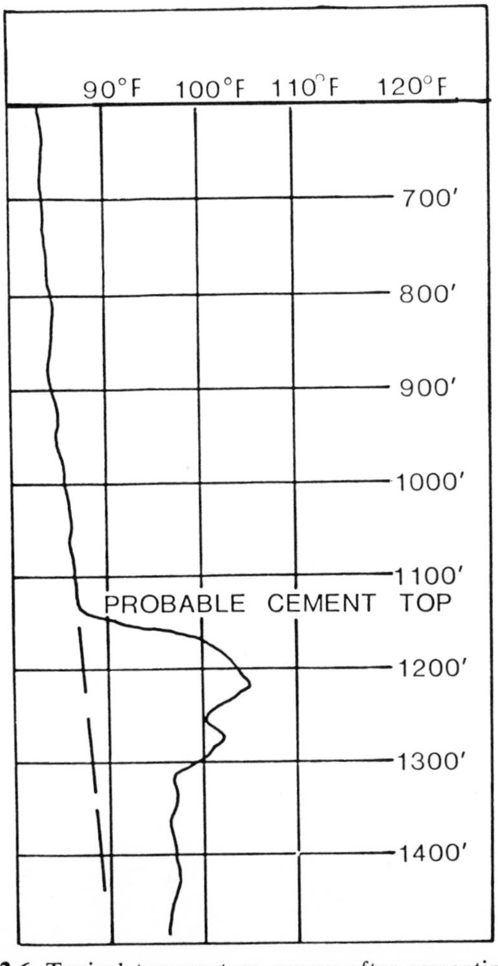

Fig. 12.6 Typical temperature survey after cementing.

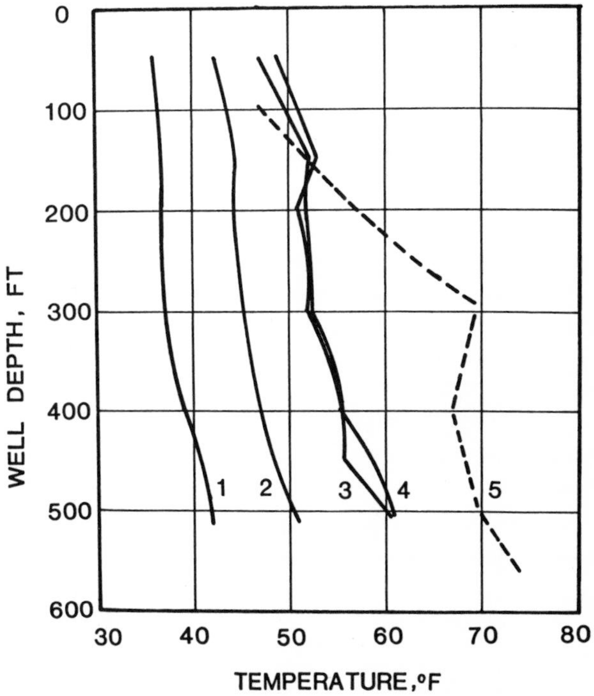

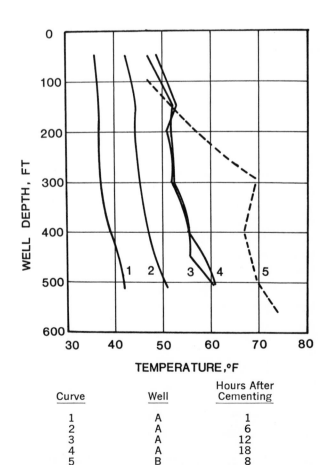

Curve	Well	Hours After Cementing
1	A	1
2	A	6
3	A	12
4	A	18
5	B	8

Fig. 12.7 Down-hole temperatures during the setting of cement.

accurate. It is not necessary to be so prompt with a tracer survey, and in fact with a long-half-life tracer, the survey can be conducted at any time in the life of the well.

The principal advantage of the radioactive-tracer method is that it positively and accurately determines the location of the tracer. Its principal disadvantages are that it requires special health precautions, it may interfere with radioactive surveys run for purposes other than locating cement tops, and it costs more than other locating methods. A base survey, before cementing, to establish the natural gamma ray emission of the formation is usually necessary to enable clear interpretation of the radioactive-tracer survey.

Fig. 12.9 illustrates how a radioactive-tracer survey (gamma ray log) distinguishes the top of good cement.

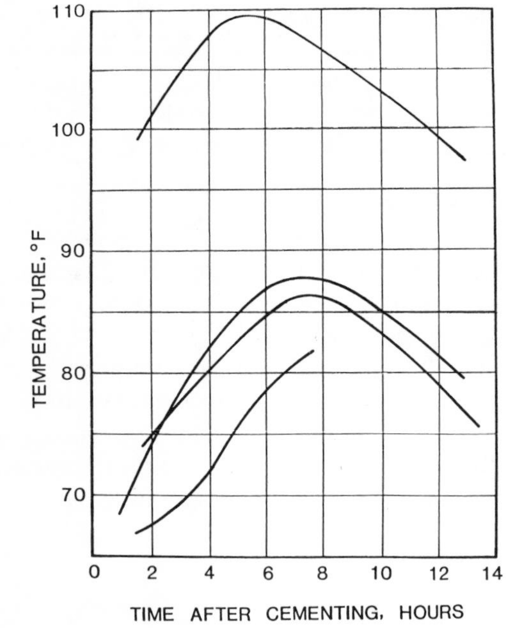

Fig. 12.8 Heat of hydration for four different wells (well depth, 400 ft).

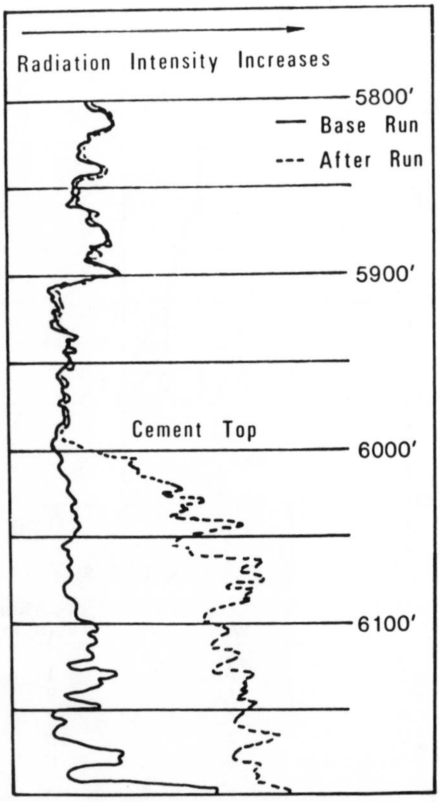

Fig. 12.9 Typical radioactive-tracer survey.

Bond Logging

The acoustic or bond log is the most widely used aid to locating cement behind pipe. Operating on an acoustic principle, it transmits a signal or a vibration and receives that signal and records its arrival time.[8-10] (See Fig. 12.10.) Both the time of arrival and the amplitude of the vibration are used to determine bonding conditions, since both the casing and the formation, when acoustically coupled, have characteristic arrival times and amplitudes (Fig. 12.11). The principle is similar to tapping on a wall to locate the position of studding behind the wallboard.[10]

If the pipe is free and not held firmly by cement, it will vibrate, creating a strong signal. If the cement is firmly bonded to the pipe and to the formation, the signal shows no pipe vibration, and the received signal is characteristic of the formation behind the pipe. In the simple curve bond log, a signal may be received when the cement is bonded to the pipe but not to the formation. Very little signal is received from the formation if mud cake interferes between the cement and the formation.

If the casing is resting against the wall of the hole so that the cement does not completely surround the pipe, and there is, in addition, a channel in the cement, both pipe signal and formation signal are present, since

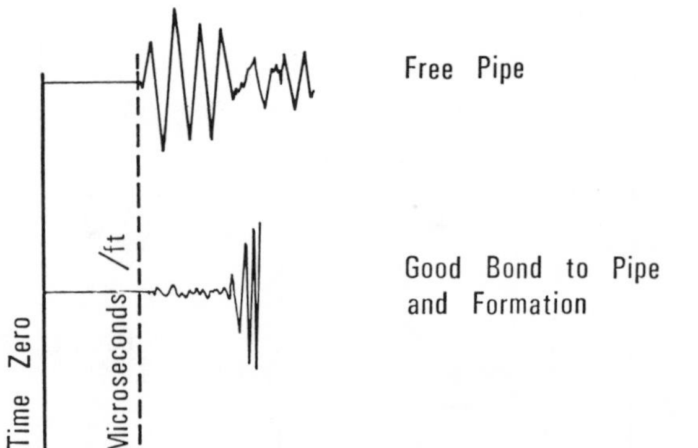

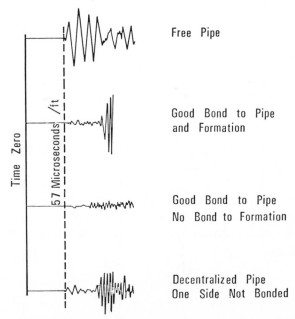

Fig. 12.10 Acoustic signals in shallow cased test holes with built-in cementing conditions.[12]

the pipe is partially free on one side and the formation is partially coupled on the other side. Cement sheath thickness and compressive strength tend to affect the amplitude of the pipe vibration.[8] However, data based on that fact cannot be used directly because a thin cement sheath bonded to both pipe and formation would look like an infinitely thick sheath so far as the vibrational amplitude of the pipe is concerned.

The amplitude and the times of arrival of the acoustic signal are determined by the path that the signal takes to reach the receiver. If the casing is not acoustically coupled to the cement, the path will be down the casing and the amplitude will be large. If they are acoustically coupled and the formation and cement are also coupled, the path will be through the formation and the time of arrival will be determined by the formation characteristics (primarily porosity). The amplitude will also reflect the formation character and may be quite small opposite an unconsolidated formation that is gas filled. In addition, if the cement sheath approaches 2 in. in thickness and is acoustically coupled to the casing, the amplitude of the signal will be extremely small.[11,12] In soft-rock country, these signals are separated in time so that the amplitude measurements during a fixed time make it possible to identify cement/formation bonding. In dense, hard-rock formations, the two signals are not necessarily separated in time and the identification of cement/formation bonding becomes more difficult.

Improvements in the interpretation of acoustic waves between transmitter and receiver have led to a technique of recording cement/pipe and cement/formation bonding in wellbores.[12-14] (See Figs. 12.12 and 12.13.)

If acoustic logging is to be successful, ample time must be allowed for the cement to develop strength before the log is run. The generally recommended time is 24 to 36 hours.

Bond logs have been greatly improved since they were first introduced. In the effort to improve them, particular emphasis has been placed on the interpretation of signals.

Following are some general conclusions concerning bond logs.[15]

1. Bonding, if adhesion is implied, of the cement to the pipe is not necessary for a satisfactory bond log recording. Intimate physical contact or adequate acoustic coupling is all that is required for minimum acoustic transmission of the pipe signal.

2. Sensitive tool calibration is necessary to locate small channels in heavy-weight cements that do not contain additives, but sensitive calibration may lead to misinterpretation in cements that do contain "filler," or light-weight additives.

3. Cementing compositions that do not contain bulking additives show, in general, better acoustic and physical properties than those that do contain them.

4. Coating on pipe has a negligible effect on the final

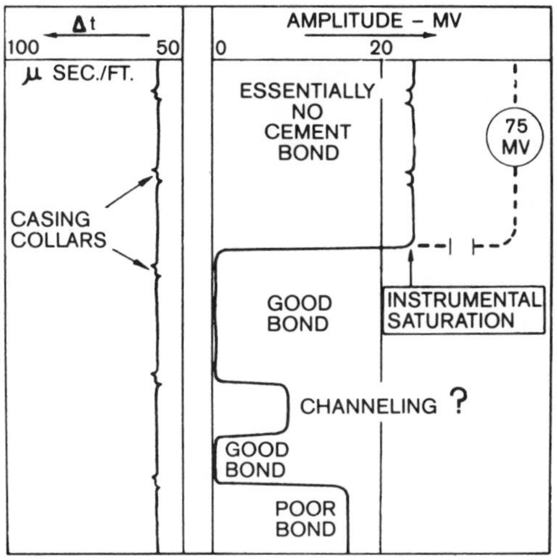

Fig. 12.11 Schematic of the cement bond log.[9]

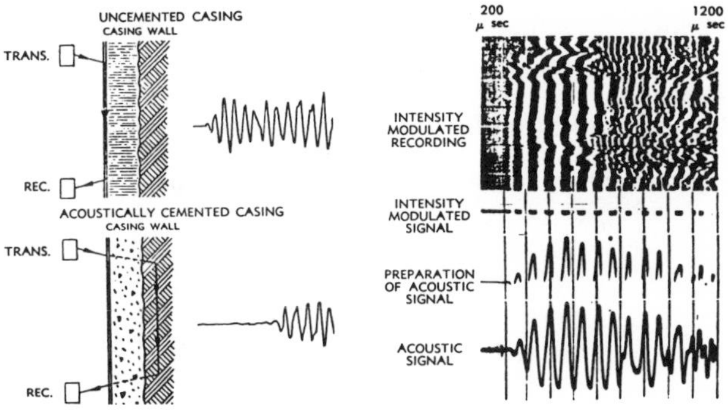

Fig. 12.12 Acoustic energy travel in cased wells (left), and recording technique (right).[15]

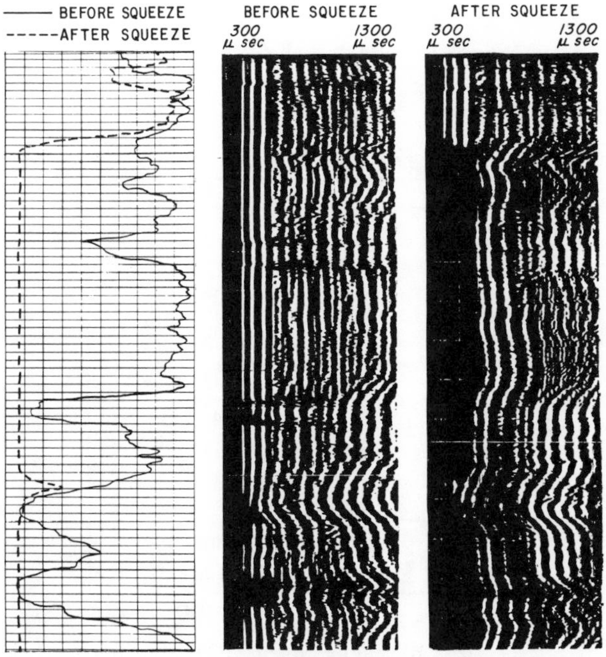

Fig. 12.13 Single-curve acoustic log (left) and intensity-modulated recordings, showing the effects of squeeze cementing.[15]

bond log reading; however, it lowers the support coefficient and the resistance to communication.

5. Excessive closed-in pressure during cement curing is detrimental to the bonding of cement to casing and a poor bond will be reflected on the log.

12.6 Perforating — Effects on the Cement Sheath

For most well completions, casing is set through the producing zone and cemented, and then perforated for production.[16-19] The primary aim in perforating is to get deep penetration with a burrless hole in the pipe and with a minimum of cement fracturing.

The factor most commonly considered with respect to the influence of perforating is cement compressive strength. Of more vital concern, however, are the effects of perforating on the hydraulic bond and the mechanical bond of the cement to the pipe and the formation. Godfrey and Methven,[20] in studies of bonding before and after perforating, found that neither the hydraulic nor the mechanical bond is affected by perforating, provided the annulus is uniformly cemented. (See Figs. 12.14 and 12.15.)

12.7 Perforating Devices and Methods

The commonly used devices for perforating, in order of popularity, are jets, bullets, hydraulic cutters, mechanical cutters, and permeators.

Jets

Jet perforation, a product of wartime research, produces a high-velocity force that penetrates casing, cement, and formation. The velocity of the jet is on the order of 30,000 ft/sec, and the impact pressure on the target is about 4 million psi.[20,21] (See Fig. 12.16.)

There are two types of jet tools commonly used for perforating — the retrievable, hollow carrier and the expendable carrier (Fig. 12.17). In the expendable gun there is a conical liner made of a friable material such as glass, cast iron, cast aluminum, or ceramics. Under the force of explosion, the liner shat-

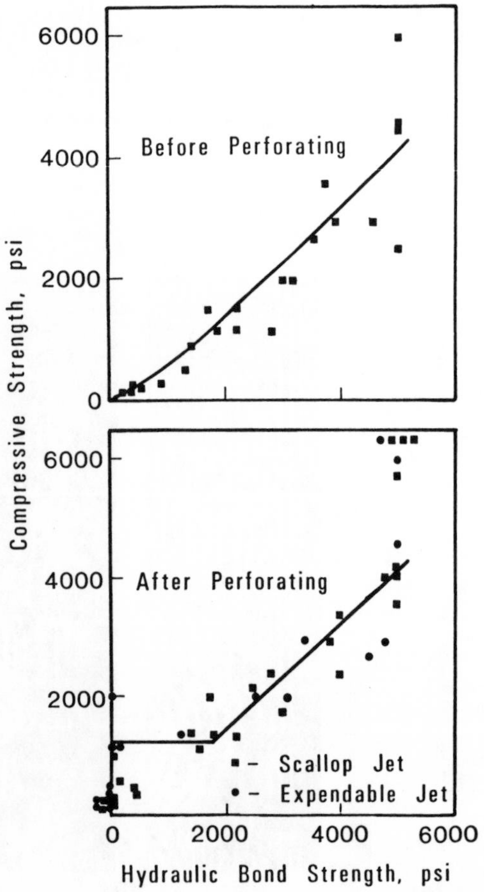

Fig. 12.14 Compressive strength vs hydraulic bond strength of cement before and after perforating.[20]

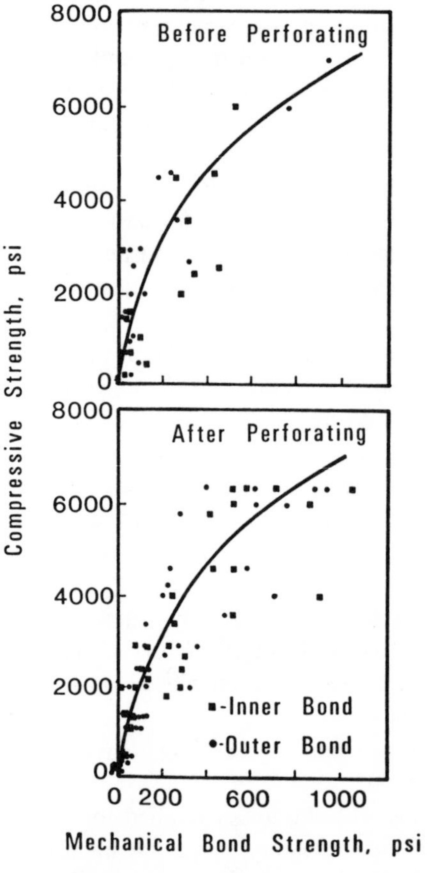

Fig. 12.15 Compressive strength vs mechanical bond strength of cement before and after perforating.[20]

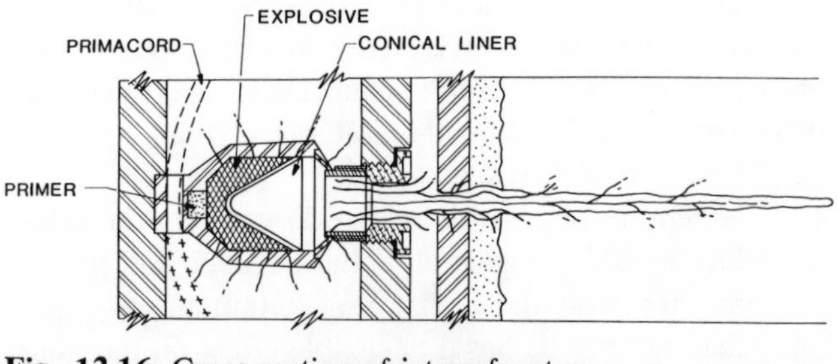

Fig. 12.16 Cross-section of jet perforator.

ters and the debris falls to the bottom of the hole. The retrievable tool is a cylindrical steel carrier that is returned to the surface and reloaded. It has the advantage of leaving no debris in the hole.

Jet guns are available in different sizes to achieve different results. The penetration achieved with jet perforating is influenced by the compressive strengths of the cement and the formation. In hard formations, jet guns usually achieve greater penetration, whereas in soft cements and soft formations, bullets may be more effective (Fig. 12.18). Penetration of jet guns is also a function of the weight of the explosive — the greater the explosive weight, the larger the shaped charge, which may create a larger hole or penetrate farther.[22]

Bullets

Because jet charges have been refined so that they can now be used under any conditions, bullet perforating has become largely obsolete. It was developed originally to perforate casing in place.[17,18] The bullet is propelled by an explosive charge and fired electrically or mechanically, depending on the design of the gun. The most commonly used bullet has an ogival profile, which penetrates well with minimum burr. Some bullets are designed especially for burrless penetration; those that are not can if necessary be equipped with caps that prevent the bullets from leaving burrs.

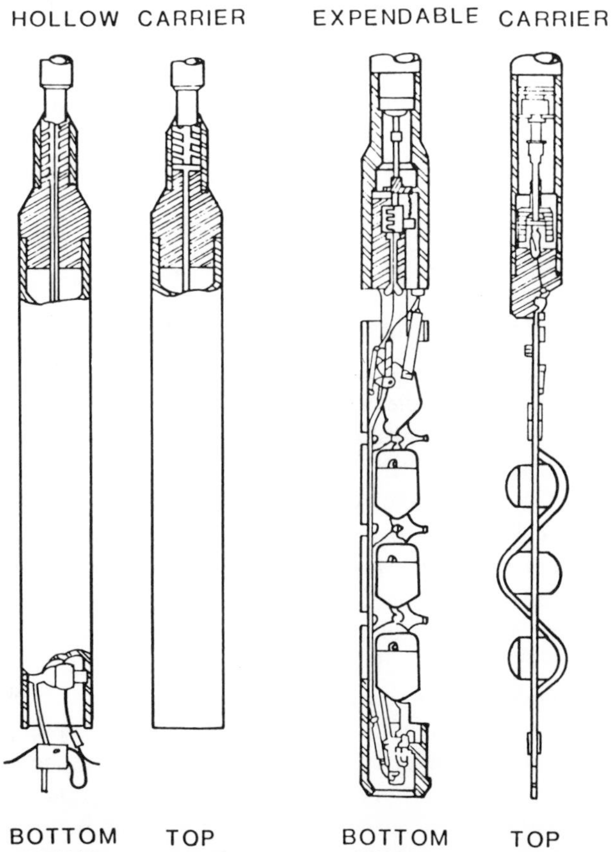

Fig. 12.17 Types of jet perforating tools.

Special-purpose bullets are available for such jobs as removing casing, fracturing formation and cement, establishing circulation in stuck tubing or drillpipe, and perforating the inner of two strings without damaging the outer string.

Bullet perforating tools, like the hollow-carrier jet perforators, are designed to hold and fire a number of bullets, so that a complete productive interval is perforated in one round trip in the hole. Perforating guns can fire the bullets singly or all at one time. When the bullets are fired singly, the tool will move in the casing as a result of recoil, and enough time must be allowed for the tool to come to rest before the next bullet is fired. When the bullets are fired simultaneously, the locations of all perforations are fixed by the configuration of the tool itself.

Hydraulic Jetting

Perforation by hydraulic jetting, or hydraulic perforating, is accomplished by forcing a high-pressure stream of sand-laden fluid through a jet aimed at the casing/formation target (Fig. 12.19).[24] The abrasive action of the sand forms a hole. This abrading jet is used mainly to prepare a formation for fracturing. Jetting a number of holes deep into the formation or

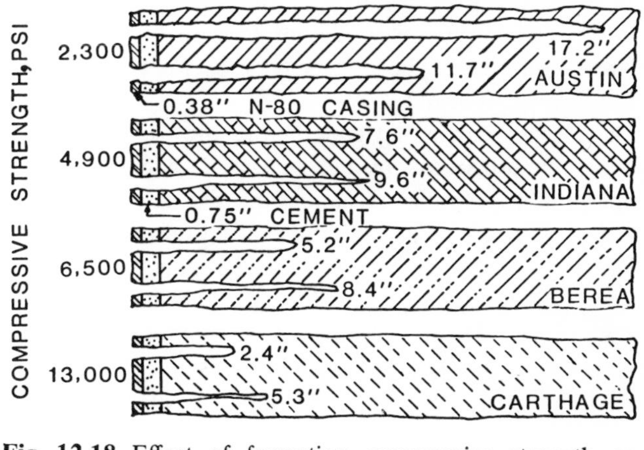

Fig. 12.18 Effect of formation compressive strength on penetrating efficiency of bullet and jet perforators.[23]

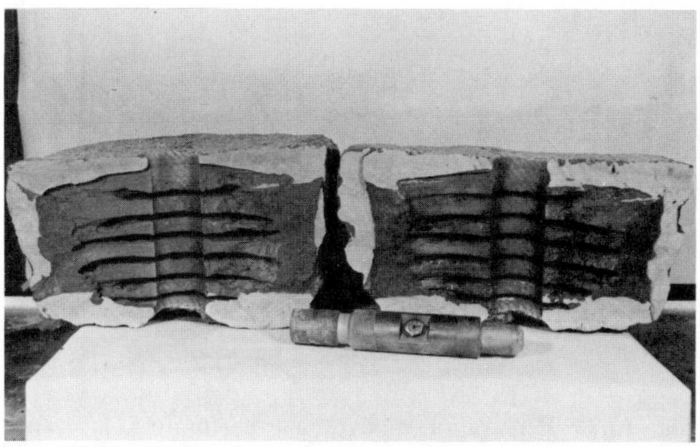
Fig. 12.19 Hydraulic jetting tool with sample of granite cut with sand-laden fluid.[24]

jetting a notch in the formation usually reduces the breakdown pressure and fixes the location of the fracture at the wellbore. The concentration of sand in the jet fluid can be varied, but usually it ranges from ½ to 1 lb per gallon of water. Hydraulic jetting causes no casing or cement damage.

Comparative penetration depths of jets, bullets, and hydraulic perforators are shown in Fig. 12.20. Of the three, hydraulic perforators are influenced the least by the strength of the formation.

Relative Performances of Bullet Guns and Jet Guns

With bullet guns the penetration is severely limited when the compressive strength of the cement or of the formation exceeds about 2,500 psi. Bullet guns do little damage to the casing except where the cement or the formation has a very high strength: there is a possibility that the high-strength cement will shatter. Optimum perforating results generally are achieved in cements of low compressive strength, ranging from 200 to 500 psi.

With jet guns, penetration is better in hard cement or in formations where the strength exceeds 3,000 psi. Only 5 to 10 percent of the explosive force creates the perforation, while the remaining 90 to 95 percent of the force creates very short term ultrahigh pressures and large shock forces on the inside walls of the shaped charge container. The effects on the two types of jet guns are as follows:

1. Hollow-carrier jet guns absorb the excess energy, which minimizes casing and cement damage. There is generally no significant damage if the cement has a compressive strength in the range of 2,000 to 6,000 psi.

2. Expendable or capsule-type jet guns exert their excess energy against the casing and may cause deformation, rupture, and splitting in smaller pipe. High-compressive-strength cement minimizes this damage by providing better support for the pipe.

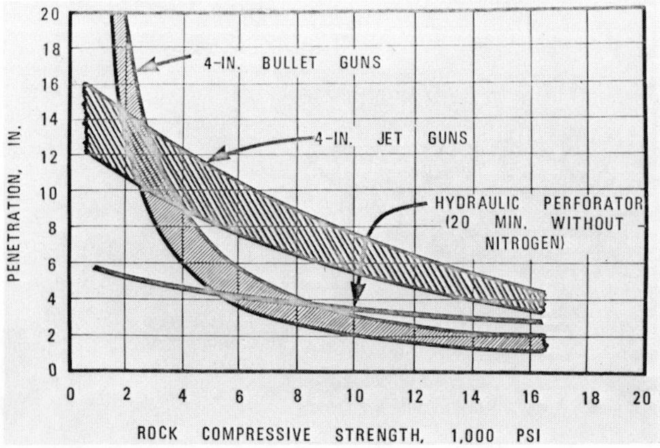

Fig. 12.20 Penetration of hydraulic perforators, jets, and bullets in rocks of various compressive strengths.

Fig. 12.21 illustrates the effect of various types and sizes of charges in several casing sizes either unsupported or well supported by cement of about 3,000-psi compressive strength. The perforating devices are denoted next to their proper explosive load with abbreviations of Jet Research Center trademarks:

S.W.	Sidewinder®
S.D.J	Super Dyna Jet®
L. J.	Link-Jet®
C. S.	Clean Strip (Kleen-Jet)®
S. J.	Swing Jet®

Mechanical Cutters

Mechnical cutters, such as knives, are used to open holes, slots, or windows to provide communication between the formation and the wellbore. They have the same advantages as the hydraulic jetting tools.

Permeators

Permeator units are welded to windows cut in sections of casing before the casing is run into the hole. The permeator units are placed in the casing to be located adjacent to the intended completion zone. Sleeves on the permeator units are extended with pump pressure to the wall of the borehole after the cement is in place but before it sets. By using acid to dissolve plugs in the units, communication is established between the inside of the casing and the borehole. An advantage of permeators is that they are extended before the cement sets and therefore do not cause damage. However, they are more expensive than jet or bullet perforators, they require more rig time, and they present more problems in establishing communication between the wellbore and the formation than do casing perforators, since the permeators do not penetrate the formation. If the casing sticks before it reaches the desired depth, there is a danger that permeator holes will be placed in the wrong interval.

12.8 Perforating in Gas-Producing Zones

Gas is not dense enough to expel debris within the perforation or to clean out a crushed zone that may form a lining around the perforation cylinder. To avoid the detrimental effect of these two factors a gas producing formation should be perforated in one of two ways. The first is to place clean fluid adjacent to the zone and reduce the hydrostatic head until the formation pressure exceeds the hydrostatic head by 500 psi. This is called a "controlled pressure completion." A second method, equally effective, is to perforate with a casing gun that is held against the casing with a rubber packer surrounding the port plug of the perforating gun. Since the interior of the gun is at atmospheric pressure all debris and usually part of the formation are expelled into the carrier, leaving a clean perforation. After perforating, a releasing valve is opened to permit the carrier to detach from the casing.

12.9 Factors Influencing Perforation

There are a great variety of possible slurries and a great number of hole variables. Those factors, as well as the following ones, influence the perforating qualities of casing-cement systems:

1. Size of the explosive charge and type of gun.
2. Cement compressive strength at the time of perforating.
3. Properties of the casing.
4. Physical support of the casing by the cement and of the cement by the formation.
5. Cement sheath thickness.
6. Nearness of the explosive to the casing (influenced by the size of the casing).
7. Perforation density.
8. Hydrostatic pressure at the time of perforating.

Perforating Time

The minimum time to wait before perforating should be based on well depth, temperature, and time for a specific type of cement to develop the correct strength for a specific type of perforator.[23] In a new well the

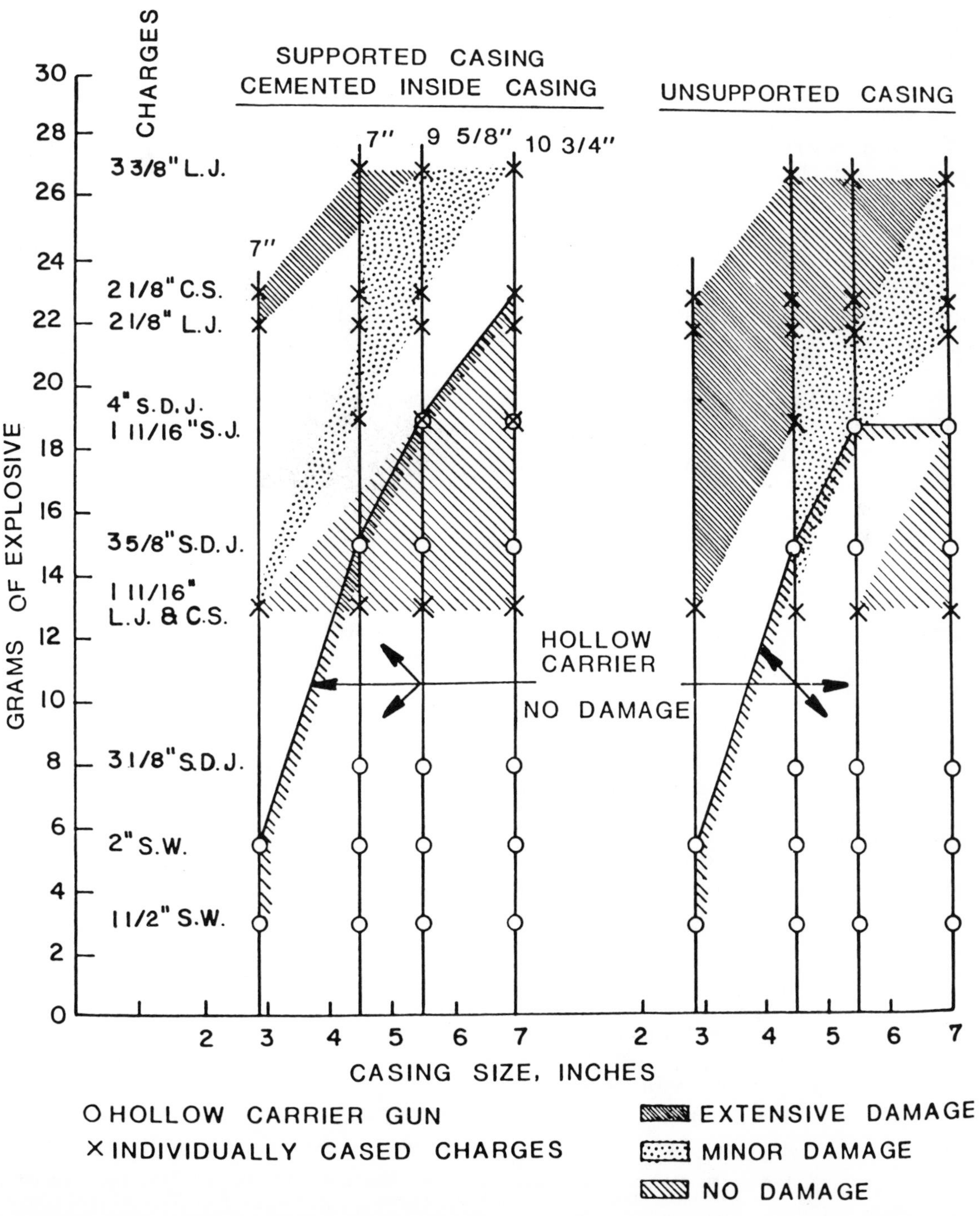

Fig. 12.21 Degree of casing damage in pipe of various sizes.[25]

strength of the set cement generally controls the time when a given zone should be perforated; down-hole pressure and temperature influence the setting time and must be considered. In old wells, however, those factors have little bearing, since the cement usually has reached its maximum strength after 30 to 60 days of curing.

Other Factors To Consider in Perforating[25]

1. Perforation Density — One or two shots per foot are usually adequate, particularly in zones to be fractured. When a sand formation is to be consolidated, a minimum number of perforations is preferred. Flow potential should always be considered for high-volume producers; however, high shot density — four or more shots per foot — tends to damage the casing and severely shatter the cement behind the pipe.

2. Economics — Perforating prices vary from area to area. In general, however, the fewer the perforations, the lower the cost. Where productive zones are separated by a number of nonproductive zones, selectively fired guns can save appreciable rig time, hence expense. Through-tubing guns can frequently save rig time if the tubing is run open-ended and set above the zone to be perforated. For new wells, tubing is usually run within a few hours after the top cement plug is pumped down. Through-tubing perforating is often carried out with no rig on the well.

12.10 Summary

Table 12.3 is a general summary of the material covered in this chapter on bonding, logging, and perforating.

References

1. Jones, P. H., and Berdine, D.: "Oil-Well Cementing—Factors Influencing Bond Between Cement and Formation," *Oil and Gas J.* (March 21, 1940) 71; *Petroleum World* (June 1940) 26; *Drill. and Prod. Prac.*, API (1940) 45-63.

2. Bearden, W. G., and Lane R. D.: "Engineered Cementing Operations To Eliminate WOC Time." *Drill. and Prod. Prac.*, API (1961) 17.

3. Maier, L. F.: "Understanding Surface Casing Waiting-on-Cement Time," paper presented at CIM 16th Annual Tech. Meeting, Calgary, Alta., Canada, May 1965.

4. Evans, G. W., and Carter, L. G.: "Bonding Studies of Cementing Compositions to Pipe and Formations," *Drill. and Prod. Prac.*, API (1962) 72.

5. Becker, H., and Peterson, G.: "Bond of Cement Composition for Cementing Wells," paper presented at Sixth World Pet. Cong., Frankfurt, June 19-26, 1963.

6. Carter, L. G., and Evans, G. W.: "A Study of Cement-Pipe Bonding," *J. Pet. Tech.* (Feb. 1964) 157-160.

7. *Halliburton Cementing Tables,* Section 230, Halliburton Co., Duncan, Okla. (March 1972).

8. Pardue, G. H., Morris, R. L., Gollwitzer, L. B., and Morgan, J. H.: "Cement Bond Log — A Study of Cement and Casing Variables," *J. Pet. Tech.* (May 1963) 545-554.

9. Grosmangin, M., Kokesh, F. P., and Majani, P.: "The

TABLE 12.3 — SUMMARY: BONDING, LOGGING, AND PERFORATING

	Remarks
Bonding	
Cement to Pipe	Cement properly placed around a centralized pipe provides a good mechanical bond. The bond can be influenced by heat of hydration, perforation, or stimulation.
Cement to Formation	This is the most critical aspect of bonding. Against a clean formation, cement will produce good hydraulic, shear, and gas bonds. Cement bonds poorly to thick mud cake. For best bond, centralize pipe, use scratchers, and move casing.
Logging	
Temperature Survey	A recording thermometer measures heat of hydration of cement. The temperature log does not yield as quantitative a cement evaluation as does a bond log. For best results, measurements should be made within 12 to 24 hours of cementing.
Radioactive-Tracer Survey	Iodine 131 or Zirconium-Niobium 95 is added to mixing water for tracer purposes. A gamma log run before and after a cement job shows where the cement is located in the annulus.
Bond Log	Bond logging is widely used to locate cement behind pipe. It is a better method of identification than temperature or radioactive-tracer logging. New logs identify both pipe and formation bond. Harder cements or longer WOC times provide the best bond logs. Acoustic waves are sometimes difficult to interpret.
Perforating	
Jets	Jets are the most widely used perforators. Retrievable hollow-carrier jets do less damage to pipe or cement. Expendable charges do some damage to small pipe. Hard cements (2,000 psi or more) are perforated best and suffer the least damage.
Bullets	Bullets perforate soft cement (500 psi) best, with very little damage. In hard cements or formations, there is less penetration.
Hydraulic Cutters	These do little damage to cement or pipe. They frequently are used to initiate fractures.
Mechanical Cutters	These are used in pipe recovery. They do little damage to cement or pipe.
Permeators	Permeators are attached to casing before the casing is run in the well. They are extended with pressure before the cement sets. The permeator plugs are opened with chemicals. The devices are not widely used.

Cement Bond Log — A Sonic Method for Analyzing the Quality of Cementation of Borehole Casings," *J. Pet. Tech.* (Feb. 1961) 165-171; *Trans.,* AIME, **222.**

10. Anderson, W. L., and Walker, T.: "Research Predicts Improved Cement Bond Evaluations With Acoustic Logs," *J. Pet. Tech.* (Nov. 1961) 1093-1097.

11. Pickett, G. R.: 'Acoustic Character Logs and Their Applications in Formation Evaluation," *J. Pet. Tech.* (June 1963) 659-667; *Trans.,* AIME, **228.**

12. Walker, T.: "A Full-Wave Display of Acoustic Signals in Cased Holes," *J. Pet. Tech.* (Aug. 1968) 811-824.

13. "Cement Quality Logging," technical report, Schlumberger Well Services (Sept. 1971).

14. Fertl, W. H., Pilkington, P. E., and Scott, J. B.: "A Look at Cement Bond Logs," *J. Pet. Tech.* (June 1974) 607-617.

15. Winn, R. H., Anderson, T. O., and Carter, L. G.: "A Preliminary Study of Factors Influencing Cement Bond Logs," *J. Pet. Tech.* (April 1962) 369-372.

16. McDowell, J. M., and Muskat, M.: "The Effect on Well Productivity of Formation Penetration Beyond Perforated Casing," *Trans.,* AIME (1950) **189,** 309-312.

17. Allen, T. O., and Atterbury, J. H.: "Effectiveness of Gun Perforating," *Trans.,* AIME (1954) **201,** 34-40.

18. Allen, T. O., and Worzel, H. C.: "Productivity Method of Evaluation Gun Perforating," *Drill. and Prod. Prac.,* API (1956) 112.

19. Krueger, R. F.: "Joint Bullet and Jet Perforation Tests — Progress Report," *Drill. and Prod. Prac.,* API (1956) 126.

20. Godfrey, W. K., and Methven, N. E.: "Casing Damage Caused by Jet Perforating," paper SPE 3043 presented at SPE-AIME 45th Annual Fall Meeting, Houston, Oct. 4-7, 1970.

21. Poulter, T. C., and Caldwell, B. M.: "The Development of Shaped Charges for Oil Well Completions," *Trans.,* AIME (1957) **210,** 11-18.

22. "Standard Procedure for Evaluation of Well Perforators," *API RP 43,* 2nd ed., API Div. of Production, Dallas (1971).

23. Thompson, G. D.: Effects of Formation Compressive Strength on Perforator Performance," *Drill. and Prod. Prac.,* API (1962) 191-197.

24. Pittman, F. C., Harriman, D. W., and St. John, J. C.: Investigation of Abrasive-Laden-Fluid Method for Perforation and Fracture Initiation," *J. Pet. Tech.* (May 1961) 489-495.

25. Robinson, R. L., Herrmann, U. O., and DeFrank, P.: "How Well Conditions Influence Perforations," paper presented at CIM 12th Annual Tech. Meeting, Calgary, Canada (1961).

Chapter 13

Regulations

13.1 Introduction

In the U. S., 43 of the 50 states have agencies regulating the drilling and cementing of wells. These regulatory bodies govern (1) the method of setting casing, (2) the volume of cement, (3) the WOC time, (4) the testing of cement jobs, (5) the squeezing, plugging, and testing of cementing plugs, and (6) the protection against pollution of fresh waters and mineral deposits.

Thirty years ago, regulations either were nonexistent or were not uniformly enforced. Through the years, industry and legislative groups have assisted each other in developing workable rules applying to the drilling and cementing of wells. Much of the technical literature that has guided these committees deals with the minimum compressive strength or bond strength required to adequately support pipe, and dates from as early as 1946.[1-5] Because they were written by a variety of groups over a long period of time, the rules are expressed in a variety of ways. Changes in technology and differences in operating environments have also contributed to the wide divergence in state regulations, particularly in regard to the volume of cement and the WOC time. There are also variations in cementing practices in fields of comparable charactertistics in the same state.

The intent of this chapter is not to cover all regulations, but to discuss only some of those that apply in the more active drilling areas of the U. S.

13.2 Regulatory Bodies Controlling the Cementing of Wells

United States — State and Federal

If one is to comply with regulations, obviously one must know that they exist. Governing bodies that control and are responsible for cementing regulations for the various states are shown in Table 13.1. The date and applicable rules are identified so that one can refer to them in seeking further information. Rules applicable to offshore drilling beyond the territories claimed by states fall under federal jurisdiction and are governed by the U. S. Geological Survey of the Dept. of the Interior.

International

Internationally, wells have been drilled in 110 countries. Less than 15 percent of those countries have rules covering casing and cementing practices. In most of the areas where there are no regulations, prudent practices are observed by the operators themselves since good practices usually lower drilling costs.

Table 13.2 lists 16 countries known to have drilling and cementing regulations. There may be others; therefore one should check locally before drilling outside the U. S.

13.3 Typical Regulations

Practically all rules cover methods of setting specific sections of casing and of plugging wells. Specifically emphasized in offshore federal regulations is conductor pipe, which is needed to start the hole in a vertical direction and to seal off badly sloughing surface material or water. Conductor pipe also serves as a connection for blowout-preventer equipment when shallow gas zones are being drilled. Although it is not always done, cement should be circulated to the ocean or Gulf floor level when conductor string is cemented.

Most state rules are very specific about setting surface casing to protect fresh-water sands from contamination, and to form a good, solid anchor to support the blowout-preventer equipment while the well is being drilled. States that have mineral deposits require that the surface pipe be cemented through these formations to protect them from oil and gas zones should they be mined at some later date.

In some areas no intermediate casing is required, while in others one or more strings may be necessary. Few states regulate the setting of production casing, even though its placement is most important in effectively isolating zones.

Specific rules applicable to active drilling areas — California, Oklahoma, Texas, and Federal Offshore —

TABLE 13.1 — REGULATORY BODIES AND RULES CONTROLLING THE CEMENTING OF WELLS IN THE UNITED STATES

State or Area	City	Regulatory Body	Date of Rule	Rule
Alabama	University	Oil & Gas Board	1967	B-9, Plug Method & Procedures; B-14, Casing Regulations
Alaska	Anchorage	Dept. of Natural Resources, Oil & Gas Div.	1969	2102 (a, b, c, d, e); 2056; 2057
Arizona	Phoenix	Oil & Gas Conservation Commission	1971	108 A, B, C; 109 A, B; 110; 202 A, B, C
Arkansas	El Dorado	Oil & Gas Commission	1970	Order No. 2-39, B-8, Para. A, B, C, E—Well Plugging
California	Sacramento	Dept. of Conservation, Div. of Oil & Gas	1968	Bulletin 82, Chap. I; Vol. 14, No. 3; Vol. 11, No. 4; Vol. 6, No. 1; Vol. 6, No. 6; Vol. 13, No. 5
Colorado	Denver	Oil & Gas Conservation Commission, Dept. of Natural Resources	1970	315 (b, c, d, e, f, i); 325; 327 (a, b)
Connecticut	Hartford	Dept. of Health (water wells only)	1971	Sec. 19-13-B5, 1f
Delaware	Dover	Dept. of Natural Resources and Environmental Control	1969	Sec. 10
Florida	Tallahassee	Dept. of Natural Resources, Bureau of Geology	1962	115-B-2.08 Plug Method & Procedures; 115-B-2.13 Casing Requirements
Georgia	Atlanta	Dept. of Mines, Mining & Geology	1945	B-9; B-14
Hawaii*				
Idaho	Moscow	Bureau of Mines & Geology	1963	XXXII
Illinois	Springfield	Dept. of Mines & Minerals, Oil & Gas Div.	1974	VIII (6)B, 2, 3, 4—Casing Cementing; XI (5); XI-A (2), (3), (4), (6), (7)
Indiana	Indianapolis	Dept. of Natural Resources, Oil & Gas Div.	1972	22-J-Surface Casing; 33-B-Plugging
Iowa	Des Moines	Natural Resources Council	1965	Chap. 84—No. 7 (1-3), No. 16 (2)
Kansas	Topeka	Kansas Corporation Commission, Conservation Div.	1966	82-2-119; 82-2-123; 82-2-401
Kentucky	Lexington	Dept. of Mines & Minerals, Oil & Gas Div.	1967	Reg. 4, Sec. 3; Reg. 5, Secs. 3, 4, 5, 6
Louisiana	Baton Rouge	Dept. of Conservation	1974	Order No. 29-B, Sec. V-D, Para. 5—Cementing Time; Sec. V—Casing Program; Sec. IX—Plugging
Maine*				
Maryland	Annapolis	Dept. of Water Resources	1969	2.36 B-8, Table II
Massachusetts*				
Michigan	Lansing	Dept. of Natural Resources, Div. of Geological Survey	1963	302 (B, C, F); 306; 802; 804; 806
Minnesota	St. Paul	Dept. of Conservation, Div. of Water, Soils & Minerals	1965	Circular No. 6
Mississippi	Jackson	Oil & Gas Board	1970	11, 12, 28, 53
Missouri	Rolla	Dept. of Public Health & Welfare, Div. of Geological Survey & Water Resources	1975	27-Plugging Application; 28-Procedures; 11-Surface Casing; 50-2-Plugging
Montana	Helena	Oil & Gas Conservation Commisision	1954	206.1 (a); 232.2 (a)
Nebraska	Sidney	Oil & Gas Conservation Commission	1969	312 (a, b, c, d, i); 331 (a, b); 404
Nevada	Reno	Dept. of Conservation & Natural Resources	1954	107, 201
New Hampshire*				
New Jersey	Trenton	Dept. of Health (water wells only)	1966	4.6
			1960	Revised Statistics
New Mexico	Santa Fe	Oil Conservation Commission	1955	107, Para. (a)
			1955	Order No. R-11-A
			1956	Order No. 63
New York	Albany	Dept. of Conservation, Oil & Gas Div.	1972	202
North Carolina	Raleigh	Dept. of Conservation & Development, Mineral & Resources Div.	1946	No. 1 (A, B, C, D)
North Dakota	Bismarck	Industrial Commission of North Dakota	1971	107; 202
Ohio	Columbus	Dept. of Natural Resources, Oil & Gas Div.	1974	Chap. 1509.15—Plugging; Chap. 1509.19—Plugging**
Oklahoma	Oklahoma City	Oklahoma Corporation Commission		Ref. No. 42-70
			1969	206 (b, c, f) Casing Cementing; 604 (e, f, g, h) Plugging
Oregon	Portland	Dept. of Geology & Mineral Industries	1971	10-198 (4); 10-014 (2)
Pennsylvania	Harrisburg	Dept. of Mines & Mineral Industries, Oil & Gas Div.	1969	Sec. 204—Casing; 206—Plugging
Rhode Island*				
South Carolina*				
South Dakota	Rapid City	Oil & Gas Board	1974	
Tennessee	Nashville	Oil & Gas Board	1968	Reg. No. 6, Sec. 3; Reg. No. 7, Secs. 3, 4, 5, 6, 10
Texas	Austin	Railroad Commission	1970	13—Casing (A) 1, 2a, b; (B) 2; (D); (E) 1a, b, c 14—Plugging (C) 15—(A), (B)
Utah	Salt Lake City	Dept. of Natural Resources, Oil & Gas Conservation Div.	1969	C-8(c); C-23(a-6); C-25(b); D-1(b, 1-7); E-3
Vermont*				
Virginia	Big Stone Gap	Dept. of Labor & Industry; Mines, Oil, & Gas Div.	1974	45.1-122; 45.1-129, -138, -140

*No recognized regulation.

**Details are not given in the Code Chapter 1509, but they do appear in various memoranda.

TABLE 13.1 — Cont'd.

State or Area	City	Regulatory Body	Date of Rule	Rule
Washington	Olympia	Dept. of Natural Resources, Div. of Mines & Geology	1954	14; 15; 24 (a, b, c, f)
West Virginia	Charleston	Dept. of Mines, Oil & Gas Div.	1973	Chap. 22-4, Sec. 2.10 (1, 2, 4); Sec. 2.11 (1, 2)
Wisconsin	Madison	Dept. of Natural Resources	1938	Appendix to Wisconsin Well Code
Wyoming	Casper	Oil & Gas Conservation Commission	1976	320 (a, b, c, d, g); 315
Federal	Washington, D.C.	Dept. of the Interior, Conservation Div.	1971	250.41 (a) 1; 250.44
Gulf of Mexico	Washington, D.C.	Dept. of the Interior, Conservation Div.	1975	OCS Order No. 2, 1(B, C, D, E); OCS Order No. 3, 1(A through I) 2
Pacific Region	Washington, D.C.	Dept. of the Interior, Conservation Div.	1969	OCS Order No. 10, 1(A1 through A7)

*No recognized regulation.

**Details are not given in the Code Chapter 1509, but they do appear in various memoranda.

are summarized below. Table 13.3 is a digest of WOC-time requirements for various states.

California

The State of California (Bulletin 82) defines general rules[7] to be followed, yet is flexible enough to allow state engineers to exercise their own judgment in some areas.

The general rules fall into two sections, depending on the type of hole.

1. In dry holes or uncased wells, any oil show must be covered with cement. The interface between fresh water and salt water must be covered by at least 100 linear ft of cement, and there must be a minimum plug of 25 ft at the surface. For offshore wells there is a requirement of a minimum of 200 ft of cemented surface pipe, in addition to other requirements defined by Federal Regulations.

2. In cased holes, all zones must be covered or protected with a minimum of 100 ft of cement above the top producing zone. The salt-water/fresh-water interface must also be protected with at least 100 ft of cement. A water shutoff test is required immediately above the top producing zone, and can be taken by wire line. In some cases, such a test is waived by the state inspector.

Oklahoma

The Oklahoma Drilling and Casing Procedures, governed by the Corporation Commission, are defined in Rule 206, dated 1976.[8] This rule applies to all wells drilled for oil, gas, or water, and to injection wells and disposal wells, whether the wells are drilled with rotary tools or with cable tools. It states that cementing of casing shall be by the pump and plug or displacement method and defines a minimum footage for surface pipe and a minimum WOC time. For wells that are plugged, Oklahoma rules define the volume of cement to be used and the location of the cement plug.

Texas

In Texas, the Railroad Commission has jurisdiction over the drilling and cementing of wells. It functions through 11 District Offices throughout the state.

Texas has rules that are applicable to the state as a whole and, in addition, has adopted specific field rules.[9] The statewide rules are general, whereas the individual field rules are specific and may apply only to that particular field.

Rules 13-14, dated Oct. 1, 1970, are applicable to casing and cementing. They define in general how much casing will be set, how much cement will be used, and how the casing is to be tested after completion. Rule 13(a) deals with surface casing and cementing, and is designed mainly to protect the fresh-water sands from contamination. The supply of fresh water in Texas, as in many other states, is very limited; consequently, every effort is exerted to protect all potential fresh-water sources.

U.S. Federal Regulation for the Outer Continental Shelf

Rule 250.41, Control of Wells, dated 1971, requires the casing and cementing of all wells with a sufficient number of strings in a manner necessary to (1) prevent release of fluids from any stratum through the wellbore (directly or indirectly) into the sea; (2) prevent communication between separate hydrocarbon-bearing strata (except such strata approved for commingling) and between hydrocarbon and water-bearing strata; (3) prevent contamination of fresh-water strata, gas, or water; (4) support unconsolidated sediments; and (5) otherwise provide a means of control of the

TABLE 13.2 — COUNTRIES, OTHER THAN U.S., KNOWN TO HAVE DRILLING AND CEMENTING REGULATIONS

Country	Agency
Abu Dhabi	Ministry of Petroleum
Australia	Department of Mines
Austria	Oberste Bergbehorde
Canada	Ontario-Dept. of Mines & Northern Affairs
	Alberta-Oil & Gas Conservation Board
	Saskatchewan-Dept. of Mineral Resources
Colombia	Minister of Mines & Petroleum
France	Direction Generale des Mines
Germany	Bureau of Mines
Ireland	Offshore Operating Committee-London
Italy	National Mining Bureau for Hydrocarbons
Japan	Bureau of Mines
	Petroleum Mine Safety Regulations
Libya	Petroleum Ministry
Mozambique	Geology & Mines Dept.
Netherlands	The Ministry of Mines
Turkey	Petroleum Administration
United Kingdom	Dept. of Trade & Industry—Petroleum Div.
Venezuela	Dept. of Hydrocarbons

formation pressures and fluids.[10,11]

When wells are abandoned, Rule 250.44 requires the lessee to submit a statement of reasons for abandonment and detailed plans for carrying out the work. No well shall be plugged and abandoned until the manner and method of plugging shall be approved or prescribed by the supervisor.

13.4 Permits

In most states, operators must file a notarized application covering details of the intention to drill a well. This application must describe casing and cementing programs.

Upon completion of the well, a notarized well-completion report on casing and cementing data must be filed. If the well is nonproductive, an application must be filed to plug and abandon it (and perhaps to pull the casing), with full details of the proposed plan. Upon completion of this work a detailed plugging record must be supplied.

The Geological Survey of the U. S. Dept. of the Interior requires similar reports for wells drilled on the Outer Continental Shelf.

Where wells are to be deepened or plugged back, the operator must file a report similar to that required when the well was first drilled.

13.5 Enforcement and Penalties

Most regulatory groups define the penalties for violating rules and the authority for enforcement. Where producing wells are involved, the establishment of production allowables may be withheld until the well owner complies with the regulation in question. In most instances, this procedure simplifies enforcement. Most of the states and federal bodies are in close agreement on the penalty for rule violation, differing only on the severity of the penalty.[12]

13.6 Summary

Regulations provide a means of controlling and protecting life, property, and petroleum reserves. They have been prepared by legislative bodies who have been guided by the experience of engineers, research organizations, and study committees throughout the oil industry. In the area of their jurisdiction these regulations represent a uniform practice for the drilling and cementing of casing. Most rules are not absolute or rigid, but are flexible within areas or fields. When proper evidence is presented, operators are often allowed to modify certain practices.

As the petroleum industry expands into deeper drilling areas and offshore marine operations, new regulations or modifications of existing regulations will be required. In planning any well, an operator should consult with the agency having local jurisdiction. New technology and changes in practices will continue to form the foundation for improvements in drilling regulations.

TABLE 13.3 — STATEWIDE WOC REQUIREMENTS FOR VARIOUS STATES

	Type of Job*									
	Surface Pipe				Intermediate String				Production String	
State	1	2	3	4	1	2	3	4	1	2
Alabama	—	16 hr	Up to 1,000 psi based on depth	16 hr	—	—	—	16 hr	1,500 psi or 0.2 psi/ft for 30 min.	24 hr
Colorado	500 psi or 8 hr	500 psi or 8 hr	—	8 hr	—	—	Properly	—	Properly	—
Kansas	300 psi and 8 hr	300 psi and 8 hr	—	24 hr	—	—	—	—	—	—
Louisiana	12 hr	0	Up to 100 psi based on depth	12 hr	—	0	800 to 1,500	12 hr	800 to 1,500	36 hr
Mississippi	12 hr	0	1 psi/ft up to 1,000 psi	12 hr	—	—	—	24 hr	0.2 psi/ft up to 1,500 psi	24 hr
Montana	8 hr	8 hr	Properly	8 hr	—	—	Properly	8 hr	Properly	—
New Mexico	8 hr	0	600 to 1,500	18 hr**	8 hr	0	600 to 1,500	18 hr**	600 to 1,500	18 hr**
North Dakota	12 hr	0	None	12 hr	—	—	None	12 hr	None	12 hr
Oklahoma	8 hr	Undefined	None	8 hr	—	—	—	8 hr	—	—
Texas	—	0	—	—	—	—	—	—	1,000 psi for 30 min.	24 hr
Wyoming	12 hr	0	Properly	12 hr	—	—	—	—	—	—
Federal Gulf of Mexico	12 hr	0	1,000	12 hr	1,500 psi or 0.2 psi/ft	—	—	12 hr	1,500 psi or 0.2 psi/ft for 30 min.	24 hr

*Key to Type of Job
Surface pipe and intermediate string:
1. WOC with surface pressure, without float.
2. WOC with surface pressure, with float.
3. Pressure test.
4. WOC to drill out.

Production string:
1. Pressure test.
2. WOC to perforate.

**Operator can choose between 18 hours and a time based on the strength of the cement.

References

1. Farris, R. F.: "Method for Determining Minimum Waiting-on-Cement Time," *Trans.*, AIME (1946) **165,** 175-188.

2. Cannon, G. E.: "Improvements in Cementing Practices and the Need for Uniform Cementing Regulations," *Drill. and Prod. Prac.*, API (1948) 126-133; *Pet. Eng.* (May 1949) B42.

3. Davis, S. H., and Faulk, J. H.: "Have Waiting-on-Cement Practices Kept Pace with Technology?" *Drill. and Prod. Prac.*, API (1957) 180.

4. Bearden, W. G., and Lane, R. D.: "You Can Engineer Cementing Operations To Eliminate Wasteful WOC Time," *Oil and Gas J.* (July 3, 1961) 104.

5. Maier, L. F.: "Understanding Surface Casing Waiting-on-Cement Time," paper presented at CIM 16th Annual Tech. Meeting, Calgary, Canada, May 1965.

6. *The Producing Industry in Your State,* IPAA Yearbook (1972).

7. *California Laws for Conservation of Petroleum and Gas,* Dept. of Conservation, Div. of Oil and Gas, Sacramento (1971).

8. *Drilling and Casing Procedures* (Rule 206b, Cementing; 604e, Methods of Plugging), Ref. No. 42-70, Oklahoma Corporation Commission, Oklahoma City (1969).

9. *Rule 13, Casing,* and *Rule 14, Plugging,* Railroad Commission of Texas, Oil and Gas Div., Austin (Oct. 1, 1970).

10. *Regulations Pertaining to Mineral Leasing, Operations, and Pipelines on the Outer Continental Shelf,* Code of Federal Regulations, U. S. Dept. of the Interior, Washington (April 1971).

11. *Regulations Pertaining to Mineral Leasing, Operations, and Pipelines in the Gulf Coast Region,* Code of Federal Regulations, U. S. Dept. of the Interior, Washington (Aug. 28, 1969).

12. McRee, B. C.: "Cementing Regulations Applied to Oil and Gas Wells," *Oil-Well Cementing Practices in the United States,* API, New York (1959) 209.

Chapter 14

Special Cementing Applications

14.1 Introduction

Cementing practices used in connection with oil wells have been adapted for use in other types of holes; for example, mine shafts, water wells, waste-disposal wells, steam-producing wells, thermal-recovery or steamflood wells, and wells in permafrost environments.

All these holes are fairly shallow, rarely exceeding 6,000 ft. Except in steam-producing wells, the formation temperature is normally less than 140°F. Because the wellbore and formation conditions for such holes are unusual, the casing and cementing programs must be carefully planned.

14.2 Large-Hole Cementing

The large holes — 48 to 144 in. in diameter — are used, for example, for testing thermonuclear devices, for venting and providing escape channels in mines, and as industrial water wells. Much of the technology for drilling them has been developed since 1960.[1-4]

Casing Programs

Large-diameter casing is normally not available commercially and must be fabricated locally. In the interest of economy in the manufacture of this heavy casing, the ratio of wall thickness to diameter is decreased as the diameter of the casing is increased. Thus, even with high-strength alloys and the thickest walls possible, there is a danger that unbalanced external hydrostatic pressures developed by long cement columns during displacement will cause the casing to collapse. Therefore the collapse pressure of the pipe must be calculated, taking into account the wall thickness, the pipe diameter, and the yield stress of the steel.[5]

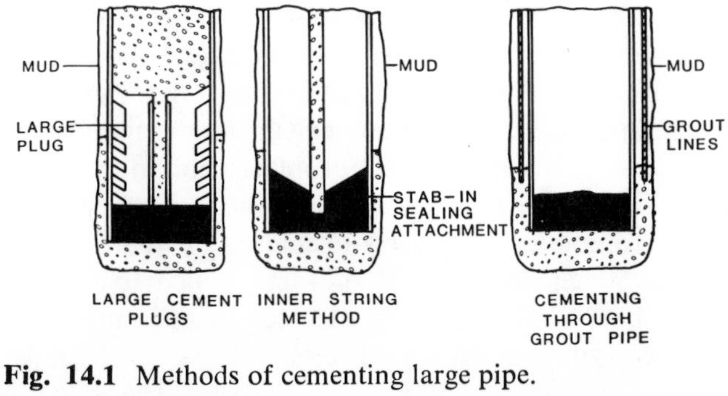

Fig. 14.1 Methods of cementing large pipe.

Cementing Techniques

There are three methods of placing cement to reduce the hazards of collapsing the pipe and still allow large volumes of cement slurry to be placed in the annulus (Fig. 14.1):

1. Filling the casing with a weighted mud that is equal to or near the density of the cement grout and then placing the cement in the conventional manner with an oversized cement plug.
2. Using the inner-string cementing method in which the drillstring is stabbed into a special, modified guide shoe. The cement is pumped down the pipe and back up the outside.
3. Pumping the cement slurry down the annulus through grout lines.

Large-diameter pipe requires a proportionately larger hole (see Table 14.1).[6] To fill the annulus that is formed requires a considerable volume of cement and a correspondingly longer mixing time (Fig. 14.2).

Where large volumes of slurries are to be mixed and placed before the cement takes an initial set, the borehole and casing must both be filled with a weighted mud having a specific gravity nearly equal to the density of the cement grout. The cement slurry is then mixed and placed in the conventional manner using special oversized cementing plugs. Objections to this

TABLE 14.1 — ESTIMATED HOLE SIZE FOR LARGE-DIAMETER PIPE

Depth (ft)	Size of Hole (in.) for Casing of Following Size (in. ID)				
	36	48	66	84	100
500	50	65	88	110	132
1,000	52	66	90	112	134
2,000	54	70	94	118	142
3,000	54	73	98	122	147
4,000	59	76	102	—	—

method are that it requires large volumes of mud and that it will leave at the bottom of the casing a large volume of cement that must be drilled out.

The inner-string technique is normally preferred. A smaller-diameter pipe is seated and sealed inside a guide shoe for slurry displacement. This allows the use of smaller latch-down cementing plugs, which function as backpressure valves after the slurry has been pumped to bottom. Also, the inner string may be withdrawn as soon as the cement takes its initial set, leaving a very small volume in the bottom to be removed when the shaft is deepened.

Cementing is commonly performed through grout lines placed between the casing and the borehole. The annulus is filled from the outside by raising the grout lines as the slurry is progressively distributed around the pipe. Placement may be in stages, allowing time for the cement slurry to set before the column is raised so high that the collapse pressure becomes critical. If it is desirable to pour the cement continuously, the rate of placement can be adjusted to allow the cement to set progressively. Continuous pouring is usually at a rate of 10 to 15 ft/hr. If the cement has a setting time of 2 hours, there will be no more than 20 to 30 ft of external head placed on the pipe at any one time during placement.

In designing for and placing large pipe, burst, tensile, and collapse pressures must be considered, just as with normal pipe.[7-10] And such variable factors as buoyancy and temperature must be studied carefully.

Cementing Materials

Any of the following compositions will normally help prevent migration of fluids in the annulus of large pipe:

1. Bentonite cement or filler compositions (slurries requiring large volumes of water to reduce slurry density).
2. Pozzolan-cement compositions (API Class B, G, or H with fly ash and bentonite).
3. Common construction cement (API Class A, B, G, or H).

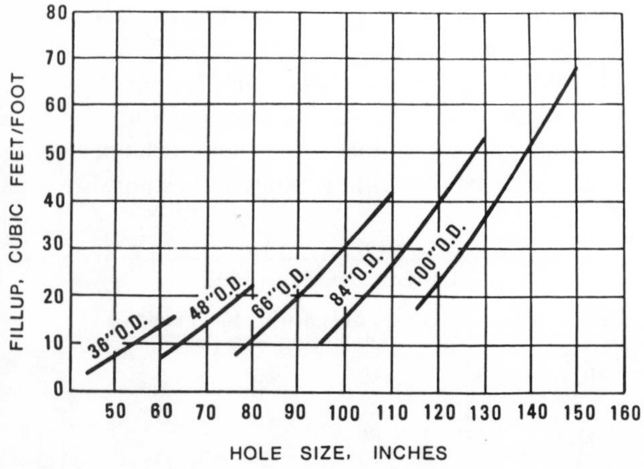

Fig. 14.2 Fillup factor — large-diameter pipe.

These cementing compositions can be designed to have variable densities and setting times and low heats of hydration, and to provide the sealing performance normally required for large-pipe hole conditions.

14.3 Water-Well Cementing

In most areas, the drilling and cementing of water wells must comply with the strictest of regulations. Therefore a fresh-water aquifer must be drilled on the basis of the best surface and subsurface information and be completed for the maximum protection and isolation of all fresh-water strata. Most water wells are completed at depths of less than 1,500 ft and without undue emphasis on casing equipment and cementing materials.[11] This does not mean, however, that water wells do not require maximum control to protect fresh-water zones from contamination and surface pollution.

Casing Programs

The National Water Well Assn. Committee on Specifications has published recommendations for water-well casing programs.[11] They state that "the casing should be new and composed of steel or other ferrous materials and shall be in accordance with the American Water Works Association Standards A100-66, ASTM or API." Normally, casing programs employed by the oil industry for surface or conductor pipe will meet these specifications and satisfy most regulations. However, in planning any casing program for water wells, the AWWA Standards should be referred to.

Completion Techniques

Water wells are normally completed in one of three ways[12,13]: (1) the water zone is cased through and the casing is perforated as it would be in an oil well; (2) the casing is set on top of the water-bearing zone and cemented, then the zone is drilled into and sand or gravel packed; or (3) casing is set at the top of the water-bearing zones and cemented and then a removable screen or liner is run through the water-bearing zone. (See Fig. 14.3.)

Cementing Techniques

In some areas, cementing may consist of pouring a small quantity of cement slurry (cement and water, or cement, sand, and water) down the annulus after the casing has been run into the well. This, however, can cause severe contamination from well fluids. A better way, especially for sealing the annulus, is to use two cementing plugs and pipe movement during placement. The most efficient method for placement in general, even though the pipe cannot be moved, is to use an inner string with a stab-in attachment on the floating device. Sometimes, instead of using a float collar or guide shoe, a pack-off device is used at the surface to force the slurry down the casing and back up the outside. For 9⅝-in. casing and larger, the latter two approaches are more economical, since the two-plug system requires additional equipment.

Cementing Materials

The pumpability, strength, and WOC time of cement are not so critical in water wells as in oil wells. Most water wells are cemented by the drilling contractor, who uses sack cement and pays little attention to weight control. Simple portland cement is used — ASTM Type I (API Class A) and water, slurried to approximately 15 lb/gal.[14] A mixture of equal portions of sand and cement is not uncommon.[15]

Water wells may be cemented using the same basic techniques used on oil wells. Persons or organizations not skilled in the art should refer to the American Water Works Assn. (Urbana, Ill.), to the National Water Well Assn. Board (Columbus, Ohio), or to service companies in the oil industry for guidance in the most effective means of water well casing and cementing.

14.4 Waste-Disposal Wells

The disposal of salt water or waste effluents is frequently accomplished by injecting them into a permeable underground formation in a depleted oil well or in a well drilled expressly for disposal purposes (Fig. 14.4). There are upwards of 40,000 salt-water disposal wells in the U. S. alone.[16,17] Well modifications are generally not required for disposing of salt water except to isolate the zone into which it is being injected. If a new well is drilled and cased expressly for this purpose, regulations should be consulted.

Wells drilled and cemented for the disposal of waste effluents require special attention, as they range in depths from 300 to 12,000 ft.[18]

Although pumping industrial waste into subsurface formations is commonly referred to as disposal, it can also be called storage, since, if the project is properly planned and executed, liquid wastes pumped into a selected stratum should remain there indefinitely without contacting fresh water or other isolated fluid-bearing formations.[19,20]

Subsurface injection is usually the most economical solution for many difficult disposal problems, but it may not solve every waste problem. Some areas have no formations suitable for injecting waste, and for some disposal problems the initial cost of properly drilling and completing a well may be too high.

In most industrial areas, local, state, or federal agencies have drawn up strict rules covering disposal wells and those rules should be taken into account in planning any well for disposal purposes. Because many of these waste solutions are corrosive, special grades of pipe and cement are required.

Completion Techniques

In the completion of wells for waste effluent disposal, a 15-in.-diameter hole is commonly drilled to approximately 200 ft below the deepest fresh-water aquifer, where 10¾-in. OD casing is set and cemented to the surface. The hole is then drilled through the potential disposal formation, where casing is set and the annulus is cemented to the surface. The most permeable sections of the disposal formation are perforated for completion. Cased-hole completions are used both in consolidated formations and in unconsolidated formations, where the wellbore tends to cave and restrict the flow of fluids into the formation. Where an open-hole completion is used in a well, the injection casing is set to the top of this disposal formation and cemented to the surface. The cement plug on bottom is then drilled out and drilling is continued into the disposal formation to the total depth. Open-hole completions are used where the formation is composed of consolidated materials such as sandstone and limestone. Unconsolidated sand formations may be gravel packed to help prevent sand cavings from filling the bottom section of the injection casing. Where the well is to be gravel packed, a large-diameter hole is reamed below the end of the injection casing and the cavity is filled with gravel. Gravel packing is also accomplished by placing either a perforated liner or a screen in the well so that it extends below the

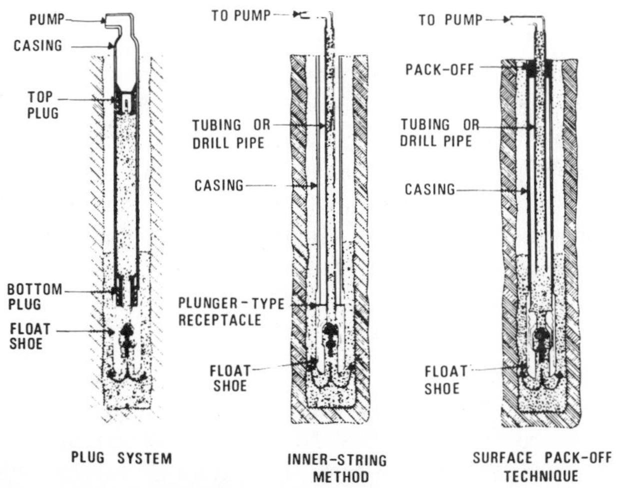

Fig. 14.3 Method of cementing water wells.[15]

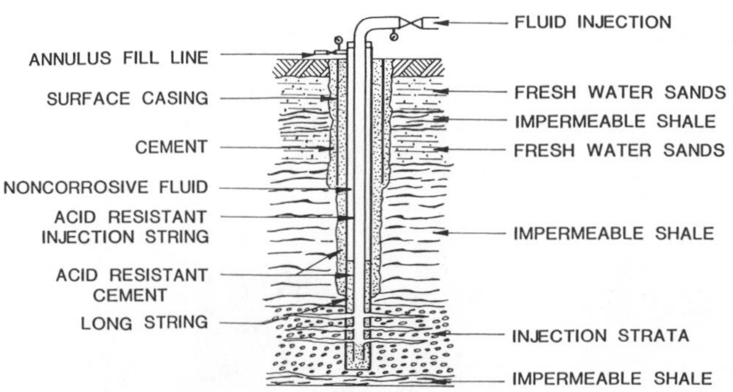

Fig. 14.4 Typical commercial waste-disposal well.[20]

casing, and gravel is introduced through the annular space and placed around the liner with the aid of a circulating fluid (Fig. 14.5).

To protect the casing, a disposal well usually employs tubing when a highly corrosive waste is injected down the hole. The practice of placing a packer below the last casing point, where feasible, helps to isolate the cement from the waste-disposal fluid, thereby allowing a greater latitude in the selection of cement as well as assuring the permeability of the well completion. The casing string exposed to a corrosive disposal fluid should be made from high-strength alloys such as Incaloy 825, Hasteloy, or Carpenter 20.

Cementing Materials

Once the casing is set to the desired depth, cement slurry should be circulated to the surface with a material that will offer maximum resistance to the disposal effluent being pumped into the well. No single cementing material will fit every disposal-well problem.[21] Cements should be tested with specific effluents to determine which works best. Some of the more commonly used compositions are zero tricalcium aluminate (C_3A) cements densified with dispersants and mixed with very little water. For highly corrosive environments, plastics, resins, or resin cementing compositions should be considered (see Section 2.7).

Some cementing compositions have been especially designed to be compatible with the disposal effluent — with radioactive or toxic waste, for example.[22,24] The waste is added through an inductor to the cement slurry and pumped into a suitable subsurface stratum for storage. Experimental work in dry shale sections with waste of low- and medium-level radioactivity indicates that the set cement remains permanently stored in a reacted state.[23]

14.5 Steam-Well Cementing

The development of drilling techniques and cementing materials for deep, hot oil wells where temperatures range to 500°F has made it possible to cement shallower geothermal steam wells having static bottom-hole temperatures in excess of 600°F. The hottest of these wells have been drilled in the Salton Sea area of California to depths of 5,000 ft for the recovery of a high-mineral-content steam containing rare metals.[26] Steam wells have also been drilled for the operation of steam generators to produce electrical power in Iceland, Italy, New Zealand, Japan, Mexico, and other parts of the world where large quantities of geothermal energy are found at shallow depths. One of the most publicized steam-recovery projects outside the U. S. is the Wairakei project in New Zealand.[27]

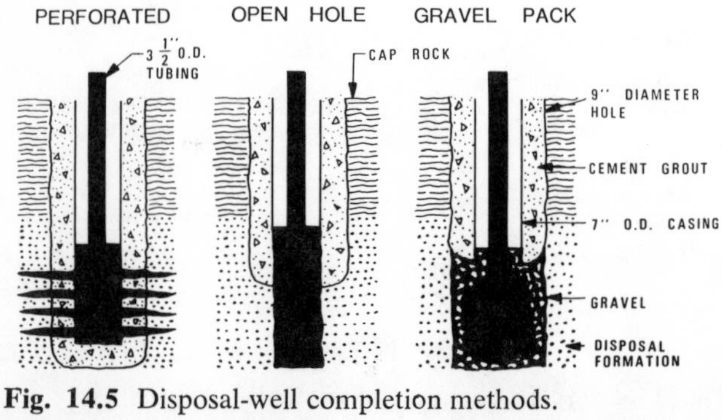

Fig. 14.5 Disposal-well completion methods.

Steam recovery projects were instituted in the U. S. as early as 1920 near San Francisco. Some 200 wells have been drilled there since 1957.[28,29] These steam wells represent some of the hottest and deepest steam deposits found anywhere in the world.[30,31] The temperature gradient — approximately 13°F for 100 ft of depth — imposes rather unusual demands on drilling muds, casing, and cement used to bond the casing to the formation.[32] Casing in steam wells is affected by temperature and undergoes creep or elongation by thermal expansion unless cemented to surface. Some of the earlier wells drilled for the recovery of geothermal steam used casing designed especially to withstand high temperatures. For later wells, however, standard oilfield casing programs were found satisfactory, particularly when casing was cemented to surface.

Because of the extremely high temperatures in steam wells, drilling mud must be circulated through a cooling system before being circulated back into the well to reduce bottom-hole circulation temperatures as much as possible before cementing. Cement should be circulated to the surface on every string of pipe to reduce buckling and minimize casing creep. Also, the cement must be placed in such a way that the pipe cannot be blown out of the hole.

Casing Programs

The main consideration in designing a casing program for steam wells is to have sufficient strength to resist longitudinal, tensile, and compressive forces and the collapse and bursting forces to which they may be subjected. Typical casing programs used in the steam wells of California are shown in Figs. 14.6A and 14.6B.

It was noted early that collapse or tension failures occurred when the casing was not properly cemented to the surface. It appeared that collapse failure was caused by heat expansion of undisplaced drilling fluid or water that had separated from the cement and become confined in pockets in the annular space between casings. In more recent wells, no failures have been reported in completely cemented wellbores, even though thermal stresses are thought to be very high.

Cementing Materials

Before cementing steam wells, it is necessary to laboratory test the specially designed cement under simulated field conditions. To achieve those conditions is very difficult. The pressure temperature thickening-time tester is limited to some 450°F; therefore, specimens

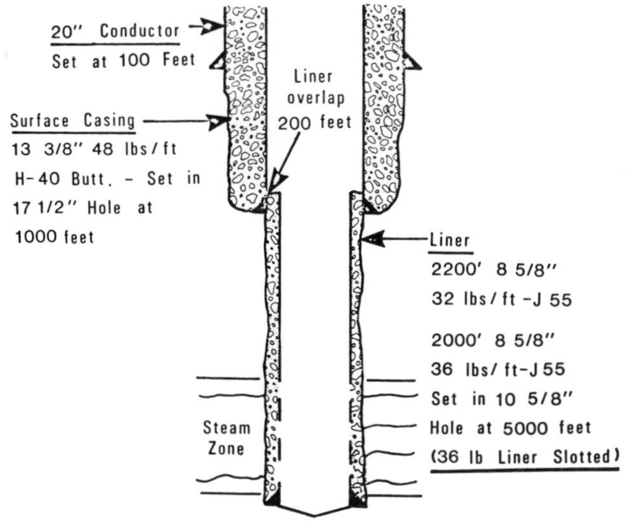

Fig. 14.6A Steam-well casing program — Geyser area, California.[30]

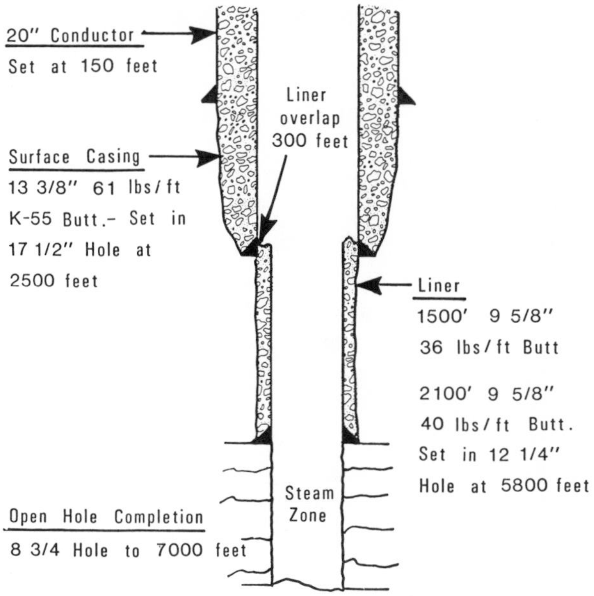

Fig. 14.6B Steam-well casing program — Imperial Valley, California.[30]

are usually preheated in the tester so that during the actual test the critical temperature gradients can be reached. In testing cements for strength, it may be necessary first to fire the cement test cubes in a furnace if there is no autoclave for the purpose. Although these procedures cannot duplicate field conditions, they do represent techniques closely related to cementing of thermal recovery wells, and this provides an excellent background for designing slurries for geothermal steam wells.

A major difficulty in curing cement at high temperatures is that of stabilizing them with sufficient silica flour to maintain relatively high strengths and low permeabilities.[33-35] Typical values of the physical properties of API Class G cement with and without bentonite and silica flour as used in the California steam wells are listed in Table 14.2.

14.6 Cementing in Permafrost Environments

Until the discovery of large oil deposits in the arctic region of Alaska in 1968, very little information had been reported on well completions in permafrost environments. One of the earliest accounts emphasizing the cementing of wells dates from 1952.[36] It describes the setting and cementing of casing in wells at the Naval Reserve area of northern Alaska at temperatures below freezing.[36-38] Before the discovery at Prudhoe Bay, not a great number of wells had been cemented in severe permafrost in northern Canada or Alaska.

Permafrost may be defined as any permanently frozen (continuous) subsurface formation. It may exist from a few feet to depths of 2,000 ft (Figs. 2.4 and 14.7).

TABLE 14.2 — PROPERTIES OF CEMENTS USED IN HIGH-TEMPERATURE STEAM WELLS[32]

API Class G Cement

Water (gal/sk)	Silica Flour (percent)	Bentonite (percent)	Perlite (cu ft/sk)	Mean Temperature (°F)	Test Interval (hours)	Heat Transfer Coefficient,* 2-in. Cement Sheath, Btu/(hr ft °F)
5.2	—	—	—	275	125	4.49
6.8	40	—	—	433	125	2.90
12.5	30	2.0	1.0	644	144	1.92
24.3	30	2.0	3.0	468	144	2.00

Silica Flour (percent)	Perlite (cu ft/sk)	Compressive Strength (psi)** After a Curing Time of: 1 day	3 days	7 days
0	0	425 (545)†	475 (545)†	555 (425)†
40	0	3,890 (7,330)	6,340 (11,025)	6,500 (10,010)
40	1	2,425 (3,690)	2,620 (3,580)	2,875 (3,975)
40	3	1,240 (1,690)	1,300 (1,735)	1,350 (1,825)

Silica Flour (percent)	Compressive Strength (psi)/Water Permeability (md) After Curing 27 Days**: API Class A Cement	API Class G Cement
40	7,525/0.065	7,875/0.036
50	9,625/0.101	9,580/0.023

*Values established on API Class A Cement. However, values using API Class G Cement should be comparable.
**Curing temperature, 460°F; curing pressure, 3,000 psi.
†Values in parentheses are compressive strengths of specimens cured the designated time at 440°F, then 3 days at 725°F.

Below this frozen section, the depth/temperature gradients are normal; however, since the surface temperature is so cold, the underlying formations, whether dry or water saturated, do not reach a temperature of 32°F until a certain depth is reached.

In the Arctic Islands, the formations are fairly well consolidated, so that thawing during drilling and cementing does not appear to be a major problem. Many frozen formations are strong enough that the holes remain in gauge despite some melting. In the Mackenzie Delta and other continental regions of Canada, the shallow formations frequently consist of ice lenses and frozen muskeg (discontinuous permafrost). Experts have noted that in Alaska, however, the thickness of permafrost varies, depending on where it is.[39,40] Permafrost thins out in the ocean and certain lake areas and disappears completely offshore in water depths of 10 ft or more. It varies in ice content — some sections are as little as 20 percent ice and others as much as 90 percent. It resembles snow. Ice lenses, which are composed of a complex growth of ice crystals and fine particles of silt and clay, also occur in permafrost. The ice lenses vary in thickness from a fraction of an inch to several feet. Measured temperatures at depths of 25 to 100 ft range from 8° to 15°F and increase steadily to the freezing point at the bottom of the permafrost. Warm spots (above 32°F) have been observed in permafrost, as well as flowing streams.

Materials for both above-ground and below-ground installations in these remote areas must be selected with care.[41,42] As temperatures drop below freezing, steel used for tubing and casing becomes brittle and loses it ductility, which is vital at low temperatures. Casing subjected to shock loads has been reported to crack.[38] Thus it must be handled with the utmost care in freezing conditions.

Holes should be drilled with a minimum of thawing around the wellbore. Melting can cause the thawed earth to subside, particularly in the upper 200 ft of the well. If the permafrost melts, drag forces can be

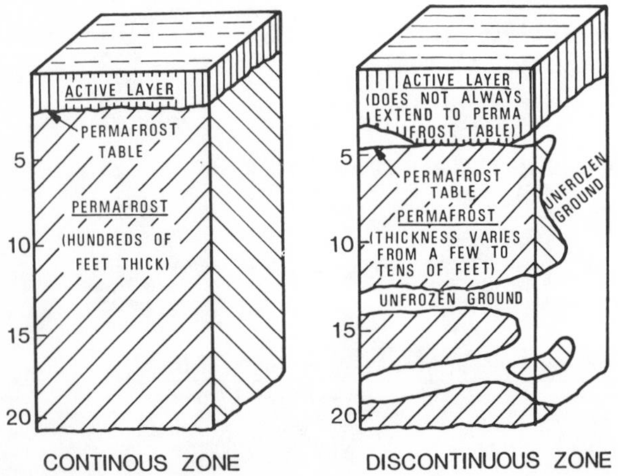

Fig. 14.7 Structures of continuous and discontinuous permafrost.[39]

transferred from the soil to the casing. The cement used should have a low heat of hydration and should develop strength without freezing at about 20°F. The annulus between the casing strings must either be cemented completely to the surface or contain a nonfreezing fluid to prevent the casing damage that can occur when freezable fluids expand as the permafrost freezes back.

Completion Techniques

In completing wells in permafrost, two distinct casing programs must be designed.[43] One involves the pipe, cements, and materials used through the permafrost; the other includes the cementing materials and casing used below the permafrost. Greater emphasis obviously must be placed on the section that goes through the permafrost. Below that, little special care is normally required.

The following precautions should be taken in dealing with permafrost:

1. Use high-strength ductile casing designed for subfreezing conditions.
2. Design the cementing and casing programs so that injection fluids or warm oil produced during the life of the well cannot harm or melt the permafrost section.
3. Leave no fluid in the annulus through the permafrost that might freeze and damage the casing.

To minimize the effects of warm fluids either during drilling or during production, some operators have insulated wells with a sheath of polyurethane foam to depths of 600 to 800 ft.[42] Polyurethane, with a thermal conductivity of 0.18 Btu/hr/ft, is an excellent insulator. In some of the earlier wells, a 16-in. casing was left uncemented between the 20-in. and the 13⅜-in. casings. This way refrigerant fluids could be circulated between the casing strings if insulation was required through the permafrost. These precautions were taken on the basis of thermal model studies that showed the amount of thawing that could result from producing warm oil over prolonged periods (see Fig. 14.8).

The need to eliminate freezable substances in the annulus between casing strings after cementing cannot be overemphasized. Bursting or collapsing of surface casing in permafrost has been attributed to ice rings formed from water or water-base drilling fluids left in the annular space; those fluids increase in volume some 9 percent upon freezing.[44] Such forces develop slowly, generally over a period of 3 to 5 months. The expanding ice usually collapses the inner string.

Various casing programs have been used to combat the ice forces. Fig. 14.9 illustrates some of the more basic casing designs. While many problems have already been encountered and solved, with prolonged production of warm oil many new ones could develop.

In designing a casing program, drag-down forces, subsidence, and freeze-back must be considered. Most

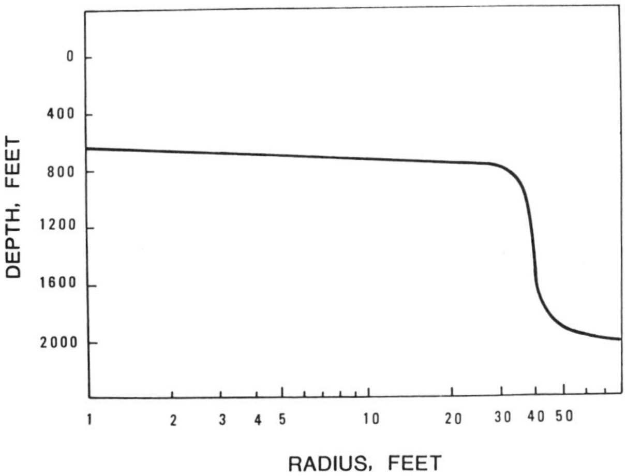

2½ in. of insulation inside 13⅜-in. casing to 700 ft.
Geothermal gradient: 0.012 °F/ft above 2,000 ft.
Thermal conductivity of earth: 20 Btu/(day ft °F) below 2,000 ft.
Thermal conductivity of earth: 30 Btu/(day ft °F) above 2,000 ft.
Flow rate: 15,000 BOPD.
Time: 20 years.

Fig. 14.8 Thaw profile for permafrost to depth of 2,000 ft.[42]

operators agree that casing must be strong enough to resist subsiding effects; therefore, 13⅜-in. (N-80, 68 lb/ft) casing is commonly set through the permafrost zone.

Cementing Compositions

Considerable research has gone into designing cement for permafrost regions.[44-50] Any composition used to bond pipe to ice or pipe to pipe should have a low heat of hydration to prevent hole enlargement due to melting or erosion and to allow the cement to set at temperatures of 15° to 32°F without requiring excessive heating of the mixing water. Also, in a minimum of time, it should develop adequate strength to support the pipe and to bond to the permafrost. The cement sheath should be thick enough to prevent further thawing of the permafrost.

Usually, with 30-in. conductor pipe and 20-in. surface casing, gypsum portland cements with freeze suppressant are used. Below the permafrost, API Class G cement with dispersant and retarder for placement is commonly used for 13⅜-, 9⅝-, and 7-in. casing programs. Because gypsum is one of the few materials that will set in subfreezing weather, gypsum cement blends are highly suitable for the permafrost regions. (Research[36] has shown that regular portland cements or high-early-strength cements — API Class C — accelerated with or without salt are not satisfactory.) In a few of the early casing programs on the Naval Reserve wells, pure gypsum cement (Cal-Seal) was successfully used to cement surface pipe. Table 14.3 lists the properties of both gypsum cement and refractory cements in these environments.

To suppress the freezing point of the cement slurry during the setting process, sodium chloride is often used. Fig. 14.10 shows how the freezing temperature of salt water varies with the concentration of salt. The slurry temperature that results when cold water

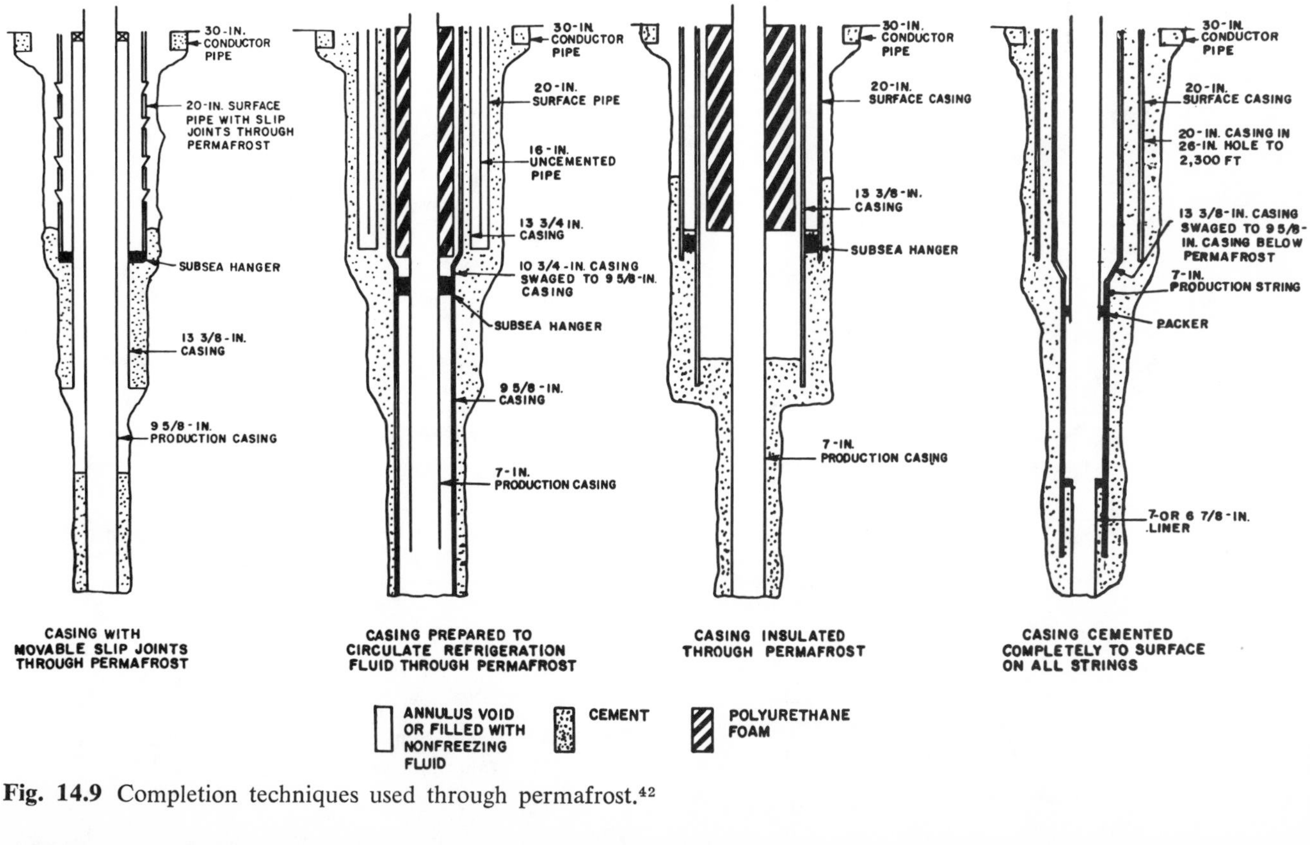

Fig. 14.9 Completion techniques used through permafrost.[42]

TABLE 14.3 — THERMAL PROPERTIES OF CEMENTS USED IN PERMAFROST

Cement	Heat of Hydration (Btu per lb of slurry)	Thermal Conductivity, Btu/(hr ft °F)	Diffusivity (ft^2/hr)	Slurry Weight (lb/cu ft)
Refractory	92 to 115	0.375	0.0085	114
Gypsum-portland	12 to 14	0.486	0.0094	90

and cold cement are mixed is shown in Fig. 14.11.

Most slurries used in the permafrost regions — regardless of composition — must have a fluid time of 2 to 4 hours because the large pipe, and possible washouts, require more cement than normal, thus more time for placement. Breakdowns and other unpredictable problems can arise, and they, too, must be allowed for.

Summary

In the short period since operations in the permafrost regions began, many obstacles have been overcome, and much has been learned about cementing in the harsh environment. To minimize damage in permafrost, the following should be considered:

1. Use drilling fluids to minimize hole enlargements due to melting or erosion. The frozen area should be disturbed as little as possible, and then only as drilling requires.
2. Displace all water-base fluids from the annulus and between casing strings with cement or a nonfreezing flush to reduce the possibility of collapsing or bursting the casing.
3. Use heavy ductile pipe designed for cold regions.
4. Use gypsum-portland cement, which sets without freezing, to provide support for the casing.
5. Insulate the casing to reduce thawing and consequent down-drag caused by the production of warm oil.
6. Below the permafrost zones, use API Class G cement with enough retarder to allow placement.

References

1. "Boring a 5-Foot Mine Shaft," *Mech. Eng.* (April 1937) 286.
2. Allen, J. H.: "Super Drills for Boring Shafts and Tunnels," *Mining World* (Dec. 1959) 20-25.
3. Bawcom, J. W.: "Rotary Drilling of Large Diameter Vertical Holes," paper presented at Northern Ohio Geol. Soc. Symposium on Salt, Cleveland, May 1962.
4. Burke, R. G.: "How Those 72-Inch Holes Are Being Drilled," *Oil and Gas J.* (April 27, 1964).
5. Timoshenko, S. P., and Gere, J.: *Theory of Elastic Stability* 2nd ed., McGraw-Hill Book Co., Inc., New York (1961).
6. Dellinger, T. B.: "Mechanics and Economics of Big Hole Drilling," *World Oil* (Oct. 1965).
7. Cain, P. A., and Folinsbee, J. R.: "Shaft Drilling at Lynn Lake With Oil Well Drilling Equipment," paper presented at CIM Annual Western Meeting, Winnipeg, Canada, Oct. 18, 1965.
8. Caldwell, H. L.: "Big Hole Drilling Technology Is Changing," *World Oil* (Sept. 1965) 103-107.
9. Crews, S. H.: "What It Takes To Drill Big Holes," *Pet. Eng.* (Oct. 1964).
10. Ball, W. W.: "How To Cement Big Pipe," *Oil and Gas J.* (Nov. 22, 1954) 86-92.
11. *AWWA Standard for Deep Wells,* American Water Works Assn., Natl. Water Well Assn., Urbana, Ill. (Sept. 1958).
12. Anderson, K. E.: *Water Well Handbook,* Missouri Water Well Drillers Assn. (1959).
13. Brown, B. D.: "Cementing Water Wells," *Public Works* (Sept. 1959) **90,** No. 9, 99-100.
14. *Proposed Specifications for Large Capacity Rotary Drilled Gravel Packed Wells,* 11th ed., NWWA Specifications Committee, Columbus, Ohio (1958).

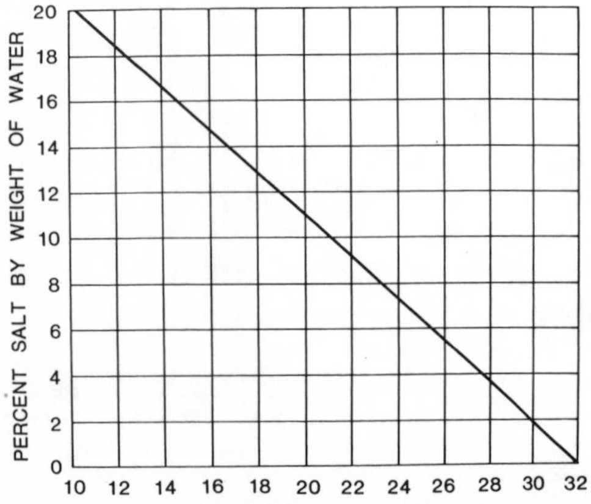

Fig. 14.10 Salt concentration vs freezing temperature.[44]

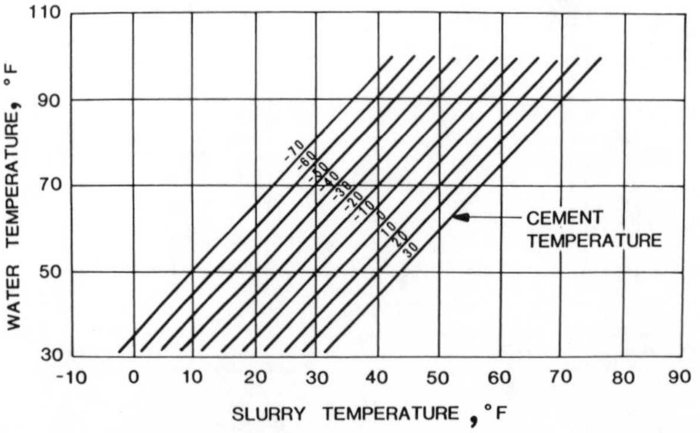

Fig. 14.11 Slurry temperature vs temperature of mixing water and of cement solids (neat Class G cement; 44 percent water by weight of cement).[44]

15. *Water Well Services,* Company Brochure, Halliburton Co., Duncan, Okla. (1961).

16. Baker, W. M.: "Waste Disposal Well Completion and Maintenance," *Industrial Water and Wastes* (Nov.-Dec. 1963) 43-47.

17. Albright, J. D.: "Subsurface Salt Water Disposal," paper presented at AIChE Southern Regional Meeting (Nov. 1965).

18. Donaldson, E. C.: "Subsurface Disposal of Industrial Wastes in the United States," Information Circular 8212, USBM (1964).

19. Caswell, C. A.: "Underground Waste Disposal Concepts and Misconceptions," *Environmental Science and Technology* (Aug. 1970) **4,** No. 8.

20. Warner, D. L.: "Deep-Well Disposal of Industrial Wastes," *Chem. Eng.* (Jan. 4, 1965) 73-78.

21. Ostroot, G. W., and Donaldson, A. L.: "Subsurface Disposal of Acidic Effluents," paper SPE 3201 presented at SPE-AIME Evangeline Section Regional Meeting, Lafayette, La., Nov. 9-10, 1970.

22. Belter, W. G.: "Radioactive Wastes," *Industrial Science and Technology* (Sept. 1962) 42-51.

23. Slagle, K. A., and Stogner, J. M.: "Techniques of Disposing of Industrial Wastes in Deep Wells," paper presented at National Pollution Control Exposition and Conference, Houston, April 1968.

24. Stogner, J. M.: "Underground Disposal of Solid Wastes," paper presented at 21st Oklahoma Industrial Waste Conference, Stillwater, March 1970.

25. *Deep Well Injection Symposium,* proceedings of Seventh Industrial Water and Waste Conference, Texas Water Pollution Control Assn. (June 1967).

26. Anderson, E. T.: "How World's Hottest Hole Was Drilled," *Pet. Eng.* (Oct. 1961).

27. "Geothermal Steam for Power in New Zealand," *Bull. 117,* L. I. Grange, Ed., New Zealand Geological Survey, Dept. of Scientific and Industrial Research, Wellington (1955).

28. Bayliss, B. P.: "Introduction to Geothermal Energy," paper SPE 4176 presented at SPE-AIME 43rd Annual California Regional Fall Meeting, Bakersfield, Nov. 8-10, 1972.

29. Koenig, J. B.: "Worldwide Status of Geothermal Exploration and Development," paper SPE 4179 presented at SPE-AIME 43rd Annual California Regional Fall Meeting, Bakersfield, Nov. 8-10, 1972.

30. Cromling, J.: "Geothermal Drilling in California," *J. Pet. Tech.* (Sept. 1973) 1033-1038.

31. Anderson, D. N.: "Geothermal Development in California," paper SPE 4180 presented at SPE-AIME 43rd Annual California Regional Fall Meeting, Bakersfield, Nov. 8-10, 1972.

32. Ostroot, G. W., and Shryock, S. H.: "Cementing Geothermal Steam Wells," *J. Pet. Tech.* (Dec. 1964) 1425-1429; *Trans.,* AIME, **231.**

33. Ludwig, N. C., and Pence, S. A.: "Properties of Portland Cement Pastes Cured at Elevated Temperatures and Pressures," *Proc.,* American Concrete Institute (1956) **52,** 673-687.

34. Carter, L. G., and Smith, D. K.: "Properties of Cementing Compositions at Elevated Temperatures and Pressures," *J. Pet. Tech.* (Feb. 1958) 20-28.

35. Ostroot, G. W., and Walker, W. A.: "Improved Composition for Cementing Wells With Extreme Temperatures," *J. Pet. Tech.* (March 1961) 277-284; *Trans.,* AIME, **222.**

36. White, F. L.: "Setting Cements in Below Freezing Conditions," Part 1, *Pet. Eng.* (Aug. 1952) B7.

37. White, F. L.: "Casing Can Be Cemented in Permafrost Area," *World Oil* (Dec. 1953) 119.

38. Von der Ahe, K. L.: "Operating Problems in Oil Exploration in the Arctic," *Pet. Eng.* (Feb. 1953) B12.

39. Brown, R. J. E.: *Permafrost in Canada,* U. of Toronto Press, Toronto (1970).

40. Stonley, R.: "Discussion of Thermal Considerations in Permafrost," *Proc.,* Geol. Seminar on the North Slope of Alaska, Pacific Section AAPG, Los Angeles (1970).

41. Thorvaldson, W. M.: "Low Temperature Cementing," paper presented at CIM 13th Annual Meeting, Calgary, Canada (1962).

42. Blount, E. M., Prueger, N. J.: "The Development of a Well Completion System for Deep Permafrost," paper SPE 3251 presented at SPE-AIME Southwestern Alaska Section Regional Meeting, Anchorage, May 5-7, 1971.

43. Larminie, F. G.: *The Arctic as an Industrial Environment — Some Aspects of Petroleum Development in Northern Alaska,* BP Alaska, Inc., Fairbanks (Jan. 1971).

44. Cunningham, W. C., Fehrenbach, J. R., and Maier, L. F.: "Arctic Cements and Cementing," *J. Cdn. Pet. Tech.* (Oct.-Dec. 1972).

45. Maier, L. F., Carter, M. A., Cunningham, W. C., and Bosley, T. G.: "Cementing Practices in Cold Environments," *J. Pet. Tech.* (Oct. 1971) 1215-1220.

46. Stude, D. L.: "High Alumina Cement Solves Permafrost Cementing Problems," *Pet. Eng.* (Sept. 1969) 64.

47. Morris, E. F.: "Evaluation of Cement Systems for Permafrost," paper presented at AIME 99th Annual Meeting, Denver, Colo., Feb. 15-19, 1970.

48. Kljucec, N. M., Telford, A. S., and Bombardieri, C. C.: "Cementing Arctic Wells Through Permafrost," paper 7257 presented at CIM 23rd Annual Tech. Meeting, Calgary, Canada, May 16-19, 1972.

49. Kljucec, N. M., and Telford, A. S.: "Well Temperature Monitoring with Thermistor Cables Through Permafrost," paper 7256 presented at CIM 23rd Annual Tech. Meeting, Calgary, Canada, May 16-19, 1972.

50. Perkins, T. K., Rochon, J. A., and Knowles, C. R.: "Studies of Pressures Generated Upon Refreezing of Thawed Permafrost Around a Wellbore," *J. Pet. Tech.* (Oct. 1974) 1159-1166; *Trans.,* AIME, **257.**

Appendix A

Examples of Primary Cementing Jobs

A.1 Job No. 1 for XYZ Oil Company

Type of job: 13⅜-in. surface pipe to be cemented at 400 ft (122 m)
Nominal bit size: 17.5 in. (444.5 mm)
Average hole size: 17.5 in. (444.5 mm)
Cement: API Class C, 450 sk
Additive: Calcium chloride (2 percent), 1.88 lb per sack of cement
Method of measuring temperature: estimate
Accessories: Guide shoe, insert float, top casing plug, one centralizer, 4 lb of thread-locking compound

Temperature, °F (°C)	70	(21)
Thickening time, hours	3	
WOC time, hours	12	
Mixing water, gal/sk (1/sk)	6.3	(23.8)
Bulk weight, lb/sk (kg/sk)	95.88	(43.5)
Slurry weight, lb/gal (kg/1)	14.81*	(1.8)
Slurry volume, cu ft/sk (cu m/sk)	1.339	(0.038)
Volume of slurry required (excluding excess), cu ft (cu m)	589	(16.7)
Excess volume, percent	100	

Estimated Cost

Cementing service charge		$ 235.00
Floating equipment charge		338.70
Cementing materials charges		
Materials	$1,047.26	
Transportation	431.46	
	$1,478.72	1,478.72
Miscellaneous charges		22.00
Total Cost of job		$2,074.42

A.2 Job No. 2 for XYZ Oil Company

Type of job: 9⅝-in. intermediate casing to be cemented at 5,300 ft
Nominal bit size: 12.25 in. (311.15 mm)
Average hole size: 12.25 in. (311.15 mm)
Method of measuring temperature: estimate

*Cement will have an approximate compressive strength of 2,865 psi at this weight.

Accessories: Down jet float shoe, float collar, stage cementing tool with free-falling plug set, four centralizers, 4 lb of thread-locking compound

Stage One

Cement: Gel cement, 475 sk
Additives per sack of cement: 6 percent bentonite, 5 lb salt, 0.25 lb Flocele, 5 lb gilsonite

Temperature, °F (°C)	110	(43)
Thickening time, hours	3	
WOC time, hours	18	
Mixing water, gal/sk (1/sk)	10.9	(41.2)
Bulk weight, lb/sk (kg/sk)	102.47	(46.5)
Slurry weight, lb/gal (kg/1)	12.41**	(1.49)
Slurry volume, cu ft/sk (cu m/sk)	2.082	(0.059)
Volume of slurry required (excluding excess), cu ft (cu m)	991	(28.1)
Excess volume, percent	50	

Tail-In

Cement: API Class C, 300 sk
Additive per sack of cement: 3 lb salt

Temperature, °F (°C)	110	(43)
Thickening time, hours	3	
WOC time, hours	18	
Mixing water, gal/sk (1/sk)	6.3	(23.8)
Bulk weight, lb/sk (kg/sk)	97	(44)
Slurry weight, lb/gal (kg/1)	14.85**	(1.78)
Slurry volume, cu ft/sk (cu m/sk)	1.346	(0.038)
Volume of slurry required (excluding excess), cu ft (cu m)	404	(11.44)
Excess volume, percent	50	

Stage Two

Cement: Gel cement, 575 sk
Additives per sack of cement: 6 percent bentonite, 5 lb salt, 0.25 lb Flocele, 5 lb gilsonite

Temperature, °F (°C)	110	(43)
Thickening time, hours	3	
WOC time, hours	18	

Mixing water, gal/sk (1/sk)	10.9	(41.2)
Bulk weight, lb/sk (kg/sk)	102.47	(46.5)
Slurry weight, lb/gal (kg/1)	12.41**	(1.49)
Slurry volume, cu ft/sk (cu m/sk)	2.082	(0.059)
Volume of slurry required (excluding excess), cu ft (cu m)	1,174	(33.2)
Excess volume, percent	50	

Estimated Cost

Cementing service charge			$1,565.40
Floating equipment charge			1,609.40
Cementing materials charges			
	Materials	Transportation	
Stage One	$1,398.26	$ 486.73	
Tail-In	702.94	291.00	
Stage Two	1,745.62	589.20	
	$3,846.82	$1,366.93	5,213.75
Miscellaneous charges			22.00
Total cost of job			$8,410.55

A.3 Job No. 3 for XYZ Oil Company

Type of job: 5½-in. production casing (tying in to 9⅝-in. casing) cemented at 13,000 ft (3,962 m)
Nominal bit size: 8.75 in. (222.25 mm)
Average hole size: 8.75 in. (222.25 mm)
Accessories: Down jet float shoe, float collar, top casing plug, five centralizers, 2 lb thread-locking compound
Method of measuring temperature: estimate

Initial Cementing

Cement: Light weight, 1,165 sk
Additives per sack of cement: 0.8 percent fluid-loss agent, 0.25 lb Flocele, 5 lb gilsonite

Temperature, °F (°C)	180	(82)
Thickening time, hours	3	
WOC time, hours	18	
Mixing water, gal/sk (1/sk)	8.7	(32.9)
Bulk weight, lb/sk (kg/sk)	80.85	(36.7)
Slurry weight, lb/gal (kg/1)	12.28†	(1.47)
Slurry volume, cu ft/sk (cu m/sk)	1.669	(0.048)
Volume of slurry required (excluding excess), cu ft (cu m)	1,945	(55.1)
Excess volume, percent	20	

Tail-In

Cement: API Class H, 350 sk
Additive per sack of cement: 0.75 percent dispersant

Temperature, °F (°C)	180	(82)
Thickening time, hours	3	
WOC time, hours	18	
Mixing water, gal/sk (1/sk)	5.2	(19.7)
Bulk weight, lb/sk (kg/sk)	94.70	(42.9)
Slurry weight, lb/gal (kg/1)	15.68†	(1.88)
Slurry volume, cu ft/sk (cu m/sk)	1.176	(0.033)
Volume of slurry required (excluding excess), cu ft (cu m)	389	(11.0)
Excess volume, percent	50	

Estimated Cost

Cementing service charge			$1,726.00
Floating equipment charge			299.25
Cementing materials charges			
	Materials	Transportation	
Initial cementing	$4,506.22	$ 941.90	
Tail-In	1,067.50	331.47	
	$5,573.72	$1,273.37	6,847.09
Miscellaneous charges			11.00
Total cost of job			$8,883.34

**At these weights, cement will have an approximate compressive strength of 400 psi (Stages One and Two); 300 psi (Tail-In).

†At these weights, cement will have an approximate compressive strength of 500 psi (12.28-lb/gal initial slurry); 3,000 psi (15.68-lb/gal tail-in).

Appendix B

Examples of Squeeze Jobs

B.1 Job No. 1 for XYZ Oil Company

Type of job: High-pressure squeeze to 2,000 psi (13 790 kPa); put 30 sk of cement in formation

Well depth, ft (m)	5,800	(1,768)
Depth of perforations, ft (m)	5,380 to 5,385	(1,639.8 to 1,641.3)
Well temperature, °F (°C)	120	(49)
Casing		
Size, in. (mm)	5½	(139.7)
Weight, lb/ft (kg/m)	17	(25.3)
Tubing		
Size, in. (mm)	2⅜	(60.3)
Weight, lb/ft (kg/m)	4.7	(6.9)
Length, ft (m)	5,300	(1,615)
Squeeze tool	Retrievable	
Displacing fluid	Fresh water	
Weight, lb/gal (kg/1)	8.34	(1.0)
Cement*		
API Class	H	
Quantity, sk	150	
Slurry weight, lb/gal (kg/1)	15.6	(1.83)
Thickening time, hours:min.	2:26	

Mixing Water*

For cement:

150 sk × 5.2 gal/sk = 780 gal, or 18.57 bbl (2.94 cu m)

To pump through perforations:

5,300 ft × 0.00387 bbl/ft = 20.51 bbl

85 ft × 0.0232 bbl/ft = 1.97 bbl

22.48 bbl (3.56 cu m)

To reverse out	22.48 bbl	(3.56 cu m)
Minimum total for job	63.53 bbl	(10.06 cu m)

Pressure To Reverse After Squeeze Job

Cement volume:*

120 sk × 1.18 cu ft/sk = 141.6 cu ft (4.0 cu m)

Casing capacity (5½ in., 17 lb/ft):

0.1305 cu ft/ft × 85 ft = 11.09 cu ft (0.31 cu m)

130.51 cu ft (3.68 cu m)

130.51 cu ft × 46.067 ft/cu ft = 6,012.2 ft

Cement slurry (15.6 lb/gal) 0.8096 psi/ft

Water (8.34 lb/gal) 0.4328 psi/ft

0.3768 psi/ft × 5,300 ft = 1,997 psi (13 769 kPa)

Maximum Hydraulic Pressure at 5,385 ft at Time of Squeeze

Cement slurry pressure

0.8096 psi/ft × 5,385 ft = 4,359 psi (30 055 kPa)

Pump pressure	2,000 psi	(13 790 kPa)
Total pressure	6,359 psi	(43 845 kPa)

B.2 Job No. 2 for XYZ Oil Company

Type of job: Low-pressure squeeze to seal perforations

Well depth, ft (m)	9,000	(2,743)
Depth of perforations, ft (m)	8,000 to 8,025	(2,438 to 2,446)
Well temperature, °F (°C)	140	(60)
Casing		
Size, in. (mm)	5½	(139.7)
Weight, lb/ft (kg/m)	17	(25.3)
Tubing		
Size, in. (mm)	2	(50.8)
Weight, lb/ft (kg/m)	4	(6.0)
Length, ft (m)	8,000	(2,438)
Squeeze tool	Retrievable	
Displacing fluid	Salt water	
Weight, lb/gal (kg/1)	9.5	(1.14)
Cement		
API Class	H	
Quantity (plus 0.75 percent fluid-loss agent), sk	75	
Slurry weight, lb/gal (kg/1)	15.3	(1.83)
Thickening time, hours:min.	3:40	

*Data from service company handbook.

Slurry Volume and Tubing Capacity

Cement slurry (75 sk), bbl (cu m)	16.4 (2.6)
Capacity of tubing:	
8,000 ft × 0.00387 (bbl/ft)	30.95 bbl (4.9 cu m)

Pressure on Formation When Tubing Is Full of Fluids

Cement slurry fillup, 16.4 bbl, or 4,240 ft		
4,240 × 15.3 lb/gal × 0.052	= 3,373 psi	(23 257 kPa)
Salt-water displacing fluid, 14.55 bbl, or 3,760 ft		
3,760 × 9.5 lb/gal × 0.052	= 1,858 psi	(12 811 kPa)
Total hydrostatic pressure	5,231 psi	(36 068 kPa)

Maximum Pressure To Avoid Fracturing the Zone

8,000 ft × 0.85	= 6,800 psi	(46 886 kPa)
Hydrostatic pressure (cement slurry plus salt water)	− 5,231 psi	(36 068 kPa)
	1,569 psi	(10 818 kPa)
Safety factor	− 300 psi	(2 069 kPa)
Safe surface pressure	1,269 psi	(8 749 kPa)

Note: With each 1,000 ft of cement slurry that is placed, surface pressure can be increased 300 psi.

Final Safe Squeeze Pressure

1,269 psi + (300 psi × 4,240 ft of cement slurry)	= 2,529 psi	(17 437 kPa)

Appendix C

Calculations for Down-Hole Plugging

C.1 Example Balanced-Cement-Plug Job for XYZ Oil Company

Place a 110-cu-ft cement plug in the bottom of 8,000 ft of 7-in., 26-lb casing, using 2½-in. tubing. Equalize the cement column and balance a column of water to the same height in the tubing/casing annulus as within the tubing.

First, find the equalization point with the following

Equalization Point Formula:

$$h_c = \frac{V}{V_a + V_i}, \quad \text{(C-1)}$$

where

h_c = height of cement column, linear ft,
V = volume of cement slurry, cu ft,
V_a = volume of annulus between tubing (or drillpipe) and casing (or hole), cu ft/linear ft,
V_i = volume inside tubing (or drillpipe or casing), cu ft/linear ft,

(V_a and V_i are found in cementing service company handbook tables, where V_a is given and C and V_i is given as T, on "Volume and Height Between Tubing and Casing" under the cu ft/linear ft column.)

$$h_c = \frac{110 \text{ cu ft}}{(0.1697 + 0.0325) \text{ cu ft/ft}},$$

$$= \frac{110}{0.2022} = 544 \text{ linear ft.}$$

Then determine the displacement volume required to equalize the cement column:

$$8{,}000 \text{ ft} - 544 \text{ ft} = 7{,}456 \text{ ft} \times 0.00579 \text{ bbl/ft*}$$
$$= 43.17 \text{ bbl.}$$

To balance the water to be placed ahead of the cement with the water that is to follow the cement, one must obtain the height that 1 bbl of water will fill in the space between the casing and the tubing, and the height that 1 bbl of water will fill inside the tubing:

Height between tubing and casing = 33.11 ft/bbl*
Height inside tubing = 172.7 ft/bbl*
172.7/33.11 = 5.22 bbl.

A ratio of 5.22 bbl of water ahead of the cement to 1 bbl of water behind it will give a balanced column of 172.7 ft of water. This ratio may be used to balance any amount of water that it may be necessary to use. Two barrels of water in the tubing behind the cement fills up 345.4 ft. To balance this in the annulus requires 2 × 5.22, or 10.44 bbl of water ahead of the cement.

C.2 Sacks of Cement for a Given Plugging Operation

Ascertain the number of sacks of API Class G cement mixed with water (5.2 gal/sk) required to place a 200-ft plug in 8¾-in. hole (assuming no washout).

$$N = \frac{(h)(C_h)}{Y}, \quad \text{(C-2)}$$

where

N = number of sacks,
h = height of fill, linear ft,
C_h = capacity of hole (from Cementing Tables), cu ft/linear ft,
Y = yield of sack cement with additives (from Cementing Tables), cu ft/sk.

$$N = \frac{200 \text{ ft} \times 0.4176 \text{ cu ft/ft}}{1.18 \text{ cu ft/linear ft}} = 70.78 \text{ sk.}$$

*From service company handbook tables.

Appendix D

Flow Calculations for Example Primary Cementing Jobs

D.1 Job No. 1 for XYZ Oil Company

It is desired to obtain the following:

1. The pumping rate required to put slurry into turbulent flow in the annulus.
2. The frictional pressure drop of slurry in the annulus and in the pipe.
3. The hydraulic horsepower required to overcome friction losses.

Conditions

Hole size: 10 in. (254 mm)
Casing size, OD: 7 in. (177.8 mm)
Casing weight: 35 lb/ft (52.1 kg/m)
Cementing depth: 5,000 ft (1,524 m)
Type of slurry: neat API Class G cement
From Handbook: ID of 7-in. (177.8 mm) 35-lb casing is 6.004 in. (152.5 mm).
From laboratory measurements on cement slurry:

$$n' = 0.30$$

$$K' = 0.195$$

$$\rho = 15.6 \text{ lb/gal } (1.87 \text{ kg/1})$$

1. Pumping Rate Required for Turbulence

$$v_c = \frac{1{,}129\ K'\ (96/d_i)^{n'\frac{1}{2-n'}}}{\rho},$$

or

$$v_c^{2-n'} = \frac{1{,}129\ K'\ (96/d)^{n'}}{\rho}, \quad . \quad . \quad . \quad . \quad . \quad (11.5)$$

For the annulus,

$$d = d_o - d_i = 10 \text{ in.} - 7 \text{ in.} = 3 \text{ in. } (254 \text{ mm} - 178.8 \text{ mm} = 76.2 \text{ mm}).$$

$$(96/d)^{n'} = 2.83,$$

$$v_c^{1.7} = \frac{1{,}129\ (0.195)\ (2.83)}{15.6} = 40.0;$$

$$v_c = 8.7 \text{ ft/sec } (2.65 \text{ m/sec}).$$

Rearranging the equation for displacement velocity (Eq. 11.3),

$$q_b = \frac{vd^2}{17.15}.$$

For the annulus, $d^2 = d_o^2 - d_i^2 = 100$ in. $- 49$ in. $= 51$ in. (2540 mm $-$ 1244.6 mm $=$ 1295.4 mm); therefore,

$$q_b = \frac{8.7\ (51)}{17.15} = 25.8 \text{ bbl/min. } (4.10 \text{ cu m/min.}).$$

2. *Frictional Pressure Drop*
In the annulus:
Reynolds number, N_{Re}, is 2,100,
Fanning friction factor, f, is 0.0074,

$$d = d_o - d_i = 10 \text{ in.} - 7 \text{ in.} = 3 \text{ in. } (254 \text{ mm} - 178.8 \text{ mm} = 76.2 \text{ mm}).$$

$$\Delta p_f = \frac{0.039\ L\rho v^2 f}{d}. \quad . \quad . \quad . \quad . \quad . \quad . \quad . \quad . \quad (11.7)$$

$$\Delta p_f = \frac{0.039\ (5{,}000)\ (15.6)\ (8.7)\ (8.7)\ (0.0074)}{3}$$

$$= 568 \text{ psi } (3916.36 \text{ kPa}).$$

In the casing:
Pumping rate, q_b, is 25.8 bbl/min. (4.10 cu m/min.),
Diameter, d, is 6.004 in. (152.5 mm).

From Eq. 11.3,

$$v_d = \frac{17.15\ q_b}{d} = \frac{17.15\ (25.8)}{6.004\ (6.004)}$$

$$= 12.19 \text{ ft/sec } (3.745 \text{ m/sec}).$$

From Eq. 11.4,

$$N_{Re} = \frac{1.86 (v^{2-n'}) \rho}{K'\ (96/d)^{n'}}$$

$$= \frac{1.86\ (71)\ (15.6)}{0.195\ (2.3)} = 4{,}590.$$

$$\Delta p_f = \frac{0.039\ L\rho v^2 f}{d},$$

$$= \frac{0.039(5{,}000)(15.6)(12.29)(12.29)(0.0062)}{6.004}$$

$$= 475 \text{ psi } (3275.13 \text{ kPa}).$$

To make a complete hydraulic analysis of the system, it is necessary to know n', K', and the density of both the mud in the well and the fluid used to displace the top plug. After calculations similar to those above have been made on these fluids, it is possible to make wellhead pressure calculations dependent upon the location of the cement slurry in the well and to estimate what the maximum wellhead pressure will be during the cementing operation. From this figure the hydraulic horsepower can be calculated as follows:

$$\text{hhp} = \frac{\text{pressure (lb/sq ft)} \times \text{pumping rate (cu ft/min.)}}{33{,}000}$$

$$\text{hhp} = 0.0244 \times \text{psi} \times \text{bbl/min.},$$

$$= (0.0244)\ (475)\ (25.8) = 299.$$

D.2 Job No. 2 for XYZ Oil Company

On this job it is desired to run a water flush, a Newtonian fluid, ahead of the cement. It is necessary, therefore, to calculate the pumping rate required to get the water flush into turbulent flow.

Conditions

Hole size: 6 in. (152.4 mm),
Casing size: 4½ in. (114.3 mm),
Casing weight: 11.6 lb/ft (17.3 kg/m),
Cementing depth: 5,000 ft (1524 m).

$$N_{\text{Re}} = \frac{928\ d_i v \rho}{\mu},$$

where

d_i = diameter, in. (m),

v = velocity, ft/sec (m/sec),

μ = viscosity, cp,

ρ = density, lb/gal (kg/1).

Turbulence begins at $N_{\text{Re}} = 2{,}100$.

$$N_{\text{Re}} = \frac{928\ d_i v \rho}{\mu};$$

$$v_c = \frac{2.26\mu}{d\rho};$$

$$v = \frac{2{,}100}{928\ d\rho} = \frac{2.26}{d\rho};$$

$$d = d_o - d_i = 6.0 \text{ in.} - 4.5 \text{ in.} = 1.5 \text{ in.}$$
$$(152.4 \text{ mm} - 114.3 \text{ mm} = 38.1 \text{ mm}).$$

For water flush: $\mu = 1$ cp; $\rho = 8.33$ lb/gal (1.0 kg/1)

$$v_c = \frac{2.26\ (1)}{(1.5)\ (8.33)} = 0.177 \text{ ft/sec } (0.0539 \text{ m/sec});$$

$$d^2 = d_o^2 - d_i^2 = 36 \text{ in.} - 20.25 \text{ in.} = 15.75 \text{ in.}$$
$$(914.4 \text{ mm} - 514.35 \text{ mm} = 400.05 \text{ mm});$$

$$q_b = \frac{vd^2}{17.15} = \frac{(0.177)\ (15.75)}{(17.15)},$$

$$= 0.162 \text{ bbl/min. required for turbulence in annulus.}$$

Nomenclature

Symbols

A = area, sq in.
C_h = capacity of hole, cu ft/linear ft
d = diameter, in.
d_e = equivalent diameter, in.
d_h = diameter of hole, in.
d_i = inside diameter of pipe, in.
d_{iop} = inside diameter of outer pipe, in.
d_o = outside diameter of casing, in.
d_{oip} = outside diameter of inner pipe, in.
D = depth, ft
$D_{\max}$ = maximum depth of casing, ft
f = Fanning friction factor, dimensionless
F_t = filtrate recovered in t minutes, ml
F_{30} = filtrate recovered in 30 minutes, ml
g_c = pressure gradient of cement slurry, psi/ft
g_m = pressure gradient of mud in annulus, psi/ft
g_{m2} = pressure gradient of mud in annulus outside casing, psi/ft
g_w = pressure gradient of water, psi/ft
h = height, ft
h_c = height of cement column, ft
h_m = height of mud column, ft
h_w = height of water column, ft
K' = fluid consistency index, $lb_f sec^{n'}/ft^2$
L = length of pipe, ft
n' = flow behavior index, dimensionless
N = number (Appendix C)
N = range extension factor of the Fann torque spring
N_{Re} = Reynolds number, dimensionless
$p_{b\max}$ = maximum annular backup pressure, psi
p_h = hydrostatic pressure, psi
p_{hs} = hydrostatic pressure of squeeze column, psi
p_s = support pressure to resist collapse, psi
$p_{s\max}$ = maximum allowable surface squeeze pressure, psi
p_{WBS} = "working burst strength" of casing, psi
p_{WCS} = "working collapse strength" of weakest casing within 1,000 ft above packer, psi
Δp_f = frictional pressure drop, psi
q = flow rate, cu ft/min.
q_b = pumping rate, bbl/min.
q_{cf} = pumping rate, cu ft/min.
q_d = displacement rate, bbl/min.
S_r = shear rate, sec^{-1}
S_s = shear stress, lb_f/ft^2
t_c = contact time, minutes
v = velocity, ft/sec
v_a = velocity in annulus, ft/sec
v_i = velocity inside pipe, ft/sec
$\dot{v}_c$ = critical velocity, ft/sec
$\overline{v_d}$ = average displacement velocity, ft/sec
V = volume, cu ft
V_a = volume of annulus, cu ft/linear ft
V_i = volume inside tubing, drillpipe, or casing, cu ft/linear ft
V_t = volume of fluid in turbulent flow, cu ft
w_m = mud weight, lb/gal
Y = yield point, lb_f/ft^2
μ = viscosity, cp
μ_p = plastic viscosity, cp
ρ = fluid or slurry density, lb/gal

Abbreviations

API	American Petroleum Institute
ASTM	American Society for Testing Materials
BHT	bottom-hole temperature
°C	degrees Celsius (or centigrade)
cm	centimetre
CMC	carboxymethyl cellulose
CMHEC	carboxymethyl hydroxyethyl cellulose
gm	gram
HEC	hydroxyethyl cellulose
hhp	hydraulic horsepower
hp	horsepower
kg	kilogram
kPa	kiloPascal
l	litre
lb_f	pounds force
lb_m	pounds mass
mev	million electron volts
mm	millimetre
ppm	parts per million
PV	plastic viscosity (μ_p)
sk	sack
Uc	units of consistency
WOC	waiting on cement
YP	yield presure (p_y)
μsec	milliseconds

Tables of Recommended SI Units and Conversion Factors for the Metric System as Taken From API Publication 2564, *Metric Practice Guide* (ASTM E 380, ANSI Z210.1)

Symbol	Name	Quantity	Type of Unit
A	ampere	electric current	Base SI unit
a	annum (year)	time	Allowable (not official SI) unit
bar	bar	pressure	Allowable (not official SI) unit, $= 10^5$Pa
°C	degree Celsius	temperature	Allowable (not official SI) unit, $=$ K $-$ 273.15
°	degree	plane angle	Allowable (not official SI) unit
d	day	time	Allowable (not official SI) unit, $=$ 24 h
g	gram	mass	Allowable (not official SI) unit, $= 10^{-3}$ kg
h	hour	time	Allowable (not official SI) unit, $= 3.6 \times 10^3$ s
ha	hectare	area	Allowable (not official SI) unit, $= 10^4$ m^2
J	joule	work, energy	Derived SI unit, $=$ 1 N•m
K	kelvin	temperature	Base SI unit
kg	kilogram	mass	Base SI unit
kn	knot	velocity	Allowable (not official SI) unit, $=$ 1.852 km/h
litre	litre	volume	Allowable (not official SI) unit, $=$ 1 dm^3
m	metre	length	Base SI unit
min	minute	time	Allowable (not official SI) unit
N	newton	force	Derived SI unit, $=$ 1 kg•m/s^2
naut. mi	nautical mile	length	Allowable (not official SI) unit, $=$ 1.852 km
Pa	pascal	pressure	Derived SI unit, $=$ 1 N/m^2
rad	radian	plane angle	Supplementary SI unit
s	second	time	Base SI unit
T	tesla	magnetic induction	Derived SI unit, $=$ 1 Wb/m^2
t	tonne	mass	Allowable (not official SI) unit, $= 10^3$kg $=$ 1 Mg
V	volt	electric potential	Derived SI unit, $=$ 1 W/A
W	watt	power	Derived SI unit, $=$ 1 J/s
Wb	weber	magnetic flux	Derived SI unit, $=$ 1V•s

Quantity and SI Unit		Customary Unit	API Preferred SI Metric Unit	Conversion Factor (Multiply value in customary unit by factor to get value in metric unit.)
FACILITY THROUGHPUT, CAPACITY				
Throughput (Mass Basis)	kg/s	US ton/d	t/d	9.071 847 E − 01
		US ton/h	t/h	9.071 847 E − 01
		lb/h	kg/h	4.535 924 E − 01
Throughput (Volume Basis)	m^3/s	bbl/d	t/a*	5.803 036 E + 01
		ft^3/d		1.179 869 E − 03
		bbl/h		1.589 873 E − 01
		ft^3/h		2.831 685 E − 02
		US gal/h	m^3/h	3.785 412 E − 03
		US gal/min	m^3/h	2.271 247 E − 01

*Based on a density of 1,000 kg/m^3.

Quantity and SI Unit		Customary Unit	API Preferred SI Metric Unit	Conversion Factor (Multiply value in customary unit by factor to get value in metric unit.)	
		SPACE, TIME			
Length	m	mi	km	1.609 344*	E + 00
		yd	m	9.144*	E — 01
		ft	m	3.048*	E — 01
		in	mm	2.54*	E + 01
		cm	mm	1.0*	E + 01
		mil	μm	2.54*	E + 01
		micron (μ)	μm	1	
Length/Length	m/m	ft/mi	m/km	1.893 939	E — 01
Length/Volume	m/m^3	ft/US gal	m/m^3	8.051 964	E + 01
		ft/ft^3	m/m^3	1.076 391	E + 01
		ft/bbl	m/m^3	1.917 134	E + 00
Area	m^2	mi^2	km^2	2.589 988	E + 00
		yd^2	m^2	8.361 274	E — 01
		ft^2	m^2	9.290 304*	E — 02
		in^2	mm^2	6.451 6*	E + 02
		cm^2	mm^2	1.0*	E + 02
Area/Volume	m^2/m^3	ft^2/in^3	m^2/cm^3	5.699 291	E — 03
Area/Mass	m^2/kg	cm^2/g	m^2/kg	1.0*	E — 01
			m^2/g	1.0*	E — 04
Volume, Capacity	m^3	m^3	m^3	1	
		yd^3	m^3	7.645 549	E — 01
		bbl (42 US gal)	m^3	1.589 873	E — 01
		ft^3	m^3	2.831 685	E — 02
			dm^3(litre)	2.831 685	E + 01
		US gal	m^3	3.785 412	E — 03
			dm^3(litre)	3.785 412	E + 00
		litre	dm^3(litre)	1	
		US qt	dm^3(litre)	9.463 529	E — 01
		US fl oz	cm^3	2.957 353	E + 01
		in^3	cm^3	1.638 706	E + 01
		ml	cm^3	1	
Volume/Length (Linear Displacement)	m^3/m	bbl/in	m^3/m	6.259 342	E + 00
		bbl/ft	m^3/m	5.216 119	E — 01
		ft^3/ft	m^3/m	9.290 304*	E — 02
		US gal/ft	m^3/m	1.241 933	E — 02
Time	s	yr	a	1	
		wk	d	7.0*	E + 00
		d	d	1	
		h	h	1	
		min	s	6.0*	E + 01

*All following digits would be zeros.

Quantity and SI Unit		Customary Unit	API Preferred SI Metric Unit	Conversion Factor (Multiply value in customary unit by factor to get value in metric unit.)
MECHANICS				
Velocity (Linear), Speed	m/s	mi/h	km/h	1.609 344* E + 00
		ft/s	m/s	3.048* E − 01
		ft/min	m/s	5.08* E − 03
		ft/h	mm/s	8.466 667 E − 02
Reciprocal Velocity	s/m	μs/ft	s/m	3.280 840 E + 00
Corrosion Rate	mm/a	in/yr (ipy)	mm/a	2.54* E + 01
Rotational Frequency	rev/s	rev/s	rev/s	1
		rev/min (rpm)	rev/s	1.666 667 E − 02
Acceleration (Linear)	m/s^2	ft/s^2	m/s^2	3.048* E − 01
		gal (cm/s^2)	m/s^2	1.0* E − 02
Acceleration (Rotational)	rad/s^2	rad/s^2	rad/s^2	1
Momentum	kg•m/s	lb•ft/s	kg•m/s	1.382 550 E − 01
Force	N	US ton_f	kN	8.896 443 E + 00
		kg_f (kp)	N	9.806 650* E + 00
		lb_f	N	4.448 222 E + 00
		N	N	1
Bending Moment, Torque	N•m	US ton_f•ft	kN•m	2.711 636 E + 00
		kg_f•m	N•m	9.806 650* E + 00
		lb_f•ft	N•m	1.355 818 E + 00
		lb_f•in	N•m	1.129 848 E − 01
		pdl•ft	N•m	4.214 011 E − 02
Bending Moment/ Length	N•m/m	lb_f•ft/in	N•m/m	5.337 866 E + 01
		kg_f•m/m	N•m/m	9.806 650* E + 00
		lb_f•in/in	N•m/m	4.448 22 E + 00
Moment of Inertia	$kg•m^2$	$lb•ft^2$	$kg•m^2$	4.214 011 E − 02
Stress	Pa	US ton_f/in^2	MPa (N/mm^2)	1.378 951 E + 01
		kg_f/mm^2	MPa (N/mm^2)	9.806 650* E + 00
		US ton_f/ft^2	MPa (N/mm^2)	9.576 052 E − 02
		lb_f/in^2 (psi)	MPa (N/mm^2)	6.894 757 E − 03
		lb_f/ft^2 (psf)	kPa	4.788 026 E − 02
		dyn/cm^2	Pa	1.0* E − 01
Yield Point, Gel Strength (Drilling Fluid)		$lb_f/100\ ft^2$	Pa	4.788 026 E + 01
Mass/Length	kg/m	lb/ft	kg/m	1.488 164 E + 00
Mass/Area Structural Loading, Bearing Capacity (Mass Basis)	kg/m^2	US ton/ft^2	t/m^2	9.764 855 E + 00
		lb/ft^2	kg/m^2	4.882 428 E + 00

*All following digits would be zeros.

Quantity and SI Unit		Customary Unit	API Preferred SI Metric Unit	Conversion Factor (Multiply value in customary unit by factor to get value in metric unit.)
FLOW RATE				
Flow Rate (Mass Basis)	kg/s	US ton/min	kg/s	1.511 974 E + 01
		US ton/h	kg/s	2.519 958 E — 01
		US ton/d	kg/s	1.049 982 E — 02
		US ton/yr	kg/s	2.876 664 E — 05
		lb/s	kg/s	4.535 924 E — 01
		lb/min	kg/s	7.559 873 E — 03
		lb/h	kg/s	1.259 979 E — 04
Flow Rate (Volume Basis)	m^3/s	bbl/d	dm^3/s	1.840 131 E — 03
		ft^3/d	dm^3/s	3.277 413 E — 04
		bbl/h	dm^3/s	4.416 314 E — 02
		ft^3/h	dm^3/s	7.865 791 E — 03
		US gal/h	dm^3/s	1.051 503 E — 03
		US gal/min	dm^3/s	6.309 020 E — 02
		ft^3/min	dm^3/s	4.719 474 E — 01
		ft^3/s	dm^3/s	2.831 685 E + 01
Flow Rate/Length (Mass Basis)	kg/s•m	lb/s•ft	kg/s•m	1.488 164 E + 00
		lb/h•ft	kg/s•m	4.133 789 E — 04
Flow Rate/Length (Volume Basis)	m^2/s	US gal/min•ft	m^2/s (m^3/s•m)	2.069 888 E — 04
		US gal/h•in	m^2/s (m^3/s•m)	4.139 776 E — 05
		US gal/h•ft	m^2/s	3.449 814 E — 06
Flow Rate /Area (Mass Basis)	kg/s•m^2	lb/s•ft^2	kg/s•m^2	4.882 428 E + 00
		lb/h•ft^2	kg/s•m^2	1.356 230 E — 03
Flow Rate /Area (Volume Basis)	m/s	ft^3/s•ft^2	m/s (m^3/s•m^2)	3.048* E — 01
		ft^3/min•ft^2	m/s (m^3/s•m^2)	5.08* E — 03
		US gal/h•in^2	m/s (m^3/s•m^2)	1.629 833 E — 03
		US gal/min•ft^2	m/s (m^3/s•m^2)	6.790 972 E — 04
		US gal/h•ft^2	m/s (m^3/s•m^2)	1.131 829 E — 05
Flow Rate/ Pressure Drop (Productivity Index)	m^3/s•Pa	bbl/d•psi	m^3/d•kPa	2.305 916 E — 02
DENSITY, SPECIFIC VOLUME, CONCENTRATION				
Density (Gases)	kg/m^3	lb/ft^3	kg/m^3	1.601 846 E + 01
			g/m^3	1.601 846 E + 04
Density (Liquids)	kg/m^3	lb/US gal	kg/dm^3	1.198 264 E — 01
		lb/UK gal	kg/dm^3	9.977 644 E — 02
		lb/ft^3	kg/dm^3	1.601 846 E — 02
		g/cm^3	kg/dm^3	1
Density (Solids)	kg/m^3	lb/ft^3	kg/dm^3	1.601 846 E — 02

*All following digits would be zeros.

Quantity and SI Unit		Customary Unit	API Preferred SI Metric Unit	Conversion Factor (Multiply value in customary unit by factor to get value in metric unit.)
Specific Volume (Gases)	m^3/kg	ft^3/lb	m^3/kg	6.242 796 E − 02
			m^3/g	6.242 796 E − 05
Specific Volume (Liquids)	m^3/kg	ft^3/lb	dm^3/kg	6.242 796 E + 01
		UK gal/lb	dm^3/kg	1.022 241 E + 01
		US gal/lb	dm^3/kg	8.345 404 E + 00
Specific Volume (Mole Basis)	m^3/mol	litre/g mol	$m^3/kmol$	1
		ft^3/lb mol	$m^3/kmol$	6.242 796 E − 02
Specific Volume (Clay Yield)	m^3/kg	bbl/US ton	m^3/t	1.752 535 E − 01
		bbl/UK ton	m^3/t	1.564 763 E − 01
Concentration (Mass/Mass)	kg/kg	wt %	kg/kg	1.0* E − 02
			g/kg	1.0* E − 05
		wt ppm	mg/kg	1
Concentration (Mass/Volume)	kg/m^3	lb/bbl	kg/m^3	2.853 010 E + 00
		lb/1000 US gal	g/m^3	1.198 264 E + 02
		lb/1000 bbl	g/m^3	2.853 010 E + 00
Concentration (Volume/Volume)	m^3/m^3	bbl/bbl	m^3/m^3	1
		ft^3/ft^3	m^3/m^3	1
		bbl/acre•ft	dm^3/m^3	1.288 931 E − 01
		US gal/ft^3	dm^3/m^3 (litre/m^3)	1.336 806 E + 02
		ml/US gal	dm^3/m^3 (litre/m^3)	2.641 720 E − 01
		Vol ppm	cm^3/m^3	1
			dm^3/m^3 (litre/m^3)	1.0* E − 03
		US gal/1000 bbl	cm^3/m^3	2.380 952 E + 01
Concentration (Mole/Volume)	mol/m^3	lb mol/US gal	$kmol/m^3$	1.198 264 E + 02
		lb mol/ft^3	$kmol/m^3$	1.601 846 E + 01
		std ft^3 (60° F, 1 atm)/bbl	$kmol/m^3$	7.518 21 E − 03
Concentration (Volume/Mole)	m^3/mol	US gal/1000 std ft^3 (60°F/60°F)	$dm^3/kmol$	3.166 91 E + 00

MASS, AMOUNT OF SUBSTANCE

Mass	kg	UK ton	t	1.016 047 E + 00
		US ton	t	9.071 847 E − 01
		US cwt	kg	4.535 924 E + 01
		kg	kg	1
		lb	kg	4.535 924 E − 01
		oz (troy)	g	3.110 348 E + 01
		oz (av)	g	2.834 952 E + 01
		g	g	1
		grain	mg	6.479 891 E + 01
Amount of Substance	mol	lb mol	kmol	4.535 924 E − 01
		g mol	kmol	1.0* E − 03

*All following digits would be zeros.

Quantity and SI Unit		Customary Unit	API Preferred SI Metric Unit	Conversion Factor (Multiply value in customary unit by factor to get value in metric unit.)

CALORIC VALUE, HEAT, ENTROPY, HEAT CAPACITY

Quantity	SI Unit	Customary Unit	API Preferred SI Metric Unit	Conversion Factor
Calorific Value (Mass Basis)	J/kg	Btu/lb	MJ/kg	2.326 000 E — 03
Calorific Value (Volume Basis — Solids and Liquids)	J/m^3	Btu/ft^3	MJ/m^3	3.725 895 E — 02
			kJ/m^3	3.725 895 E + 01
		$kcal/m^3$	MJ/m^3	4.184* E — 03
			kJ/m^3	4.184* E + 00
		cal/ml	MJ/m^3	4.184* E + 00
Calorific Value (Volume Basis — Gases)	J/m^3	cal/ml	kJ/m^3	4.184* E + 03
		$kcal/m^3$	kJ/m^3	4.184* E + 00
		Btu/ft^3	kJ/m^3	3.725 895 E + 01
Specific Heat Capacity (Mass Basis)	J/ kg•K	kW•h/kg•°C	kJ/kg•°C	3.6* E + 03
		Btu/lb•°F	kJ/kg•°C	4.186 8* E + 00

TEMPERATURE, PRESSURE, VACUUM

Quantity	SI Unit	Customary Unit	API Preferred SI Metric Unit	Conversion Factor
Temperature (Absolute)	K	°R	K	5/9
		°K	K	1
Temperature (Traditional)	K	°F	°C	5/9(°F - 32)
		°C	°C	1
Temperature/Length (Geothermal Gradient)	K/m	°F/100 ft	mK/m	1.822 689 E + 01
Length/Temperature (Geothermal Step)	m/K	ft/°F	m/K	5.486 4* E — 01
Pressure	Pa	atm	MPa	1.013 250* E — 01
			kPa	1.013 250* E + 02
			(bar**)	1.013 250* E + 00
		bar	MPa	1.0* E — 01
			kPa	1.0* E + 02
			(bar**)	1
		lb_f/in^2 (psi)	MPa	6.894 757 E — 03
			kPa	6.894 757 E + 00
			(bar**)	6.894 757 E — 02
		dyn/cm^2	Pa	1.0* E — 01
Liquid Head	m	ft	m	3.048* E — 01
		in	mm	2.54* E + 01
Pressure Drop/Length	Pa/m	psi/ft	kPa/m	2.262 059 E + 01
		psi/100 ft	kPa/m	2.262 059 E — 01

*All following digits would be zeros.

**In the U. S., the kilopascal is preferred and the use of the bar should be deprecated. The bar is shown here as an alternative allowable unit because some Western European countries prefer to use the bar. Where the bar is used, it should be limited to *physical measurement* (e.g., pressure gauges). For *calculations*, only the pascal or standard multiples thereof (kPa, MPa), should be used.

Quantity and SI Unit		Customary Unit	API Preferred SI Metric Unit	Conversion Factor (Multiply value in customary unit by factor to get value in metric unit.)
TRANSPORT PROPERTIES				
Diffusivity	m^2/s	ft^2/s	mm^2/s	9.290 304* E + 04
		cm^2/s	mm^2/s	1.0* E + 02
Thermal Conductivity	W/m·K	cal/s·cm²·°C/cm	W/m·°C	4.184* E + 02
		Btu/h·ft²·°F/ft	W/m·°C	1.730 735 E + 00
		kcal/h·m²·°C/m	W/m·°C	1.162 222 E + 00
Heat Transfer Coefficient	W/m²·K	cal/s·cm²·°C	kW/m²·°C	4.184* E + 01
		Btu/s·ft²·°F	kW/m²·°C	2.044 175 E + 01
Volumetric Heat Transfer Coefficient	W/m³·K	Btu/s·ft³·°F	kW/m³·°C	6.706 611 E + 01
		Btu/h·ft³·°F	kW/m³·°C	1.862 947 E − 02
Surface Tension	N/m	dyn/cm	mN/m	1
Viscosity (Dynamic)	Pa·s	lb_f·s/in²	Pa·s	6.894 757 E + 03
		lb_f·s/ft²	Pa·s	4.788 026 E + 01
		kg_f·s/m²	Pa·s	9.806 650* E + 00
		lb/ft·s	Pa·s	1.488 164 E + 00
Viscosity (Kinematic)	m^2/s	ft^2/s	mm^2/s	9.290 304* E + 04
		in^2/s	mm^2/s	6.451 6* E + 02
		ft^2/h	mm^2/s	2.580 64* E + 01
Permeability	m^2	darcy	μm^2	9.869 233 E − 01
		millidarcy	μm^2	9.869 233 E − 04
ENERGY, WORK, POWER				
Energy, Work	J	Btu	kJ	1.055 056 E + 00
		kcal	kJ	4.184* E + 00
		cal	kJ	4.184* E − 03
		ft·lb_f	kJ	1.355 818 E − 03
		lb·ft	kJ	1.355 818 E − 03
		lb·ft²/s²	kJ	4.214 011 E − 05
Impact Energy	J	kg·m	J	9.806 650* E + 00
		lb·ft	J	1.355 818 E + 00
Specific Impact Energy	J/m²	kg·m/cm²	J/cm²	9.806 650* E − 02
		lb·ft/in²	J/cm²	2.101 522 E − 03
Power	W	erg/a	TW	3.170 979 E − 27
			GW	3.170 979 E − 24
		Btu/s	kW	1.055 056 E + 00
		hydraulic horsepower — hhp	kW	7.460 43 E − 01
		Btu/min	kW	1.758 427 E − 02
		ft·lb_f/s	kW	1.355 818 E − 03
		Btu/h	W	2.930 711 E − 01
Power/Area	W/m²	Btu/s·ft²	kW/m²	1.135 653 E + 01
		Btu/h·ft²	kW/m²	3.154 591 E − 03

*All following digits would be zeros.

Quantity and SI Unit		Customary Unit	API Preferred SI Metric Unit	Conversion Factor (Multiply value in customary unit by factor to get value in metric unit.)
Heat Release Rate, Mixing Power	W/m^3	hp/ft^3	kW/m^3	2.633 414 E + 01
		$cal/h \cdot cm^3$	kW/m^3	1.162 222 E + 00
		$Btu/s \cdot ft^3$	kW/m^3	3.725 895 E + 01
		$Btu/h \cdot ft^3$	kW/m^3	1.034 971 E — 02
Specific Fuel Consumption (Volume Basis)	m^3/J	$m^3/kW \cdot h$	dm^3/MJ	2.777 778 E + 02
		US gal/hp·h	dm^3/MJ	1.410 089 E + 00
Fuel Consumption (Automotive)	m^3/m	US gal/mi	dm^3/ 100 km (litre/100 km)	2.352 146 E + 02
		mi/US gal	km/dm^3 (km/litre)	4.251 437 E — 01

*All following digits would be zeros.

Bibliography

Aagaard, P. M., and Besse, C. P.: "A Review of the Offshore Environment — 25 Years of Progress," *J. Pet. Tech.* (Dec. 1973) 1355-1360.

Abadie, H. G.: "Oil Well Repair by Scabbing Methods," *Oil Weekly* (Dec. 2, 1940) **99,** No. 13, 18-28.

"A Basic Oil-Well Cement," paper 701-61-A, a report from the API Committee on Standardization of Oil-Well Cements (Feb. 1965) preprint.

"Acid-Soluble Admixtures With Portland Cement," *Report No. 69,* Halliburton Oil Well Cementing Co., Duncan, Okla. (Aug. 1942).

"Acid Soluble Cement," *The Acidizer* (1941) **5,** No. 4, 1-8.

"Additive Lightens Weight of Cement Slurry," *Drilling* (July 1958) **19,** No. 9, 131.

Adkins, J. E.: "Permanent Well Completions Save Time and Money in Workovers," *World Petroleum* (April 1954) **25,** No. 4, 58.

"Advantages of Aquagel Cement for Use in Oil Wells," *Drilling Mud* (Nov. 1943) **10,** No. 1, 3-19.

"Advantages of Rapid-Setting Cement in Oil Wells," *Engineering and Mining J.* (Feb. 9, 1924) **117,** No. 6, 239.

"A High Early Strength Cement," *Chem. and Eng. News* (Oct. 10, 1946) **24,** No. 9, 2684.

Akins, D. W., Jr.: "Liner Job Well Remedial Work in East Texas Field," *Pet. Eng.* (Jan. 1947) 96.

Albright, J. D.: "Subsurface Salt Water Disposal," paper presented at AIChE Southern Regional Meeting (Nov. 1965).

Allen, J. H.: "Super Drills for Boring Shafts and Tunnels," *Mining World* (Dec. 1959) 20-25.

Allen, T. O., and Atterbury, J. H.: "Effectiveness of Gun Perforating," *Trans.,* AIME (1954) **201,** 34-40.

Allen, T. O., and Worzel, H. C.: "Productivity Method of Evaluating Gun Perforating," *Drill. and Prod. Prac.,* API (1956) 112.

"All-Metal Cementing Basket," *Pet. Eng.* (July 1937) 91.

"A Look at Deep Drilling," *World Oil* (May 1968) 57.

Alquist, F. N., and Miller, H. H.: "Effect of Calcium Chloride in Oil-Well Cements," *Oil and Gas J.* (July 17, 1941) 42.

"Ancient Chinese Drilled Deep Gas Wells by Primitive Methods," *Crane Valve World* (Aug. 1942) 128.

Anderson, D. N.: "Geothermal Development in California," paper SPE 4180 presented at SPE-AIME 43rd Annual California Regional Fall Meeting, Bakersfield, Nov. 8-10, 1972.

Anderson, E. T.: "How World's Hottest Hole was Drilled," *Pet. Eng.* (Oct. 1961).

Anderson, F. M.: "A Study of Surface Casing and Open-Hole Plug-Back Cementing Practices in the Mid-Continent District," *Drill. and Prod. Prac.,* API (1955) 312-325.

Anderson, F. M.: "Effects of Mud-Treating Chemicals on Oil-Well Cements," *Oil and Gas J.* (Sept. 29, 1952) 283-284.

Anderson, F. M.: "Modern Oil Well Cementing Operations," paper presented at Symposium on Oilwell Cements, Mid-Year Meeting, API, Pittsburgh, June 16, 1953; *Calif. Oil World* (Aug. 1953) **49,** No. 16, 1.

Anderson, F. M.: "Use of Activated Charcoal in Cement to Combat Effects of Contamination by Drilling Muds," *Trans.,* AIME (1952) **195,** 314-315.

Anderson, K. E.: *Water Well Handbook,* Missouri Water Well Drillers Assn. (1959).

Anderson, T. O., Winn, R. H., and Walker, T.: "A Qualitative Cement Bond Evaluation Method," *Drill. and Prod. Prac.,* API (1964) 24-32.

Anderson, W. L., and Walker, T.: "Research Predicts Improved Cement Bond Evaluations With Acoustic Logs," *J. Pet. Tech.* (Nov. 1961) 1093-1097.

Annis, M. R., and Monaghan, P. H.: "Differential Pressure Sticking — Laboratory Studies of Friction Between Steel and Mud Cake," *J. Pet. Tech.* (May 1962) 537-543; *Trans.,* AIME, **225.**

A Primer of Oil Well Drilling, 3rd ed., Pet. Extension Service, U. of Texas, Austin (1957).

Arnold, R., and Garfias, R.: "The Cementing Process of Excluding Water from Oil Wells as Practiced in California," Tech. Paper 32, USBM (1913) 5.

Aspdin, J.: "An Improvement in the Modes of Producing Artificial Stone," British Patent No. 5022 (1824).

ASTM Standards, Part III, American Society for Testing Materials, Philadelphia, Pa. (1970).

A Symposium on Plugging To Abandon, minutes and summaries of talks presented to Geol. Survey Div., Mich. Dept. of Natural Resources, Mt. Pleasant, Sept. 25, 1973.

AWWA Standard for Deep Wells, American Water Works Assn., Natl. Water Well Assn., Urbana, Ill. (Sept. 1958).

Bachman, L. T.: "Discussion of Cements and Their Application in Shutting Off Water in Oil Wells," *Petroleum World* (Sept. 1927) **12,** No. 9, 82.

Bachman, L. T.: "Influence Related to Use of Cement in Wells," *Oil Weekly,* Part I (Jan. 26, 1924) 25; Part II (Feb. 2, 1924) 27.

Bade, J. F.: "Cement Bond Logging Techniques — How They Compare and Some Variables Affecting Interpretation," *J. Pet. Tech.* (Jan 1963) 17-22.

Bailey, J. E., and Dimit, C. E.: "Plug Back Work With Plastics in East Texas Field," *Drill. and Prod. Prac.,* API (1943) 82-86.

"Baker Cement Equipment Designed for Efficiency," *Oil and Gas J.* (June 22, 1933) 84.

"Baker Cement Retainer and Special Equipment," Baker Oil Tools, Inc., Los Angeles (1938) 25 pp.

Baker, M. L., and Kirby, C. E.: "Gilsonite — An Effective Cement Additive," *Drilling* (May 15, 1964) 140.

Baker, M. L., and Kirby, C. E.: "Gilsonite — A Slurry Loss Preventive," *The Drilling Contractor* (Jan.-Feb. 1965).

Baker, R. C.: "Cement Retainer," U. S. Patent No. 1,035,674 (Aug. 13, 1912), filed Jan. 29, 1912.

Baker, R. C.: "Plug for Well Casing," U. S. Patent No. 1,392,619 (Nov. 18, 1913), filed Nov. 20, 1911.

Baker, R. T.: "Cementing Important Factor in Drilling," *Oil and Gas J.* (June 14, 1923) 22.

Baker, W. L.: "Deep Drilling Casing and Cementing Practices," *Pet. Eng.* (Feb. 1932) 28.

Baker, W. M.: "Waste Disposal Well Completion and Maintenance," *Industrial Water and Wastes* (Nov.-Dec. 1963) 43-47.

Ball, W. F., and Slagle, K. A.: "Pozzolanic Slurries for Injection Wells," *World Oil* (Feb. 1, 1958) 79-82.

Ball, W. W.: "How to Cement Big Pipe," *Oil and Gas J.* (Nov. 22, 1954) 86-92.

Banister, J. A.: "Methods and Materials for Placing Cement Plugs in Open Holes," paper presented at meeting of Interstate Oil Compact Commission, Yellowstone, Wyo., June 10-12, 1957.

Barkis, B.: "Cementing," *Petroleum World (London)* (Dec. 1931) **28,** 411.

Barkis, B.: "Cementing Surface Casing," *Pet. Eng.* (Nov. 1951) B-54.

Barkis, B.: "Cements and Cementing Practice," *API Proc. Sec. IV (Prod. Bull. 208)* (1931) **17,** 157.

Barkis, B.: "Open-Hole Cement Plugs," *World Oil* (June 1952) 134.

Barkis, B.: "Time Factors in Cementing," *Oil and Gas J.* (Jan. 12, 1950) 74; *Pet. Eng. (Ref. Annual)* (1950) 870.

Barkis, B., and Wright, K. A.: "Casing Movement While Cementing," *International Oil* (Oct. 1941) **2,** 21-25.

Barkis, B., and Wright, K. A.: "Effect of Hole Enlargement on Cementing and Completion," *World Oil* (Dec. 1947) 109.

Barnes, J.: "Influence of Controlled Cement Slurry Weights in Squeeze Jobs," *Pet. Eng.* (Aug. 1937) 30.

Barnes, K. B.: "Dowell Develops Brand New Cement," *Oil and Gas J.* (Nov. 7, 1955) 80.

Barnes, K. B.: "New Type Casing Shoes Pack Off Bottom Hole During Cementing," *Oil and Gas J.* (March 3, 1945) 47.

Bawcom, J. W.: "Rotary Drilling of Large Diameter Vertical Holes," paper presented at Northern Ohio Geol. Soc. Symposium on Salt, Cleveland, May 1962.

Bayliss, B. P.: "Introduction to Geothermal Energy," paper SPE 4176 presented at SPE-AIME 43rd Annual California Regional Fall Meeting, Bakersfield, Nov. 8-10, 1972.

Bayliss, J.: "Good Drilling Practice: Chapter 7 — Cementing," *Oil Weekly* (May 23, 1938) **89,** No. 11, 34.

Beach, H. J.: "Improved Bentonite Cements Through Partial Acceleration," *J. Pet. Tech.* (Sept. 1961) 923-926; *Trans.,* AIME, **222.**

Beach, H. J., Frawley, F. E., Jr., EnDean, H. J., and Yates, D.: "Causes and Prevention of Failures in Cement Pipe Linings," *J. Pet. Tech.* (Jan 1970) 51-57.

Beach, H. J., and Goins, W. C., Jr.: "A Method of Protecting Cements Against the Harmful Effects of Mud Contamination," *Trans.,* AIME (1957) **210,** 148-152.

Beach, H. J., O'Brien, T. B., and Goins, W. C., Jr.: "Controlled Filtration Rate Improves Cement Squeezing," *World Oil* (May 1961) 87-93.

Beach, H. J., O'Brien, T. B., and Goins, W. C., Jr.: "Formation Cement Squeezed By Using Low-Water-Loss Cements," *Oil and Gas J.* (May 29 and June 12, 1961).

Beach, H. J., O'Brien, T. B., and Goins, W. C., Jr.: "Low-Water-Loss Cements Improve Formation Cement Squeezes," *Oil and Gas J.,* Part I (May 29, 1961) 155-158; Part II (June 12, 1961) 154-156.

Beach, H. J., O'Brien, T. B., and Goins, W. C., Jr.: "The Role of Filtration in Cement Squeezing," *Drill. and Prod. Prac.* API (1961) 27-35.

Beale, A. F., and Eilers, L. H.: "A New Oil-Well Cementing Material," *Drill. and Prod. Prac.,* API (1956) 515 (abstr.); complete paper presented at API Southwestern Dist. Spring Meeting, Fort Worth, Tex., March 21-23, 1956.

Bearden, W. G.: "Effect of Temperature and Pressure on the Physical Properties of Cement," *Oil-Well Cementing Practices in the United States,* API, New York (1959) 56.

Bearden, W. G., and Lane, R. D.: "Engineered Cementing Operations To Eliminate WOC Time," *Drill. and Prod. Prac.,* API (1961) 17.

Bearden, W. G., and Lane, R. D.: "You Can Engineer Cementing Operations To Eliminate Wasteful WOC Time," *Oil and Gas J.* (July 3, 1961) 104.

Bearden, W. G., Spurlock, J. W., and Howard, G. C.: "Control and Prevention of Interzonal Flow," *J. Pet. Tech.* (May 1965) 579-584; *Trans.,* AIME, **234.**

Beaupre, C. J.: "Drilling and Completion Programs and Problems, Rojo Caballos Penn Field, Pecos County, Texas," paper presented at API Div. of Production Spring Meeting, Fort Worth, Tex., March 1963.

Becker, H., and Peterson, G.: "Bond of Cement Compositions for Cementing Wells," paper presented at Sixth World Pet. Cong., Frankfurt, June 19-26, 1963.

Beck, R. W., Nuss, W. F., and Dunn, T. H.: "The Flow Properties of Drilling Muds," *Drill. and Prod. Prac.,* API (1947) 9-22.

Beckstrom, R. C.: "Open-Hole Diameter Changes Located and Measured by Recording Calipers," *Oil Weekly* (May 27, 1935) **77,** No. 11, 19-20.

Beeson, C. M.: "Well Completion Practice (Part I)," *World Oil* (Nov. 1949) 87.

Beirute, R., and Tragesser, A.: "Expansive and Shrinkage Characteristics of Cements Under Actual Well Conditions," *J. Pet. Tech.* (Aug. 1973) 905-909.

Bell, H. W.: "Cement in Oil and Gas Wells," *Conservation News* (Nov. 1923) **1,** No. 10; *Mining and Metallurgy* (Jan. 1924).

Bell, H. W., and Grimm, M. W.: "'Tests on Cement and Accelerators," *Bull.,* AAPG (March 1928) **12,** No. 3, 279-281.

Bell, W. T., and Auberlinder, G. A.: "Perforating High-Temperature Wells," *J. Pet. Tech.* (March 1961) 211-216.

Belter, W. G.: "Radioactive Wastes," *Industrial Science and Technology* (Sept. 1962) 42-51.

Bender, A.: "Plugging Wells," *Azerbaidzhanskoe Neftyanoe Khoz* (March 1934) **14,** No. 3, 63-68.

Benge, J.: "Venezuela Cementing Problems," *World Oil* (Feb. 1948) 202-203.

Bergman, W. E., Hurley, J. R., and Shell, F. J.: "Low-Water-Loss, Low-Density Cement," *Oil and Gas J.* (Sept. 1955) 107-110.

Bergman, W. E., Hurley, J. R., and Shell, F. J.: "New Oil Well Cement Blend," *World Oil,* Part I (Sept. 1955) 99; Part II (Nov. 1955) 150.

Best, W. E.: "Squeeze Cementing," *World Oil* (Sept. 1948) 132.

Bignell, L. G. E.: "Cementing Casing," *Oil and Gas J.* (June 9, 1938) 63.

Bignell, L. G. E.: "Deep Wells Create Casing Problems," *Oil and Gas J.* (Oct. 3, 1929) 69.

Bignell, L. G. E.: "Oklahoma Well Takes More Than Two Million Pounds of Cement," *Oil and Gas J.* (Aug. 18, 1938) 46-47.

Bignell, L. G. E.: "Sulfate-Resisting Cement," *Oil and Gas J.* (Oct. 27, 1938) 188.

"Big Well-Cementing Job Completed in Record Time," *Oil and Gas J.* (May 29, 1967) 82-86.

Billington, S. A.: "Practical Approach to Lost Circulation Problems," *The Drilling Contractor* (July-Aug. 1963) 52.

Bingham, I. F.: "Cementing Hot Wells," paper presented at AIME meeting, San Antonio, Tex., Oct. 6-7, 1938.

Bingham, I. F.: "Characteristics Common to Various Types of Cements and How To Use These Cements for Best Results in Oil Field Work," paper presented at AIME meeting, San Antonio, Tex., Oct. 6-7, 1938.

Bingham, I. F.: "Present Trends in Rotary Well Drilling and Completion," *The Science of Petroleum,* Oxford U. Press, London (1935) 406-429.

Binkley, G. W., Dumbauld, G. K., and Collins, R. E.: "Factors Affecting the Rate of Deposition of Cement in Unfractured Perforations During Squeeze-Cementing Operations," *Trans.,* AIME (1958) **213,** 51-58.

Blanks, R. F.: "The Use of Portland-Pozzolan Cement by the Bureau of Reclamation," paper presented at Spring Meeting, Portland Cement Assn., New York, May 11, 1948.

Blanks, R. F., and Kennedy, H. B.: *The Technology of Cement and Concrete,* John Wiley & Sons, Inc., New York (1955) 1-64.

Bleakley, W. B.: "A Really Engineered Cement Job (Slurry-Flo Plan)," *Oil and Gas J.* (Feb. 12, 1961).

Bleakley, W. B.: "Cut Lost Circulation While Cementing," *Oil and Gas J.* (Aug. 26, 1963).

Bleakley, W. B.: "North Slope Operators Tackle Production Problems," *Oil and Gas J.* (Oct. 25, 1971) 89.

Bleakley, W. B.: "What It Takes To Make a Good Well Completion," *Oil and Gas J.* (June 11, 1962) 125.

Blount, E. M., and Prueger, N. J.: "The Development of a Well Completion System for Deep Permafrost," paper SPE 3251 presented at SPE-AIME Southwestern Alaska Section Regional Meeting, Anchorage, May 5-7, 1971.

Bogue, R. H.: "Calculation of Compounds in Portland Cement," *Ind. and Eng. Chem.* (Analytical Ed.) (1929) **1,** 192-196.

Bogue, R. H.: *Chemistry of Portland Cement,* 2nd ed., Reinhold Publishing Co., New York (1955).

Bogue, R. H.: "The Constitution of Portland Cement Clinker," *Concrete* (1926) **29,** No. 1, 14-16.

Boice, D., and Diller, J.: "A Better Way To Squeeze Fractured Carbonates," *Pet. Eng.* (May 1970) 79-82.

Book of ASTM Standards, Part 3, American Society for Testing Materials (1970).

Boon, A. R.: "Cement Aids Oil Production," *Scientific American* (June 1942) **166,** No. 6, 272-273.

Boone, D. E.: "A Guide to Effective Cementing," *Oil and Gas J.* Part 1 (May 26, 1969) 72; Part 2 (June 9, 1969) 81; Part 3 (June 30, 1969) 95.

"Boring a 5-Foot Mine Shaft," *Mech. Eng.* (April 1937) 286.

Born, S.: "Comparative Data Concerning the Use of Neat and Aquagel Cement in Cementing Casing," Baroid Sales Report, Baroid, Houston (June 1937) 15 pp.

Boughton, L. D., and Dellinger, T. B.: "New Cementing Process for Big Pipe in a Salt Plug," *World Oil* (Jan. 1965) 105-108.

Broughton, L. D., Pavlich, J. P., and Wahl, W. W.: "The Use of Dispersants in Cement Slurries To Improve Placement Techniques," paper SPE 412 presented at SPE-AIME 33rd Annual Fall Meeting, Los Angeles, Oct. 8-10, 1962.

Bouman, C. A.: "Buckling of Oil Well Piping," *Pet. Eng.* (June 1956) B60.

Bowden, C. B.: "Hesitation Squeeze: Cementing Technique . . . as Used in Cementing for an Initial Water Shut-Off," *Oil and Gas J.* (July 13, 1953) 111.

Bowden, C. B.: "Squeeze Cementing Techniques in Cementing Casing for Initial Water Shut-Off," *Drill. and Prod. Prac.*, API (1953) 68-74.

Bowers, C. N.: "Design of Casing Strings," paper 514-G presented at SPE-AIME 30th Annual Fall Meeting, New Orleans, La., Oct. 2-5, 1955.

Bradshaw, J. R.: "Established Method Adapted to Cementing of Wells Drilled Through Mine Tunnels," *Pet. Eng.* (July 1938) 25-26.

Brice, J. W., Jr., and Holmes, B. C.: "Engineered Casing Cementing Programs Using Turbulent Flow Techniques," *J. Pet. Tech.* (May 1964) 503-508.

Briggs, G. E.: "Squeeze Cement Job Cuts Off Undesired Water in TXL Field," *Drilling* (Oct. 1954) **15,** No. 12, 103.

Briggs, G. E., and Porter, W.: "Improved Wireline Cementer Developed to Cope With Water Shut-Off and Channelling Problems," *J. Pet. Tech.* (Nov. 1958) 17-20.

Brighton, J. A., and Jones, J. B.: "Fully Developed Turbulent Flow in Annuli," *J. of Basic Engineering Trans.*, ASME (Sept. 1964) **86,** Series D, No. 4, 835.

Broussard, P. J., and McNatt, J. C.: "An Improved High Pressure Squeeze Cementing Technique," *Drilling* (July 1958) **19,** No. 9, 100.

Brown, B. D.: "Cementing Water Wells, *Public Works* (Sept. 1959) **90,** No. 9, 99-100.

Brown, R. J. E.: *Permafrost in Canada,* U. of Toronto Press, Toronto (1970).

Brown, R. W., Bucy, B. D., Pruitt, G. T., and Crawford, H. R.: "Cement Rheology — A Tool for Better Completions," *Pet. Eng.* (Feb. 1963) 70.

Bruist, E. H.: "A New Approach in Relief Well Drilling," *J. Pet. Tech.* (June 1972) 713-722.

Bryan, B.: "Deep Well Cementing Problems," *Oil Bull.* (June 1930) 610.

Bugbee, J. M.: "Lost Circulation — A Major Problem in Exploration and Development," *Drill. and Prod. Prac.*, API (1953) 14-27.

Bugbee, J. M.: "Lost Circulation Can Be Avoided or Minimized," *World Oil* (May 1953) 150.

Bugbee, J. M.: "Lost Circulation," *Oil and Gas J.*, Part 1 (March 11, 1954) 228; Part 2 (April 5, 1954) 125; Part 3 (April 26, 1954) 257; Part 4 (May 3, 1954) 112.

Bugbee, J. M.: "Procedures To Be Used in Fighting Lost Circulation," *World Oil* (June 1953) 132.

"Bulk Type Oil-Well Cementing Introduced in Illinois and California," *Oil Weekly* (Sept. 30, 1940) 99, No. 4, 20-21.

Burch, D. D.: "Casing Shoe," U. S. Patent No. 1,603,447 (Oct. 19, 1926), filed Feb. 25, 1926.

Burke, R. G.: "How Those 72-Inch Holes Are Being Drilled," *Oil and Gas J.* (April 27, 1964).

Burke, R. H.: "Gulf Coast Cementing Practice," *J. Pet. Tech.* (Oct. 1952) 26.

Burkhardt, J. A.: "Wellbore Pressure Surges Produced by Pipe Movement," *J. Pet. Tech.* (June 1961) 595-605; *Trans.*, AIME, **222.**

Buster, J. L.: "A Short-Cut Method of Calculating Your Cement Slurry," *Oil and Gas J.* (Jan 6, 1958) 112-115.

Buster, J. L.: "Cementing Multiple Tubingless Completions," *Oil and Gas J.*, Part 1 (June 8, 1964) 121-125; Part 2 (June 15, 1964) 89-91.

Buster, J. L.: "Plan Turbulence Into Your Cement Job," *Pet. Eng.* (May 1962).

Byrd, E. E., and Jessen, F. W.: "Thermal Characteristics of High-Temperature Oil Well Cements," *Ind. and Eng. Chem.* (Oct. 1942) **34,** No. 10, 1142-1148.

Cain, J. E., Shryock, S. H., and Carter, L. G.: "Cementing Steam Injection Wells in California," *J. Pet. Tech.* (April 1966) 431-436.

Cain, P. A., and Folinsbee, J. R.: "Shaft Drilling at Lynn Lake with Oil Well Drilling Equipment," paper presented at CIM Annual Western Meeting, Winnipeg, Man., Canada, Oct. 18, 1965.

Caldwell, B. M., and Griggs, G. E., Jr.: "A New Retrievable Wire Line Cementing Tool," paper 530-G presented at SPE-AIME 30th Annual Fall Meeting, New Orleans, La., Oct. 2-5, 1955.

Caldwell, B. M., and Owen, H. D.: "A New Tool for Perforating Casing Below Tubing," *Trans.,* AIME (1954) **201,** 29-36.

Caldwell, H. L.: "Big Hole Drilling Technology Is Changing," *World Oil* (Sept. 1965) 103-107.

California Laws for Conservation of Petroleum and Gas, Dept. of Conservation, Div. of Oil and Gas, Sacramento (1971).

"California's Oil," API, 12 (1948).

Campbell, J. L. P.: "Checking Cement Jobs With Radioactivity," *Tomorrow's Tools Today* (1944) **10,** 1st Quarter, 24-26.

Campbell, J. L. P.: "Tracing Cement Travel With Carnotite," *Oil Weekly* (Jan 21. 1946) **120,** No. 8, 47-49.

Cannon, G. E.: "Changes in Hydrostatic Pressure Due to Withdrawing Drill Pipe from the Hole," *Drill. and Prod. Prac.,* API (1934) 42.

Cannon, G. E.: "Improvements in Cementing Practices and the Need for Uniform Cementing Regulations," *Drill. and Prod. Prac.*, API (1948) 126-133; *Pet. Eng.* (May 1949) B42.

Cannon, G. E., and Sullins, R. S.: "Problems Encountered in Drilling Abnormal-Pressure Formations," *Drill. and Prod. Prac.*, API (1946) 29.

Cardwell, W. T., Jr.: "Pressure Changes in Drilling Wells Caused by Pipe Movement," *Drill. and Prod. Prac.*, API (1953) 97.

Carnahan, D. A.: "Whys of Casing Failures in Steam Wells," *Pet. Eng.* (Sept. 1966) 98-105.

Carney, L. L.: "Cement Spacer Fluid," *J. Pet. Tech.* (Aug. 1974) 856-858.

Carter, L. G., Cook, C., and Snelson, L.: "Cementing Research in Directional Gas Well Completions," paper SPE 4313 presented at SPE-AIME European Spring Meeting, London, April 2-3, 1973.

Carter, L. G., and Evans, G. W.: "A Study of Cement-Pipe Bonding," *J. Pet. Tech.* (Feb. 1964) 157-160.

Carter, L. G., and Harris, F.: "How To Squeeze Those Perforations," *Drilling International* (Jan. 1964) 62.

Carter, L. G., Harris, F., and Smith, D. K.: "Remedial Cementing of Plugged Perforations," paper SPE 759 presented at SPE-AIME 34th Annual California Regional Fall Meeting, Santa Barbara, Oct. 23-25, 1963.

Carter, L. G., and Slagle, K. A.: "A Study of Completion Practices To Minimize Gas Communication," *J. Pet. Tech.* (Sept. 1972) 1170-1174.

Carter, L. G., Slagle, K. A., and Smith, D. K.: "Resilient Cement Decreases Perforating Damage," paper presented at API Mid-Continent Dist. Div. of Production Spring Meeting, Amarillo, Tex., April 1968.

Carter, L. G., and Smith, D. K.: "Properties of Cementing Compositions at Elevated Temperatures and Pressures," *J. Pet. Tech.* (Feb. 1958) 20-28.

Carter, L. G., Waggoner, H. F., and George, C. R.: "Expanding Cements for Primary Cementing," *J. Pet. Tech.* (May 1966) 551-558.

Case, J. B.: 'Cementing and Cementing Projects," *Petroleum Engineer's Handbook No. 1,* The Petroleum Engineer Publishing Co., Dallas (1930) 98-106.

Case, J. B.: "Landing and Cementing Long Water Strings," *Petroleum Times,* Part I (March 30, 1929) 572; Part II (April 13, 1929) 668; Part III (May 11, 1929) 851.

Casewell, C. A.: "Underground Waste Disposal Concepts and Misconceptions," *Environmental Science and Technology* (Aug. 1970) **4,** No. 8.

"Casing Centralizers," *API Spec. 10D,* API, Dallas, 1st ed. (April 1971); 2nd ed. (Feb. 1973).

"Casing-Landing Recommendations," *API Bull. D7,* API, Dallas (June 1955).

"Casing Shoe Eliminated Plug Drillout," *Oil and Gas J.* (Sept. 2, 1957) 160.

"Casing String Design Factors," paper 851-29-I, report of API Mid-Continent Dist. Study Committee on Casing Programs (March 1955).

"Casing the World's Deepest Well," *Oil and Gas J.* (June 23, 1958) 77.

"Casing, Tubing, Drill Pipe and Line Pipe Properties," *API Bull. 5C3,* 1st ed. (Dec. 1971); 2nd ed. (Nov. 1974).

"Cement Additive Reduces Weight," *Petroleum Week,* (June 13, 1958) **6,** No. 11, 32.

"Cementing Collar," *Oil Weekly* (July 18, 1932) **66,** No. 5, 61.

"Cementing Equipment: Cementing Float Shoes; Gilsonite Cement for Light Weight Slurry," *Pet. Eng.* (July 15, 1958) B55.

"Cement Quality Logging," Tech. Report, Schlumberger Well Services (Sept. 1971).

"Cement Record Set," *Oil and Gas J.* (May 7, 1956) 96.

Cement Standards of the World — Portland Cement and Its Derivatives, CEMBUREAU, Paris (1967).

Cheatham, J. B., Jr., and McEver, J. W.: "Behavior of Casing Subjected to Salt Loading," *J. Pet. Tech.* (Sept. 1964) 1069-1076.

"Checking Cement Behind Pipe With Radioactivity Log," *Calif. Oil World* (July 15, 1947) **40,** No. 13, 9-13.

Childers, M. A.: "Primary Cementing of Multiple Casing," *J. Pet. Tech.* (July 1968) 751-762; *Trans.,* AIME, **243.**

Christian, W. W., Chatterji, J., and Ostroot, G. W.: "Gas Leakage in Primary Cementing — A Field Study and Laboratory Investigation," paper SPE 5517 presented at SPE-AIME 50th Annual Fall Technical Conference and Exhibition, Dallas, Sept. 28-Oct. 1, 1975.

Christman, S. A.: "Offshore Fracture Gradients," *J. Pet. Tech.* (Aug. 1973) 910-914.

"Chronic Emergencies Put Oil-Well Cementing in the Category of Rugged Operations," *The Bearing Engineer* (May-June 1954) **14,** No. 3, 4.

Clark, C. R., and Carter, L. G.: "Mud Displacement With Cement Slurries," *J. Pet. Tech.* (July 1973) 775-783.

Clark, C. R., and Jenkins, R. C.: "Cementing Practices for Tubingless Completions," paper SPE 4609 presented at SPE-AIME 48th Annual Fall Meeting, Las Vegas, Nev., Sept. 30-Oct. 3, 1973.

Clark, C. R., Steel, J. H., and Gidley, J. L.: "Coarse-Grind Cement in Oil Well Cementing," *Oil and Gas J.* (Jan. 24, 1972) 50-53.

Clark, E. H., Jr.: "A Graphic View of Pressure Surges and Lost Circulation," *Drill. and Prod. Prac.,* API (1956) 424-438.

Clark, E. H., Jr.: "Bottom Hole Pressure Surges While Running Pipe," *Pet. Eng.* (Jan. 1955) B68.

Clark, E. H., Jr., and Murray, A. S.: "A Study of Primary Cementing," paper presented at CIM Meeting, Calgary, Canada, May 28-30, 1958.

Clark, E. H., Jr., and Murray, A. S.: "Primary Cementing," *Oil and Gas J.,* Part 1 — Cementing Procedures (Aug. 25, 1958) 70-73; Part 2 — Analyzing Cementing Problems (Sept. 1, 1958) 179-181; Part 3 — Sound Casing Procedures (Sept. 15, 1958) 226-230.

Clark, R. C., Jr.: "A Note on Well Cementing at Low Temperatures," *Trans.*, AIME (1949) **186,** 327.

Clark, R. C., Jr.: "A Ten-Pound Cement Slurry for Oil Wells," *J. Pet. Tech.* (Sept. 1953) 93-94.

Clark, R. C., Jr.: "Influence of Different Types of Formation Waters on Disintegration of Cements," *Trans.*, AIME (1950) **189,** 367-369.

Clason, C. E.: "Cement Substitutes for Elimination of Water," *Oil Weekly* (Dec. 10, 1945) **120,** No. 2, 54-57; *Chem. Abstr.* (1946) **40,** 2963.

Clason, C. E.: "Evolution and Use of Gypsum Cement for Oil Wells," *World Oil* (Aug. 1949) 119-126.

Clegg. J. D.: "Casing Failure Study — Cedar Creek Anticline," *J. Pet. Tech.* (June 1971) 676-684.

"Co-Axial Spiral Used Successfully in Cementing of Oil Wells," *International Oil* (Jan. 1941) **2,** No. 1, 79.

"Code for Testing Cement Used in Wells," *API Code 32,* 1st ed., API, Dallas (1948).

Coffer, H. F., Reynolds, J. J., and Clark, R. C., Jr.: "A Ten-Pound Cement Slurry for Oil Wells," *Trans.*, AIME (1954) **201,** 146-148.

"Cold Weather Ready Mixed Concrete," National Ready Mixed Concrete Assn. Pub. 130 (Sept. 1968).

Coleman, J. R., and Corrigan, G. L.: "Fineness and Water-Cement Ratio in Relation to Volume and Permeability of Cement," *Trans.*, AIME (1941) **142,** 205-215.

"Committee Report on Oil-Well Cements," *API Proc. Sec. IV (Prod. Bull. 226)* (1940) **21,** 120-121.

"Continuous Indicator Rotary Mud Density," *Calif. Oil World* (May 9, 1940) **33,** No. 9, 30.

Copland, G. V., and Smith, D. K.: "Improved Cementing Through Automation," *Pet. Eng.* (Nov. 1958) B113-B115.

Coppinger, J. E., and Goode, J. M.: "Improved Primary Cementing Practices Through Hydraulic Well-Bore Analysis," *Drill. and Prod. Prac.*, API (1963) 37-41.

Corley, C. B., Jr., and Rike, J. L.: "Tubingless Completions," *Drill. and Prod. Prac.*, API (1959) 7.

Corrigan, G. L.: "Effects of Bentonite on Well Cement," Report, U. of Texas, Austin (Nov. 1941) 21 pp.

Courtney, F. G.: "Permanent Completions Pay Off for Gulf in South Louisiana," *Pet. Eng.* (Dec. 1955) B46.

Covell, K. A., and Parson, C. P.: "Drilling Practices Improved and Many Economies Are Effected in Oklahoma City Operations," *Oil and Gas J.* (Nov. 17, 1932) 48.

Covlin, R., Jr.: "How To Get Better Liner Cement Jobs," *World Oil* (June 1968) 55.

Cox, W. R.: "Here Are the Down-Hole Factors That Influence Casing-Landing Procedures," *Oil and Gas J.* (July 22, 1957) 86-90.

Craft, B. C.: "Tensile and Compressive Strength of Formations," *Oil and Gas J.* (July 1, 1957) 138-139.

Craft, B. C., and Hawkins, M. F.: *Applied Petroleum Reservoir Engineering,* Prentice-Hall, Inc., Englewood Cliffs, N. J. (1959) 319.

Craft, B. C., Holden, W. R., and Graves, E. D., Jr.: *Well Design: Drilling and Production,* Prentice-Hall, Inc., Englewood Cliffs, N. J. (1962) 43-48, 55-79, and 212-213.

Craft, B. C., Johnson, T. J., and Kirkpatrick, H. L.: "Effects of Temperature, Pressure, and Water-Cement Ratio on the Setting Time and Strength of Cement," *Trans.*, AIME (1935) **114,** 62-68.

Craft, B. C., and Kail, R. S.: "Well Cementing," *Oil Weekly* (May 15, 1955) **113,** No. 11, 44; paper presented at AIME Meeting, Houston, May 8-10, 1944.

Craft, B. C., and Stephenson, A. H.: "Effect of Calcium Chloride on High Early Strength Cement," *J. Pet. Tech.* (June 1952) 11-12.

Crawford, P. B., and Kennedy, H. T.: "Perlite Additive Improves Cement Characteristics," *World Oil* (Dec. 1957) 117-123.

Crews, S. H.: "What It Takes To Drill Big Holes," *Pet. Eng.* (Oct. 1964).

Cromling, J.: "Geothermal Drilling in California," *J. Pet. Tech.* (Sept. 1973) 1033-1038.

Crone, B. L., and White, G. L.: "A New Cement-in-Oil Slurry," paper 952-G presented at SPE-AIME 28th Annual California Regional Fall Meeting, Los Angeles, Oct. 17-18, 1957.

Cunningham, W. C., Fehrenbach, J. R., and Maier, L. F.: "Arctic Cements and Cementing," *J. Cdn. Pet. Tech.* (Oct.-Dec. 1972).

Cunningham, W. C., and Smith, D. K.: '"Effect of Salt Cement Filtrate on Subsurface Formations," *J. Pet. Tech.* (March 1968) 259-264.

"Current Cementing Practices — 1957," *Drilling* (Dec. 2, 1957) 87.

Curtis, L. B., Nance, G. E., Keefover, S. A., and Osborne, J. H.: "Progress in Determining Cementing Practices in the Rocky Mountain District," *Drill. and Prod. Prac.* API (1956) 520 (abstr.); *Oil and Gas J.* (July 23, 1956) 101.

Dale, O. O.: "Review of Oil Well Cementing Practices in California," *Petroleum World (Calif.)* (Feb. 1946) **43,** No. 2, 41-46.

Dale, O. O.: "The Effects of Some Additives on the Physical properties of Portland Cement," *Oil-Well Cementing Practices in the United States,* API, New York (1959) 62.

Daneshy, A. A.: "Experimental Investigation of Hydraulic Fracturing Through Perforations," *J. Pet. Tech.* (Oct. 1973) 1201-1206; *Trans.*, AIME, **255.**

D'Arcy, N. A., Jr.: "Heavy Duty Cementing Equipment Performs Wide Variety of Operations," *Pet. Eng.* (Aug. 1947) 74.

David, H.: "Standard Equipment Successfully Used in Squeeze Cementing at Record Depth," *Oil and Gas J.* (Sept. 27, 1947) 94.

Davie, F. E.: "Report of Special Subcommittee on Oil-Well Cements," *API Proc. Sec. IV (Prod. Bull. 223)* (1939) **20,** 80-81.

Davies, B. F., and Boorman, R. D.: "Field Investigation of Effect of Thawing Permafrost Around Wellbores at Prudhoe Bay," paper SPE 4591 presented at SPE-AIME 48th Annual Fall Meeting, Las Vegas, Nev., Sept. 30-Oct. 3, 1973.

Davis, E. L.: "Cement Properties for Well Use Discussed," *Calif. Oil World* (July 5, 1938) **31,** No. 13, 14.

Davis, E. L.: "Specifications for Oil-Well Cement," *Drill. and Prod. Prac.,* API (1938) 372-376.

Davis, L. F.: "Primary Cementing Practices," *J. Pet. Tech.* (May 1951) 21-22.

Davis, L. F.: "Results of a Survey on Primary Cementing Practices," *Drill. and Prod. Prac.,* API (1951) 191-204.

Davis, R. E.: "Pozzolanic Materials and Their Use in Concrete," *ASTM Special Tech. Pub. No. 99* (1949) 3-15.

Davis, R. E.: "What You Should Know About Pozzolans," *Engineering News Record* (April 5, 1951) **146,** No. 14, 37-40.

Davis, S. H.: "Cementing Liners," *Oil-Well Cementing Practices in the United States,* API, New York (1959) 187-188.

Davis, S. H., and Faulk, J. H.: "Are We Waiting Too Long on Cement?" *Oil and Gas J.* (April 8, 1957) 99-101.

Davis, S. H., and Faulk, J. H: "Have Waiting-on-Cement Practices Kept Pace With Technology?" *Drill. and Prod. Prac.,* API (1957) 180.

Davis, S. H., and Faulk, J. H.: "Reducing Waiting-on-Cement Intervals Can Save Rig Time," *Drilling* (Dec. (1957) 87-88.

Dawson, C. W.: "Accomplishments and Objectives of the API Standardization Committee on Oil-Well Cements," *API Proc. Sec. IV (Prod. Bull. 239)* (1953) 76-78.

Dawson, D. D., Jr.: "A Study of Lost Circulation in the Gulf Coast," *World Petroleum* (Oct. 1953) 58.

Dawson, D. D., Jr., and Goins, W. C., Jr.: "Bentonite-Diesel Oil Squeeze," *World Oil* (Oct. 1953) 222.

Dean, C. J.: "Multiple Zone Wells and New Casing and Cementing Practices as Developed at the Wilmington Field, California," paper presented at AIME meeting, Los Angeles, Oct. 20, 1938.

DeBlanc, F. X.: "Sea-Going Cementing Service for Wells in Open Waters," *Oil and Gas J.* (June 16, 1952) 368-369.

"Deep Gas Well Completion Practices Manual," Texas Iron Works, Inc., Houston (1972)

"Deep Well Cementing in Pecos County, Texas," paper 906-4-E presented at API Southwestern Dist. Div. of Production Spring Meeting, Midland, Tex. (March 1959) preprint.

Deep Well Injection Symposium, proceedings of Seventh Industrial Water and Waste Conference, Texas Water Pollution Control Assn. (June 1967).

DeGreer, D.: "Running, Setting, and Cementing Casing in Deep Wells," paper SPE 3910 presented at SPE-AIME Deep Drilling and Production Symposium, Amarillo, Tex., Sept. 11-12, 1972.

DeHetre, J. P.: "Deeper Completions Emphasize Need for Improved Casing Landing Practice," *Oil and Gas J.* (May 18, 1946) 139; *Pet. Eng.* (July 1946) 104.

Dellinger, T. B.: "Mechanics and Economics of Big Hole Drilling," *World Oil* (Oct. 1965).

Dellinger, T. B., and McLean, J. C.: "Preventing Instability in Partially Cemented Intermediate Casing Strings," paper SPE 4606 presented at SPE-AIME 48th Annual Fall Meeting, Las Vegas, Nev., Sept. 30-Oct. 3, 1973.

Dellinger, T. B., and Presley, C. K.: "Large Diameter Holes — A Growing Field for Big Rigs," *World Oil* (June 1964) 109.

Denning, R. A.: "Calibrating Nuclear Density Gages for Slurry Density," *Control Engineering* (Feb. 1965) 79-81.

"Densified Cements Help Solve Acute Deep-Well Casing Problems," *Oil and Gas J.* (May 2, 1966) 52.

"Densometer — New Tool for Drillers," *Oil and Gas J.* (Sept. 1, 1958) 116.

"Design and Control of Concrete Mixtures," Portland Cement Assn., Skokie, Ill. (July 1968).

DeWitt, G. B.: "Art of Oil Well Cementing Discussed by Technologists," *Oil and Gas J.* (Dec. 6, 1928) 186.

DeWitt, G. B.: "Mixing Device Removes Guess Work From Well Cementing," *Oil Weekly* (Dec. 21, 1928) **52,** No. 1, 39-43.

"Diacel Cement Systems," *Technical Data Booklet D-22,* Drilling Specialties Co., Bartlesville, Okla. (April 1, 1959).

Dietrich, W. K., and Willhite, G. P.: "The Casing Failure Problem in Steam Injection Wells, Cat Canyon Oil Field, Santa Barbara, California," paper 66-PET-38 presented at ASME meeting, New Orleans, La., Sept. 18-21, 1966.

Doherty, W. T.: "Oil-Well Cementing in the Gulf Coast Area," *API Proc. Sec. IV (Prod. Bull. 212)* (1933) 60.

Doherty, W. T., and Manning, M.: "Cementing Problems on Gulf Coast," *Oil and Gas J.* (Sept. 11, 1930) 30; *The Cementer* (Sept. 1930) **2,** No. 7, 1.

Doherty, W. T., and Manning, M.: "Drilling Mud Contamination Blamed for Coastal Cement Failures," *Natl. Petroleum News* (Sept. 17, 1930) **22,** No. 38, 54-55.

Doherty, W. T., and Manning, M.: "Gulf Coast Cementing Problems," *Oil and Gas J.* (April 4, 1929) 48; *Oil Weekly* (April 12, 1929) **53,** No. 4, 47-48.

Donaldson, E. C.: "Subsurface Disposal of Industrial Wastes in the United States," Information Circular 8212, USBM (1964).

Donaldson, F.: "The Injection of Cement Grout into Water Bearing Fissures," *Trans.,* AIME (1914) **48,** 136-140.

Dorsch, K. E.: "Hardening and Corrosion of Cement," *Cement and Cement Manuf.* (Feb. 1933) **6,** No. 2, 45-53.

Drilling and Casing Procedures, (Rule 206b, Cementing; 604e, Methods of Plugging) Ref. No. 42-70, Oklahoma Corporation Commission, Oklahoma City (1969).

Dubrow, M. H.: "Deep-Well Cementing," *Oil-Well Cementing Practices in the United States,* API, New York (1959) 177.

Dumbauld, G. K., Brooks, F. A., Jr., Morgan, B. E., and Binkley, G. W.: "A Lightweight, Low-Water-Loss, Oil-Emulsion Cement for Use in Oil Wells," *J. Pet. Tech.* (May 1956) 99-104.

Dumbauld, G. K., Perry, D., Binkley, G. W., and Brooks, F. A., Jr.: "An Accelerated Squeeze-Cementing Technique," *Trans.,* AIME (1956) **207,** 25-29.

Dunlap, I. R., and Patchen, F. D.: "A High-Temperature Oil Well Cement," *Pet. Eng.* (Nov. 1957) B60.

Ebel, R. E.: "Gas in the Soviet Union," paper SPE 4561 presented at SPE-AIME 48th Annual Fall Meeting, Las Vegas, Nev., Sept. 30-Oct. 3, 1973.

Edison, J. E.: "Radioactive Tracers Used To Locate Lost Circulation," *World Oil* (June 1954) 197-199.

Edwards, R. E.: "Mathematics of Cementing," *Oil Weekly,* Part I (May 27, 1946) 38; Part II (June 3, 1946) 33; Part III (June 10, 1946) 42; Part IV (June 17, 1946) 45.

Edwards, R. E.: "Rotation of Casing During Cementing," *Oil Weekly* (Feb. 17, 1947) **124,** No. 12, 24-26.

Edwards, R. E., and Nussbaumer, F. W.: "The D. V. Multiple Stage Cementer," *Pet. Eng.* (July 1947) 80-82.

Einarsen, C. A.: "High Strength Granular Sealing Material Increases Efficiency of Primary Cementing and Squeeze Cementing," *J. Pet. Tech.* (Aug. 1955) 15-18.

Elkins, L. E.: "Overview — 25 Years of Professional Contribution to Petroleum Development and Production," *J. Pet. Tech.* (Dec. 1973) 1337-1341.

"Elusive Standard Oil Well Cement Near," *Oil and Gas J.* (June 24, 1968) 55.

Enloe, J. R.: "Amerada Finds Using Multiple Casing Strings Can Cut Costs," *Oil and Gas J.* (June 12, 1967) 76-78.

Evans, G. W., and Carter, L. G.: "Bonding Studies of Cementing Compositions to Pipe and Formations," *Drill. and Prod. Prac.,* API (1962) 72.

Evans, G. W., and Carter, L. G.: "New Technique for Improving Cement Bond," *Drill. and Prod. Prac.,* API (1964) 33-38.

Evans, G. W., and Harriman, D. W.: "Laboratory Tests on Collapse Resistance of Cemented Casing," paper SPE 4088 presented at SPE-AIME 47th Annual Fall Meeting, San Antonio, Tex., Oct. 8-11, 1972.

"Expansive Cement Concretes — Present State of Knowledge," *J. American Concrete Institute* (Aug. 1970) 583.

Fagin, K. M.: "New Technique Designed for Cementing Casing," *Pet. Eng.* (Jan. 1948) 156.

Failures in Primary Cement Jobs Noted," *Drilling (10th Annual Exposition-in-Print)* (Dec. 2, 1957) 87.

Farris, R. F.: "A Practical Evaluation of Cements for Oil Wells," *Drill. and Prod. Prac.,* API (1941) 283-292.

Farris, R. F.: "Effects of Temperature and Pressure on Properties of Oil Well Cements," *Oil and Gas J.* (Feb. 22, 1940) 78.

Farris, R. F.: "Effects of Temperature and Pressure on Rheological Properties of Cement Slurries," *Trans.,* AIME (1941) **142,** 117-130 (reprint, page 306).

Farris, R. F.: "Method for Determining Minimum Waiting-on-Cement Time," *Trans.,* AIME (1946) **165,** 175-188.

Federal Specifications: Cements, Portland, SS-C-192a, Superintendent of Documents, Washington (April 1952).

"Federal Specification for Cements," *Federal Standard Stock Catalog,* Sec. IV, Part 5, U. S. Govt. Printing Office, Washington (May 20, 1946).

Fertl, W. H.: "What to Remember When Interpreting Mud Gas Cutting," *World Oil* (Sept. 1973) 67.

Fertl, W. H., Pilkington, P. E., and Scott, J. B.: "A Look at Cement Bond Logs," *J. Pet. Tech.* (June 1974) 607-617.

"15,400-Foot String of 9⅝-Inch Casing Set in Louisiana Well," *Drilling* (March 1957) **18,** No. 5, 78.

Flumerfelt, R. W.: "Laminar Displacement of Non-Newtonian Fluids in Parallel Plate and Narrow Gap Annular Geometries," *Soc. Pet. Eng. J.* (April 1975) 169-180; *Trans.,* AIME, **259.**

Folmer, L. W.: Methods of Detecting Top of Cement Behind Casing," *Oil-Well Cementing Practices in the United States,* API, New York (1959) 133-135.

Frank, W. J., Jr.: "Improved Concentric Workover Techniques," *J. Pet. Tech.* (April 1969) 401-408.

Garvin, T. R., and Moore, P. L.: "A Rheometer for Evaluating Drilling Fluids at Elevated Temperatures," paper SPE 3062 presented at SPE-AIME 45th Annual Fall Meeting, Houston, Oct. 4-7, 1970.

Garvin, T. R., and Slagle, K. A.: "Scale-Model Displacement Studies To Predict Flow Behavior During Cementing," *J. Pet. Tech.* (Sept. 1971) 1081-1088.

"Geothermal Steam for Power in New Zealand," *Bull. 117,* L. I. Grange, Ed., New Zealand Geological Survey, Dept. of Scientific and Industrial Research, Wellington (1955).

Gibbs, M. A.: "Delaware Basin Cementing — Problems and Solutions," *J. Pet. Tech.* (Oct. 1966) 1281-1265.

Gibbs, M. A.: "Improved Primary Cementing by Use of Hydraulic Analysis," *Oil and Gas J.* (Feb. 15, 1965).

Gibbs, M. A.: "Primary and Remedial Cementing in Fractured Formations," paper presented at Southwestern Pet. Short Course, Texas Technological College, Lubbock, April 22-23, 1965.

"Gilsonite Cement — For Light Weight Slurry," *Calif. Oil World* (June 1958) **51,** No. 11 (1st Issue) 30.

Glenn, E. N.: "Liner Cementing — Long Life Technique," paper presented at Southwestern Pet. Short Course, Texas Technological College, Lubbock, April 20-21, 1967.

Glenn, W. E.: "Management and Economics in the Oil Industry — A Review," *J. Pet. Tech.* (Dec. 1973) 1342-1346.

Goddard, R. D., Guest, R. J., and Anderson, T. O.: "High Resolution Fluid Measurements Improve Drilling," *The Drilling Contractor* (March-April 1973).

Godfrey, W. K.: "Effect of Jet Perforating on Bond Strength of Cement," *J. Pet. Tech.* (Nov. 1968) 1301-1314; *Trans.*, AIME, **243.**

Godfrey, W. K., and Methven, N. E.: "Casing Damage Caused by Jet Perforating," paper SPE 3043 presented at SPE-AIME 45th Annual Fall Meeting, Houston, Oct. 4-7, 1970.

Goins, W. C., Jr.: "General Considerations of Lost Circulation," *The Drilling Contractor* (Nov.-Dec. 1963) 66.

Goins, W. C., Jr.: "How To Combat Circulation Loss," *Oil and Gas J.* (June 9, 1952) 71.

Goins, W. C., Jr.: "Lost Circulation Problems Whipped With BDO (Bentonite Diesel Oil) Squeeze," *Drilling* (Sept. 1954) **15,** No. 11, 83.

Goins, W. C., Jr.: "Open-Hole Plugback Operations," *Oil-Well Cementing Practices in the United States,* API, New York (1959) 193.

Goins, W. C., Jr.: "Selected Items of Interest in Drilling Technology," *J. Pet. Tech.* (July 1971) 857-862.

Goins, W. C., Jr., Collings, B. J., and O'Brien, T. B.: "New Approach to Tubular String Design," *World Oil,* Part I (Nov. 1965) 1-6; Part II (Dec. 1965) 7-12; Part III (Jan. 1966) 13-18; Part IV (Feb. 1966) 19-24.

Goins, W. C., Jr., and Nash, F., Jr.: "A Study of Nonprimary Cementing Operations in the Pierce Junction Zone," *Report F260,* Monthly Progress Report, Tech. Services Laboratory, Gulf Oil Corp. (1957).

Goins, W. C., Jr., Weichert, J. P., Burba, J. L., Jr., Dawson, D. D., and Teplitz, A. J.: "Down-the-Hole Pressure Surges and Their Effect on Loss of Circulation," *Drill. and Prod. Prac.*, API (1951) 125-131.

Goldsmith, R. G.: "Here is a Practical View of Oil Field Hydraulics Today," *World Oil* (June 1969) 140-144.

Gonnerman, H. F., and Lerch, W.: "Changes in Characteristics of Portland Cement as Exhibited by Laboratory Tests Over the Period 1904 to 1950," *Portland Cement Assn. Research Dept. Bull. 39; ASTM Spec. Pub. 127* (1952).

Goode, J. M.: "Gas and Water Permeability Data for Some Common Oilwell Cements," *J. Pet. Tech.* (Aug. 1962) 851-854.

Gooding, P., and Halstead, P. E.: "The Early History of Cement in England," *Proc.*, Third Intl. Symposium on the Chemistry of Cement, London (1952) 1-27.

Goodman, M. A., and Wood, D. B.: "A Mechanical Model for Permafrost Freeze-Back Pressure Behavior," *Soc. Pet. Eng. J.* (Aug. 1975) 287-301; *Trans.*, AIME, **259.**

Goolsby, J. L.: "A Proven Squeeze Cementing Technique in a Dolomite Reservoir," *J. Pet. Tech.* (Oct. 1969) 1341-1346.

Gordon, G. M.: "What 1,000 Tests Revealed About Oil Well Cements," *Petroleum World* (Feb. 1930) **27,** No. 2, 84.

Graham, C. R.: "World's Deepest Well Plugged and Abandoned, Phillips No. 1-EE University," *Pet. Eng.* (March 1959) B121-B123.

Graham, H. L.: "Rheology — Balanced Cementing Improves Primary Success," *Oil and Gas J.* (Dec. 18, 1972) 53.

Gray, G. R., and Young, F. S., Jr.: "25 Years of Drilling Technology — A Review of Significant Accomplishments," *J. Pet. Tech.* (Dec. 1973) 1347-1354.

Greer, J. B.: "Effects of Metal Thickness and Temperature on Casing and Tubing Design for Deep, Sour Wells," *J. Pet. Tech.* (April 1973) 499-510; *Trans.*, AIME, **255.**

Greer, R. C., and Shryock, S. H.: "New Technique Improves Steam Stimulation Completions," *J. Pet. Tech.* (July 1968) 691-699.

Gretener, P. E.: "Temperature Anomalies in Wells Due to Cementing of Casing," *J. Pet. Tech.* (Feb. 1968) 147-151.

Griffin, W. H.: "Analysis of Present Day Oil Well Casing Cementing," *Pet. Eng.* (Nov. 1948) 117.

Grosmangin, M., Kokesh, F. P., and Majani, P.: "The Cement Bond Log — A Sonic Method for Analyzing the Quality of Cementation of Borehole Casings," *J. Pet. Tech.* (Feb. 1961) 165-171; *Trans.*, AIME, **222.**

Grossman, W.: "Air Entrainment During Primary Cement Job," *World Oil* (Jan. 1950) 110-111.

Grossman, W.: "Primary Cementing of Casing in Rotary-Drilled Wells," *World Oil* (April 1951) 153.

Guest, R. J., and Zimmerman, C. W.: "Compensated Gamma Ray Densimeter Measures Slurry Densities in Flow," *Pet. Eng.* (Sept. 1973).

"Gulf Sets 13⅜-Inch Casing to Record Depth of 11,853 Feet," *World Oil* (July 1967) 92-93.

"Gypsum Blend Cements Show Promise for North Slope Use," *Oil and Gas J.* (May 19, 1969) 58.

Hall, D.: "Bridging Effectiveness of Perlite for Light Weight Cements and Lost Circulation," paper 141-G presented at SPE-AIME Mid-Continent Meeting, Oklahoma City, Oct. 3-5, 1951.

Halliburton Cementing Tables, Section 230, Halliburton Co., Duncan, Okla. (March 1972).

Halliburton, E. P.: "Method and Means for Cementing Oil Wells," U. S. Patent No. 1,369,891 (March 1, 1921), filed June 26, 1920.

Halliburton, E. P.: "Method of Hydrating Cement and the Like," U. S. Patent No. 1,486,883 (March 18, 1924), filed June 20, 1922.

Halliburton Oil Well Cement Manual, Halliburton Co., Duncan, Okla. (1970).

Hall, J. A.: "Gulf Engineer Tallest Shoe-to-Surface Cement Job," *World Oil* (May 1949) 220-222.

Handbook of Oil Well Cementing, Halliburton Oil Well Cementing Co., Duncan, Okla. (1928).

Handin, J. O.: "Strength of Oilwell Cements at Down-Hole Pressure-Temperature Conditions," *Soc. Pet. Eng. J.* (Dec. 1965) 341-347; *Trans.*, AIME, **234.**

Hansen, W. C.: "Crystal Growth as a Source of Expansion in Portland-Cement Concrete," *Proc.,* ASTM (1963) **63,** 932-945.

Hansen, W. C.: "Oil Well Cements," *Proc.,* Third Intl. Symposium on Chemistry of Cement, London (1952).

Hansen, W. C., and Hunt, J. O.: "The Use of Natural Asphalt in Portland Cement," *Bull. 161,* ASTM (Oct. 1949) 50-58.

Harcourt, G., Walker, T., and Anderson, T. O.: "Use of the Micro-Seismogram and Acoustic Cement Bond Log To Evaluate Cementing Techniques," paper SPE 798 presented at SPE-AIME Symposium on Mechanical Engineering Aspects of Well Completions, Fort Worth, Tex., March 23-24, 1964.

Harmon, J. A., and Woodard, H. D.: "Use of Oil-Cement Slurries for Decreasing Water Production," *Drill. and Prod. Prac.,* API (1955) 451-457.

Harris, F., and Carter, L. G.: "Use a Chemical Wash and a Low Fluid Loss Cement," *Drilling* (Jan. 1964) **25,** No. 3.

Harris, F., and Carter, L. G.: "Effectiveness of Chemical Washes Ahead of Squeeze Cementing," paper 851-37-H presented at API Mid-Continent Dist. Meeting (April 1963) preprint.

Hartweg, D. G.: "Radioactive Measurement of Fluid Density," paper 60-PET-48 presented at ASME Petroleum Mechanical Engineering Conference, New Orleans, La., Sept. 1960.

Hassebroek, W. E.: "Rotating-Type Wall Cleaners and Casing Centralizers," Company Report, Halliburton Oil Well Cementing Co., Duncan, Okla. (Jan. 1948).

Hedstrom, B. O. A.: "Flow of Plastic Materials in Pipes," *Ind. and Eng. Chem.* (1952) **44,** 651.

Heise, H.: "Canada Leads in Research on Arctic Cement Operations," *Canadian Petroleum* (Aug. 1973) 30.

Helmick, W. E., and Longley, A. J.: "Pressure-Differential Sticking of Drill Pipe and How It Can Be Avoided or Relieved," *Drill. and Prod. Prac.,* API (1957) 55.

Henderson, R. A.: "Hot Wire Tool Locates Lost Circulation," *World Oil* (Oct. 1952) 187.

Henderson, R. K.: "New Fast-Setting Cement Is Valuable Off-Job Adjunct," *Pet. Eng.* (Sept. 1940) 93.

Hendrickson, J. F.: "How To Design and Run Casing Strings," *Pet. Eng.* (July 1961).

Herrington, C. G.: "Rotation of Casing While Cementing," *API Proc. Sec. IV (Prod. Bull. 226)* (1940) 109-110.

Hewitt, C. H.: "Analytical Techniques for Recognizing Water-Sensitive Reservoir Rocks," *J. Pet. Tech.* (Aug. 1963) 813-818.

"High Expanding Cement for Primary Jobs," *Pet. Eng.* (Feb. 1970) 80.

Hill, F. F.: "Cementing Deep Wells," Company Bulletin, Union Oil Co., Los Angeles (Dec. 1924).

Hill, F. F.: "Methods of Cementing Deep Oil Wells as Accomplished in California," *Oil Weekly* (Dec. 26, 1924) **36,** No. 1, 27.

Hill, F. F.: "Progress in Cementing Deep Wells," *Oil and Gas J.* (Dec. 18, 1924) 158.

Hill, F. F.: "Progress in Cementing Deep Wells as Accomplished in California," *Natl. Petroleum News* (Dec. 24, 1924) **16,** No. 52, 53-55.

Hills, J. O.: "A Review of Casing-String Design Principles and Practices," *Drill. and Prod. Prac.,* API (1951) 91.

Hills, J. O.: "Design of Casing Programs," *Drill. and Prod. Prac.,* API (1939) 369-382; *Oil and Gas J.* (May 25, 1939) 60.

Hilton, A. G.: "Mechanical Aids and Practices for the Improvement of Primary Cementing," *Oil-Well Cementing Practices in the United States,* API, New York (1959) 123.

Hilton, P. E.: "Modern Technique in Oil-Well Cementing," *Oil and Gas J.* (Dec. 3, 1942) 57.

"Historic Oklahoma Test Bests 30,000 Feet," *Oil and Gas J.* (March 6, 1972) 63.

History of Petroleum Engineering, API Div. of Production, Dallas (1961).

Hnatiuk, J.: "Environmental Conditions in the Canadian Arctic and Related Research Activities," paper SPE 4586 presented at SPE-AIME 48th Annual Fall Meeting, Las Vegas, Nev., Sept. 30-Oct. 3, 1973.

Hoch, R. S.: "Cementing Techniques Used for High Angle, S-Type Directional Wells," *Oil and Gas J.* (June 22, 1970) 88-93.

Hodges, C.: "Quick-Setting Gypsum Cements," *Oil and Gas J.* (May 9, 1940) 19 (abstr.).

Hodges, J. W.: "Squeeze Cementing Methods and Materials," *Oil-Well Cementing Practices in the United States,* API, New York (1959) 149-159.

Hoisington, G. T.: "Oil Well Cementing Practice in Kansas," paper 841-151 presented at API Spring Meeting, Amarillo, Tex., March 21-22, 1941; *Drilling* (April 1941) 10.

Holliday, G. H.: "Calculation of Allowable Maximum Casing Temperature To Prevent Tension Failures in Thermal Wells," ASME (June 11, 1969).

Holmes, C. S., and Swift, S. C.: "Calculation of Circulating Mud Temperatures," *J. Pet. Tech.* (June 1970) 670-674.

Holmquist, D.: "Casing Buckling Determined by Stress Diagram," *Pet. Eng.* (Dec. 1969).

Holmquist, J. L., and Nadai, A.: "A Theoretical and Experimental Approach to the Problem of Collapse of Deep-Well Casing," *Drill. and Prod. Prac.,* API (1939) 392.

Hopkins, H. F.: "Integrating Well Completion Practices," *J. Pet. Tech.* (Nov. 1953) 13-15.

Hopkins, H. F.: "Use of Bottom Plug in Cementing Operations," *Oil and Gas J.* (June 22, 1953) 308-309.

Horton, H. L., Morris, E. F., and Wahl, W. W.: "Improved Cement Slurries by Reduction of Water Content," paper 906-9-D presented at API Southwestern Dist. Div. of Production Spring Meeting, Midland, Tex. (March 1964) preprint.

"Hot Foot Frees Cemented Pipe," *World Oil* (Sept. 1949) 103.

Hough, J. F.: "An API Committee To Draft Standard Specifications for Portland Cement Concrete," *API Proc. (Prod. Bull. 202)* (1928) 8-15.

Hough, J. F.: "Portland Cement for Shutting Off Water in Oil Wells," *API Proc. (Prod. Bull. 201)* (1927) 11-17.

Howard, G. C., and Clark, J. B.: "Factors To Be Considered in Obtaining Proper Cementing of Casing," *Drill. and Prod. Prac.*, API (1948) 257-272; *Oil and Gas J.* (Nov. 11, 1948) 243.

Howard, G. C., and Fast, C. R.: "Significance of Pumping Rates in Squeeze Cementing," *Oil and Gas J.* (May 25, 1950) 120.

Howard, G. C., and Fast, C. R.: "Squeeze Cementing Operations," *Trans.*, AIME (1950) **189,** 53-64.

Howard, G. C., and Scott, P. P., Jr.: "An Analysis and the Control of Lost Circulation," *Trans.*, AIME (1951) **192,** 171-182.

Howard, G. C., and Scott, P. P., Jr.: "Plugging Off Water in Fractured Formations," *Trans.*, AIME (1954) **201,** 132-137.

Howell, E. P., Perkins, T. K., and Seth, M. S.: "Calculating Temperatures for Permafrost Completions," *Pet. Eng.* (April 1973) 69.

Howell, L. G., and Frosch, A.: "Detection of Radioactive Cement in Cased Wells," *Trans.*, AIME (1940) **136,** 71-78.

Hower, W. F., McLaughlin, C., Ramos, J., and Land, J.: "Water Can Be Controlled in Air and Gas Drilling," paper 1099-G presented at SPE-AIME 33rd Annual Fall Meeting, Houston, Oct. 5-8, 1958.

Hower, W. F., and Montgomery, P. C.: "New Slurry Effective for Control of Unwanted Water," *Oil and Gas J.* (Oct. 19, 1953).

Hower, W. F., and Montgomery, P. C.: "Squeeze-Cementing Mixture," *Oil and Gas J.* (Oct. 19, 1953) 136.

Hower, W. F., Ramos, J., McLaughlin, C., and Land, J.: "Fluid Grout for Water Control," *Pet. Eng.* (June 1958) B26-B29.

"How To Pre-Plan a Deep Cement Job," *Drilling-DCW* (April 20, 1972) 12.

"How 20,000-Foot Ellenburger Gas Wells Were Drilled," *World Oil* (May 1968) 58.

"How Water Is Shut Off Effectively by Slurry Oil Squeeze," *Drilling* (July 1955) **16,** No. 9, 118.

Huber, F. W.: "Cementing Oil Wells as a Problem in Cement Chemistry," *Oil Field Eng.* (April 1923) **25,** No. 4, 77-78.

Huber, T. A., and Corley, C. B., Jr.: "Permanent-Type Multiple and Tubingless Completions," *Pet. Eng.* (Feb. and March 1961).

Huber, T. A., and Tausch, G. H.: "Permanent-Type Well Completion," *Trans.*, AIME (1953) **198,** 11-16.

Huber, T. A., Tausch, G. H., and Dublin, J. R., III: "A Simplified Cementing Technique for Recompletion Operations," *Trans.*, AIME (1954) **201,** 1-7.

"Humble Has New Completion Method," *Oil and Gas J.* (Nov. 17, 1958) 136.

"Humble Runs Three Strings of 4½-Inch Pipe in One Hole," *World Oil* (March 1966) 56-59.

Ihrig, H. K.: "Fibrous Materials in Cement and Mud Open New Field in Production Practices," *Oil and Gas J.* (May 21, 1936) 98.

"Improve Cementing of Drilling Liners in Deep Wells," *World Oil* (Oct. 1973).

"Improved Primary Cementing Technique," *International Oil* (Jan. 1941) **2,** No. 1, 78.

Ingram, R.: "Amerada Sets Long String of Casing in Wyoming," *Oil and Gas J.* (Dec. 17, 1936) 44-45.

Jahns, D. F.: "Production Fundamentals: Principles and Practices of Cementing," *Pet. Eng.* (March 1957) B64.

Jennings, H. Y., Jr., and Timur, A.: "Significant Contributions in Formation Evaluation and Well Testing," *J. Pet. Tech.* (Dec. 1973) 1432-1446.

Jessen, F. W., and Webber, C. E.: "Setting of Portland Cement: Thermal Characteristics During Setting at Elevated Temperatures," *Ind. and Eng. Chem.* (June 1939) **31,** No. 6, 745-749.

Jones, E. N.: "Mud Acidization May Be Helpful Before Squeeze Cementing," *Oil and Gas J.* (April 26, 1947) 185.

Jones, P. H., and Berdine, D.: "Oil-Well Cementing — Factors Influencing Bond Between Cement and Formation," *Oil and Gas J.* (March 21, 1940) 71; *Petroleum World* (June 1940) 26; *Drill. and Prod. Prac.*, API (1940) 45-63.

Jones, R. W.: "Chemical Treatment of Cement-Contaminated Drilling Muds," *World Oil* (May 1949) 106.

Jones, W. M.: "Casing and Cementing," School of Prod. Tech., Extension Dept., U. of Texas, Austin (Sept. 19, 1955).

Kail, R. S.: "Squeeze Cementing Before Production To Reduce Gas-Oil Ratios," Company Report, Baker Oil Tools, Inc., Los Angeles (1940).

Kalousek, G. L.: Discussion of "The Reactions and Thermochemistry of Cement Hydration at Ordinary Temperatures," *Proc.*, Third Intl. Symposium on the Chemistry of Cement, London (1952) 296-311.

Kalousek, G. L.: "The Reactions of Cement Hydration at Elevated Temperatures," *Proc.*, Third Intl. Symposium on the Chemistry of Cement, London (1952) 334-354.

Kastrop, J. E.: "Cemented Liner Solves Problems of Testing Formations in Open Holes," *World Oil* (Sept. 29, 1947) 30-31.

Kastrop, J. E.: "Handling Bulk Mud and Cement on Drilling Tenders," *World Oil* (Sept. 1949) 98.

Kastrop, J. E.: "Liner Hanging Cuts Deep Well Casing Costs," *Pet. Eng.* (Jan. 1962) 104.

Kastrop, J. E.: "Technology Optimized in World's Second Deepest Well," *Pet. Eng.* (March 1974) 45.

Kastrop, J. E.: "True Pozzolan Cement Developed for Oil Field Use," *Pet. Eng.* (Nov. 1955) B33-B36.

Keller, H. H., Couch, E. J., and Berry, P. M.: "Temperature Distribution in Circulating Mud Columns," *Soc. Pet. Eng. J.* (Feb. 1973) 23-30.

Kellermann, W. F.: "Effect of Use of Blended Cements and Vinsol Resin-Treated Cements on Durability of Concrete," *J. American Concrete Institute* (1946) **17,** No. 6, 681-687, (Proc. Vol. 42); *Chem. Abstr.* (1946) **40,** 6778.

Kelly, E. F.: "Better Completions Realized From Use of Mud Acid Ahead of Cement," *Oil and Gas J.* (Jan. 6, 1949) 62.

Kennedy, H. T.: "Chemical Methods for Shutting Off Water in Oil and Gas Wells," *Trans.,* AIME (1936) **118,** 177-186.

Kennedy, H. T.: "The Control of Gas-Oil and Water-Oil Ratios by Chemical Treatment," *Drill. and Prod. Prac.,* API (1941) 214-220.

Kennedy, J. L.: "Automated Mud System Does Well in Tests," *Oil and Gas J.* (Dec. 1, 1969).

Kettenburg, R. J., and Schmieder, F. R.: "Oil-Well Casing Failures," *Drill. and Prod. Prac.,* API (1945) 185-198.

King, R. H.: "Cementing of Wells in East Texas Field To Reduce Volume of Water Produced," *Pet. Eng.* (June 1938) 56.

King, T. G.: "The Application of Permanent Completion Techniques to Offshore Operations," *J. Pet. Tech.* (Sept. 1966) 1031-1040.

Kirchner, W.: "Accelerating Hardening of Cement," *Chem. Abstr.* (1931) **25,** 392.

Kirk, W. L.: "Deep Drilling Practices in Mississippi," *J. Pet. Tech.* (June 1972) 633-642.

Klein, A., and Troxell, G. E.: "Studies of Calcium Sulfoaluminate Admixtures for Expansive Cements," *Proc.,* ASTM (1958) **58,** 986-1008.

Klinkenberg, A.: "The Neutral Zones in Drill Pipe and Casing and Their Significance in Relation to Buckling and Collapse," *Drill. and Prod. Prac.,* API (1951) 64-79.

Kljucec, N. M., and Telford, A. S.: "Well Temperature Monitoring With Thermistor Cables Through Permafrost," paper 7256 presented at CIM 23rd Annual Tech. Meeting, Calgary, Canada, May 16-19, 1972.

Kljucec, N. M., Telford, A. S., and Bombardieri, C. C.: "Cementing Arctic Wells Through Permafrost," paper 7257 presented at CIM 23rd Annual Tech. Meeting Calgary, Canada, May 16-19, 1972.

Kljucec, N. M., Telford, A. S., and Bombardieri, C. C.: "Gypsum-Cement Blend Works Well in Permafrost Areas," *World Oil* (March 1973) 49.

Knapp, I. N.: "Cementing Oil and Gas Wells," *Bull.,* AIME (March 1914) 471-488; *Trans.,* AIME (1914) **48,** 651-675.

Koch, R. D.: "The Design of Alaskan North Slope Production Wells," *Proc.,* Inst. Mech. Eng. (1971) **185,** No. 73, 989.

Koenig, J. B.: "Worldwide Status of Geothermal Exploration and Development," paper SPE 4179 presented at SPE-AIME 43rd Annual California Regional Fall Meeting, Bakersfield, Nov. 8-10, 1972.

Korepanov, V. N., and Machinskii, E. K.: "Cementation of Oil Wells with the Use of an Accelerator," *J. Inst. Pet. Tech.* (March 1932) **18,** No. 101, 197-212.

Krueger, R. F.: "Advances in Well Completion and Stimulation During JPT's First Quarter Century," *J. Pet. Tech.* (Dec. 1973) 1447-1462.

Krueger, R. F.: "Joint Bullet and Jet Perforation Tests — Progress Report," *Drill. and Prod. Prac.,* API (1956) 126.

"Lab Studies Show Why and How Gun Perforators Damage Casing," *Oil and Gas J.* (June 14, 1965).

Lafuma, H.: "Expansive Cement," *Proc.,* Third Intl. Symposium on Chemistry of Cement, London (1952) 581.

Larminie, F. G.: *The Arctic as an Industrial Environment — Some Aspects of Petroleum Development in Northern Alaska,* BP Alaska, Inc., Fairbanks (Jan. 1971).

"Latex-Cement Shows High Success Ratio," *Drilling (19th Annual Exposition-in-Print)* (Dec. 2, 1957) 33.

Laudermilk, J. L.: "Filtration Formation Cementing of Low Gas-Oil Ratio Wells," *Oil and Gas J.* (Oct. 23, 1941) 56.

Lawrence, D. K., and Toland, T.: "Preplanning Deep Holes Pays Off for Sun," *Pet. Eng.* (March 1967) 63.

Lea, F. M., and Desch, C. H.: *The Chemistry of Cement and Concrete,* Arnold & Co., London (1935; reprinted 1937, 1940).

Leon, L., Hathorn, D. H., and Saunders, C. D.: "Completion Techniques in Very Deep Wells," *Proc.,* Eighth World Pet. Cong., Moscow (1971) **3,** 159-166.

Leonardon, E. G.: "The Economic Utility of Thermometric Measurements in Drill Holes in Connection With Drilling and Cementing Problems," *Geophysics* (Jan. 1936) **1,** No. 1, 115.

Leutwyler, K.: "Casing Temperature Studies in Steam Injection Wells," *J. Pet. Tech.* (Sept. 1966) 1157-1163.

Lindsey, H. E., Jr.: "How To Cement a Liner Above Open Hole Play," *World Oil* (Feb. 1, 1972) 32-33.

Lindsey, H. E., Jr.: "Running and Cementing Deep Well Liners," *World Oil,* Part 1 (Nov. 1974); Part 2 (Dec. 1974); Part 3 (Jan. 1975).

Lindsey, H. E., Jr.: "Techniques for Liner Tie-Back Cementing," *Pet. Eng.* (July 1973) 40.

Lindsey, H. E., Jr., and Bateman, S. J.: "Improved Cementing of Drilling Liners in Deep Wells," *World Oil* (Oct. 1973) 65.

"Lone Star Introduced New Deep-Well Cement," *Oil and Gas J.* (May 22, 1941) 204.

Lord, D. L., Hulsey, B. W., and Melton, L. L.: "General Turbulent Pipe Flow Scale-Up Correlation for Rheologically Complex Fluids," *Soc. Pet. Eng. J.* (Sept. 1967) 252-258; *Trans.,* AIME, **240.**

Lubinski, A.: "Influence of Tension and Compression on Straightness and Buckling of Tubular Goods in Oil Wells," *API Proc. (Prod. Bull. 237)* (1951).

Lubinski, A., and Blenkarn, K. A.: "Buckling of Tubing in Pumping Wells, Its Effects and Means for Controlling It," *Trans.*, AIME (1957) **210,** 73-78.

Ludwig, N. C.: "Chemistry of Portland Cement Used in Oil Wells," *Oil-Well Cementing Practices in the United States,* API, New York (1959) 27-33.

Ludwig, N. C.: "Effects of Sodium Chloride on Setting Properties of Oil-Well Cements," *Drill. and Prod. Prac.*, API (1951) 20-27.

Ludwig, N. C.: "Portland Cements and Their Application in the Oil Industry," *Drill. and Prod. Prac.*, API (1953) 183-209.

Ludwig, N. C.: "Preparation of Oil Well Cement Slurries," *Oil and Gas J.* (March 29, 1951) 164-165.

Ludwig, N. C., and Pence, S. A.: "Properties of Portland Cement Pastes Cured at Elevated Temperatures and Pressures," *Proc.*, American Concrete Institute (1956) **52,** 673-687.

Lummus, J. L.: '"A New Look at Lost Circulation," *Pet. Eng.* (Nov. 1967) 69-73.

Lummus, J. L.: "Squeeze Slurries for Lost Circulation Control," *Pet. Eng.* (Sept. 1968) 59-64.

MacClain, C.: "External Casing Packer Solves Completion Problems," *Pet. Eng.* (April 1966) 83-86.

MacDonald, G. C.: "What Causes Primary Cementing Failures," *World Oil* (Aug. 1, 1956) 115.

Mahoney, B. J., and Barrios, J. R.: "Cementing Liners Through Deep High Pressure Zones," *Pet. Eng.* (March 1974) 61.

Mahony, B. J.: "New Techniques Cut Drilling Costs in the Delaware Basin," *World Oil* (Nov. 1966) 117.

Maier, L. F.: "Understanding Surface Casing Waiting-on-Cement Time," paper presented at CIM 16th Annual Tech. Meeting, Calgary, Canada, May 1965.

Maier, L. F., Carter, M. A., Cunningham, W. C., and Bosley, T. G.: "Cementing Practices in Cold Environments," *J. Pet. Tech.* (Oct. 1971) 1215-1220.

Maier, L. F., Pollock, R. W., and Heimlick, W. H.: "Problems in Cementing Salt Caverns," *Canadian Pet. Eng.* (1964) 24.

Maier, L. F., and Porteous, W. R.: "An Analysis of Cementing Problems," *Canadian Oil and Gas Ind.* (July 1957) **10,** No. 7, 102-107.

Mallinger, M. A.: "Some Controlling Factors Regarding Variable Weighting of Cement Slurries," *Trans.*, AIME (1950) **189,** 374-376.

Mangold, G. B.: "A New Material for Oil Well Cementing Problems," *Oil World* (Jan. 1951) **45,** No. 1, 2.

Mangold, G. B.: "Minimizing Oil Well Cementing Failures," *World Oil* (May 1952) 112.

Marquaire, R. R., and Brisac, J.: "Primary Cementing by Reverse Circulation Solves Critical Problem in the North Hassi-Messaoud Field, Algeria," *J. Pet. Tech.* (Feb. 1966) 146-150.

Massey, J. F., and Scott, J.: "World's Deepest Producer, Drilling/Completion Landmark," *Pet. Eng.* (Aug. 1965) 86.

Mater, B. E.: "A Study of the Relative Applicability of Retrievable and Retainer Type Production Packers to Specific Oil Well Problems," *Proc.*, Third World Pet. Cong., The Hague (1951) **II,** 250-263.

"Maximum Technology Applied to New Deep Oklahoma Well," *Pet. Eng.* (July 1973) 28.

McCray, A. W., and Cole, F. W.: "Oil Well Drilling Technology," *Cementing Operations,* U. of Oklahoma Press, Norman (1959).

McDowell, J. M., and Muskat, M.: "The Effect on Well Productivity of Formation Penetration Beyond Perforated Casing," *Trans.*, AIME (1950) **189,** 309-312.

McGhee, E.: "Equipment for Permanent Type Well Completions," *Oil and Gas J.* (Dec. 7, 1953) 111.

McGhee, E.: "How Pan American Drilled Texas' Deepest Well . . . World's Second Deepest," *Oil and Gas J.* (Feb. 3, 1958) 80-84.

McKinley, R. M., Bower, F. M., and Rumble, R. C.: "The Structure and Interpretation of Noise From Flow Behind Cemented Casing," *J. Pet. Tech.* (March 1973) 329-338; *Trans.*, AIME, **255.**

McLean, R. H., Manry, C. W., and Whitaker, W. W.: "Displacement Mechanics in Primary Cementing," *J. Pet. Tech.* (Feb. 1967) 251-260.

McLeod, H. O.: "Here Are the Special Cements," *Oil, Gas and Petrochemical Equipment* (Feb. 1971) 5-6.

McRee, B. C.: "Cementing Regulations Applied to Oil and Gas Wells," *Oil-Well Cementing Practices in the United States,* API, New York (1959) 209.

Melmick, W. E., and Longley, A. J.: "Pressure-Differential Sticking of Drill Pipe and How It Can Be Avoided or Relieved," *Drill. and Prod. Prac.*, API (1957) 55-61.

Melrose, J. C., Savins, J. G., Foster, W. R., and Parish, E. R.: "Practical Utilization of the Theory of Bingham Plastic Flow in Sationary Pipes and Annuli," *Trans.*, AIME (1958) **213,** 316-324.

Melton, L. L., and Saunders, C. D.: "Rheological Measurements of Non-Newtonian Fluids," *Trans.*, AIME (1957) **210,** 196-210.

Menzel, D.: "A New Weighting Material for Drilling Fluids Based on Synthetic Iron Oxide," paper SPE 4517 presented at SPE-AIME 48th Annual Fall Meeting, Las Vegas, Nev., Sept. 30-Oct. 3, 1973.

Merriam, R., Wechsler, O., Boorman, R., and Davies, B.: "Insulated Hot Oil Producing Wells in Permafrost — Part I, Thermal Aanalysis; Part II, Thermal Performance," *J. Pet. Tech.* (March 1975) 357-365; *Trans.*, AIME, **259.**

Messenger, J. U.: "Barite Plugs Simplify Well Control," *World Oil* (June 1969) 83.

Messenger, J. U.: "Common Rig Materials Combat Severe Lost Circulation," *Oil and Gas J.* (June 18, 1973) 57.

Messenger, J. U., and McNeil, J. S., Jr.: "Lost Circulation Corrective: Time Setting Clay Cement," *Trans.*, AIME (1952) **195,** 59-64.

Messer, P. H., Raghavan, R., and Ramey, H. J., Jr.: "Calculation of Bottom-Hole Pressures for Deep, Hot, Sour Gas Wells," *J. Pet. Tech.* (Jan. 1974) 85-92.

Metzner, A. B., and Reed, J. C.: "Flow of Non-Newtonian Fluids — Correlation of the Laminar, Transition, and Turbulent Flow Regions," *AIChE Jour.* (1955) **1,** No. 4.

Miller, D. G., Manson, P. W., and Chen, R. T. H.: "Bibliography on Sulfate Resistance of Portland Cements, Concretes and Mortars," *Paper 708, Misc. Journal Series,* U. of Minnesota, Minneapolis (1952) 133 pp; *Chem. Abstr.* (1953) **47,** 11689-i.

Miller, H. H.: "Durability of Cements in Contact With Brines and Chemical Solutions," *Pet. Eng.* (Aug. 1942) 50.

Millikan, C. V.: "Cementing," *History of Petroleum Engineering,* API Div. of Production, Dallas (1961) Chap. 7.

Millikan, C. V.: "History of Cementing," *Oil-Well Cementing Practices in the United States,* API, New York (1959) 3-25.

Mills, B.: "Rotating While Cementing Proves Economical," *Oil Weekly* (Dec. 4, 1939) **95,** No. 13, 14-15.

"Mississippi Wildcat Gets Record Casing, Cement Job," *World Oil* (Feb. 1, 1974).

"Modern Cementing Practice," *Calif. Oil World,* Part I — Cement and Its Behavior in Oil Wells (Dec. 1944) 3; Part II — Preparation of the Well for Cementing (Feb. 1945) 14; Part III — Cementing the Casing in the Well (June 1945) 11.

"Modified Herc. Shuttles North Slope Cement," *Oil and Gas J.* (Aug. 31, 1970) 85.

Moeller, R. E., and Roberts, H.: "Oil Well Cementing Practice," *Proc.,* American Concrete Institute (1947) 43-97.

Montgomery, P. C.: "Casing Cementing for Water Wells," Halliburton Oil Well Cementing Co. report presented at Missouri Water Well Drillers Assn. Meeting, Jefferson City, April 18, 1955.

Montgomery, P. C.: "Cement Accelerators Cut Rig Downtime," *Drilling* (June 1965) **26,** No. 9, 76.

Montgomery, P. C.: "Equipment, Materials, Tools for Cementing Deep Wells," *Pet. Eng.* (Jan. 1953) B40.

Montgomery, P. C.: "Multiple-Stage Cementing Uses 2nd and 3rd Stage Methods," *World Oil* (Dec. 1954) 176.

Montgomery, P. C., and Smith, D. K.: "Oil Well Cementing Practices and Materials," *Pet. Eng.* (May and June 1961).

Montgomery, P. C., and Westbrook, S. S. S.: "Casing and Cementing," Course in Natural Gas, U. of Kansas, Lawrence (1955).

Moore, J. E.: "Clay Mineralogy Problems in Oil Recovery," *Pet. Eng.,* Part 1 (Feb. 1960); Part 2 (March 1960).

Moore, P. L., and Cole, F. W.: *Drilling Operations Manual,* Petroleum Publishing Co., Tulsa, Okla. (1965).

Moore, W.: "Experimental Tests With Hydraulic Pressure Applied to Cement Plugs in Oil Well Casing," *Pet. Eng.* (Feb. 1948) 198.

Moran, J. P., and Hartweg, G. D.: "Continuous Density Recordings During Cementing Operations," *Oil and Gas J.* (April 28, 1958) 88; *Pet. Eng.* (May 1958) B40.

Morgan, B. E.: "API Specification for Oil Well Cements," *Drill. and Prod. Prac.,* API (1958) 83.

Morgan, B. E.: "The Right Cement for the Right Job," *Oil and Gas J.* (July 21, 1958) 77-81.

Morgan, B. E., and Dumbauld, G. K.: "A Modified Low-Strength Cement," *Trans.,* AIME (1951) **192,** 165-170.

Morgan, B. E., and Dumbauld, G. K.: "Bentonite Cement Proving Successful in Permanent-Type Squeeze Operations," *World Oil* (Nov. 1954) 220.

Morgan, B. E., and Dumbauld, G. K.: "Measurement of Permeability of Set Cement," *Trans.,* AIME (1952) **195,** 323-324.

Morgan, B. E., and Dumbauld, G. K.: "Recent Developments in the Use of Bentonite in Cements," *Drill. and Prod. Prac.,* API (1953) 163.

Morgan, B. E., and Dumbauld, G. K.: "Use of Activated Charcoal in Cement To Combat Effects of Contamination by Drilling Muds," *Trans.,* AIME (1952) **195,** 225-232.

Morris, E. F.: "Evaluation of Cement Systems for Permafrost," paper presented at AIME 99th Annual Meeting, Denver, Colo., Feb. 15-19, 1970.

Morris, E. F., and Motley, H. R.: "Oil Base Spacer System for Use in Cementing Wells Containing Oil Base Drilling Muds," paper SPE 4610 presented at SPE-AIME 48th Annual Fall Meeting, Las Vegas, Nev., Sept. 30-Oct. 3, 1973.

Morris, E. F., Stude, D. L., and Cameron, R. D.: "Evaluation of Cement Systems for Permafrost," *J. Cdn. Pet. Tech.* (Jan.-March 1974) 19.

Moses, P. L.: "Geothermal Gradients," *Drill. and Prod. Prac.,* API (1961) 57.

Murphy, B.: "The Use of Acid-Soluble Cement in Injection-Well Completions," *Drill. and Prod. Prac.,* API (1955) 459 (abstr.); *World Oil* (April 1956) 278.

Murphy, W. C., and Smith, D. K.: "A Critique of Filler Cements," *J. Pet. Tech.* (Aug. 1967) 1011-1016.

Neighbors, G. R., and Cromer, S.: "Some Factors Controlling the Pumping Time of Oilwell Cement," *Petroleum Technology* (Nov. 1941) **4,** No. 6, 6 pp.

Nelson, T. W., and McNiel, J. S., Jr.: "Oil Recovery by Thermal Methods," *Pet. Eng.,* Part I (Feb. 1959) B27; Part II (March 1959) B75.

"New Cement Compound Seals Off Gas Zone," *Drilling* (March 1956) **17,** No. 5, 152.

"New Cementer for All High Pressure Cementing," *Calif. Oil World* (Jan. 10, 1939) **32,** No. 1, 39.

"New Cementing Method Stops Inter-Zone Communication," *World Oil* (Aug. 1, 1969) 39.

"New Cement System Developed," *Drilling* (Dec. 1955) **17,** No. 2, 78.

Newman, K.: "The Design of Concrete Mixes With High Alumina Cement," *The Reinforced Concrete Review* (March 1960) **5,** No. 5.

"New Record Set," *Oil and Gas J.* (Jan. 23, 1956) 43.

"New Squeeze Costs Less" *Oil and Gas J.* (Nov. 12, 1956) 136.

"New Tool Improves Cement Jobs on Multiples," *Oil and Gas J.* (April 1, 1968) 95.

"New Unit Developed for 10,000-Pound Cement Operations," *Drilling* (Aug. 1947) **8,** No. 10, 58.

Nickles, S. K.: "An Instrument for Measuring the Density of Air-Entrained Fluids," paper SPE 4092 presented at SPE-AIME 47th Annual Fall Meeting, San Antonio, Tex., Oct. 8-11, 1972.

Notes on Casing and Cementing, 2nd ed., Pet. Industry Training Service, Dept. of Extension, U. of Alberta, Edmonton, Canada (1956).

"Now a Basic Oil Well Cement: A Report from the API Standardization Committee 10," *Oil and Gas J.* (May 31, 1965).

Oatman, F. W.: "Water Intrusion and Methods of Prevention in California Oil Fields," *Trans.,* AIME (1915) **48,** 627-650.

Oberg, C. H., and Master, R. W.: "The Determination of Stresses in Oil Well Casing In Place," *Pet. Eng.* (June 1947) 149.

O'Brien, T. B.: "What It Takes To Drill Ultra-Deep Wells Successfully," *World Oil* (Aug. 1, 1973) 30-34.

"Oil Well Cement — Dowell Launches a Luminous Cement," *Chem. and Eng. News* (Jan. 9, 1956) **34,** No. 2, 178.

"Oil-Well Cementing," *Finding and Producing Oil,* API, Dallas (1939) 103.

"Oil Well Cementing Practice in California," *Petroleum World (Calif.)* (Dec. 1928) **13,** 76.

Oil-Well Cementing Practices in the United States, Committee on Drilling and Production, API, New York (1959).

"Oil Well Cementing Problems Made Easy," *Petroleum World* (April 1941) **38,** No. 4, 39-40.

Oil Well Cement Manual, Halliburton Oil Well Cementing Co., Duncan, Okla. (1970).

Oliphant, S. C., and Farris, R. F.: "A Study of Some Factors Affecting Gun Perforating," *Trans.,* AIME (1947) **170,** 225-242.

O'Neal, J. E.: "The Effect of High Temperatures and Pressures on the Setting Time of Oil Well Cements," MS thesis, U. of Oklahoma, Norman (1952); *Prod. Monthly* (abstr.) (May 1954) **18,** No. 7, 4.

"On Tour," Union Oil Co. of California (Nov.-Dec. 1952).

Ostroot, G. W., and Donaldson, A. L.: "Subsurface Disposal of Acidic Effluents," paper SPE 3201 presented at SPE-AIME Evangeline Section Regional Meeting, Lafayette, La., Nov. 9-10, 1970.

Ostroot, G. W., and Shryock, S. H.: "Cementing Geothermal Steam Wells," *J. Pet. Tech.* (Dec. 1964) 1425-1429; *Trans.,* AIME, **231.**

Ostroot, G. W., and Walker, W. A.: "Improved Compositions for Cementing Wells With Extreme Temperatures," *J. Pet. Tech.* (March 1961) 277-284; *Trans.,* AIME, **222.**

Outmans, H. D.: "Mechanics of Differential Pressure Sticking of Drill Collars," *Trans.,* AIME (1958) **213,** 265-274.

Owsley, W. D.: "Cementing Practices in the Completion of High-Pressure Wells," Halliburton Oil Well Cementing Co. Report, Duncan, Okla. (1936) 4 pp.

Owsley, W. D.: "High-Pressure Pump Designed for Cementing Operations," *Pet. Eng.* (June 1938) 37-38.

Owsley, W. D.: "High-Pressure Pump for Oil Well Cementing," *Steel* (Nov. 20, 1939) **105,** 70-71; *Nickel Steel Topics* (Feb. 1939) **8,** No. 1, 9.

Owsley, W. D.: "Improved Casing Cementing Practices in the United States," *Oil and Gas J.* (Dec. 15, 1949) 76.

Owsley, W. D.: "Pumping Equipment for Oil-Well Cementing." *Mech. Eng.* (May 1942) **64,** No. 5, 388-390.

Owsley, W. D.: "Twenty Years of Oilwell Cementing," *J. Pet. Tech.* (Sept. 1953) 17-18.

Owsley, W. D., and Moeller, R. E.: "Technique of Pressure Cementing in the Petroleum, Mining, and Construction Industries," *Coal Technology* (Aug. 1948) 9-15.

Pardue, G. H., Morris, R. L., Gollwitzer, L. B., and Morgan, J. H.: "Cement Bond Log — A Study of Cement and Casing Variables," *J. Pet. Tech.* (May 1963) 545-554.

Parker, P. N.: "Cementing Successful at Low Displacement Rates," *World Oil* (Jan. 1969) 93-95.

Parker, P. N., Ladd, B. J., Ross, W. N., and Wahl, W. W.: "An Evaluation of a Primary Cementing Technique Using Low Displacement Rates," paper SPE 1234 presented at SPE-AIME 40th Annual Fall Meeting, Denver, Colo., Oct. 3-6, 1965.

Parker, P. N., and Wahl, W. W.: "Expanding Cement — A New Development in Well Cementing," *J. Pet. Tech.* (April 1966) 559-564.

Parsons, C. P.: "Condition of Hole Is Important in Running Cement Plugs," *Oil Field Eng.* (Dec. 1, 1928) **4,** No. 6, 25-29.

Parsons, C. P.: "Fibrous Material in Cement and Drilling Fluids To Prevent Loss of Cement or Mud Into Formations," *Pet. Eng.* (Nov. 1935) 30-31.

Parsons, C. P.: "Plug-Back Cementing Methods," *Trans.,* AIME (1936) **118,** 187-194.

Parsons, C. P.: "Relation of Drilling Mud and Cementing in Setting Long String of Casing in Rotary Drilled Holes," *Oil and Gas J.* (Jan. 3, 1935) 57.

Patchen, F. D.: "Reaction and Properties of Silica-Portland Cement Mixtures Cured at Elevated Temperatures," *Trans.,* AIME (1960) **219,** 281-287.

Patchen, F. D., Burdyn, R. F., and Dunlap, I. R.: "Water-in-Oil Emulsion Cements," *Trans.,* AIME (1959) **216,** 252-257.

Pavlich, J. P., and Wahl, W. W.: "Field Results of Cementing Operations Using Slurries Containing a Fluid-Loss Additive for Cement," *J. Pet. Tech.* (May 1962) 477-482.

Peret, J. W.: "Casing String Design, Handling, and Usage," *Fundamentals of Oil and Gas Production,* 3rd ed., The Petroleum Engineer Publishing Co., Dallas (June 1959).

Peret, J. W.: "Comparison of Casing Landing Methods," *Pet. Eng.* (Oct. 1953) B101.

"Performance Properties of Casing and Tubing," *API Bull. 5C2,* 16th ed., API, Dallas (March 1975).

Perkins, A. A., and Double, E.: "Method of Cementing Oil Wells," U. S. Patent No. 1,011,484 (Dec. 12, 1911), filed Oct. 27, 1909.

Perkins, T. K., Rochon, J. A., and Knowles, C. R: "Studies of Pressures Generated Upon Refreezing of Thawed Permafrost Around a Wellbore," *J. Pet. Tech.* (Oct. 1974) 1159-1166; *Trans.,* AIME, **257.**

"Permanent Type Well Completions," *World Oil* (March 1954) 168.

Pettiette, R., and Goode, J.: "How To Cement Multiple, Tubingless Wells Successfully," *Oil and Gas J.* (March 12, 1962) 110-112.

Pickett, G. R.: "Acoustic Character Logs and Their Applications in Formation Evaluation," *J. Pet. Tech.* (June 1963) 659-667; *Trans.,* AIME, **228.**

Piercy, N. A. V., Hooper, M. S., and Winney, H. F.: "Viscous Flow Through Pipes with Cores," *Phil. Mag.* (1933) **15,** No. 99, 674.

Pittman, F. C., Harriman, D. W., and St. John, J. C.: "Investigation of Abrasive-Laden-Fluid Method for Perforation and Fracture Initiation," *J. Pet. Tech.* (May 1961) 489-495; *Trans.,* AIME, **222.**

Pitts, C. A.: "Oil Well Cementing," *Pet. Eng.,* Part I (Feb. 1944) 55-56; Part II (March 1944) 87.

Pollock, R. W., Beecroft, W. H., and Carter, L. G.: "Cementing Practices for Thermal Wells," paper presented at CIM 17th Annual Tech. Meeting, Edmonton, Canada, June 1966.

Porter, E. W.: "A Low Water Loss-Low Density Cement," *Drill. and Prod. Prac.,* API (1955) 465 (abstr.); paper 875-9-I presented at API Rocky Mountain Dist. Div. of Production Spring Meeting, Denver, Colo., April 12-15, 1955.

Potter, A. R., and Louthan, J. H.: "The Application of Low Water Loss Cement in Squeeze Cementing," *Bull. D-23,* Drilling Specialties Co., Bartlesville, Okla. (March 1959).

Poulter, T. C., and Caldwell, B. M.: "The Development of Shaped Charges for Oil Well Completion," *Trans.,* AIME (1957) **210,** 11-18.

Prats, M., and Miller, W. C.: "The Role of Technical Publications in the Advancement of Fluid Injection Processes for Oil Recovery," *J. Pet. Tech.* (Dec. 1973) 1361-1370.

"Primary Cementing of Water Strings in California (Panel Discussion)," excerpt from Official Stenographic Record, 27th Annual API Spring Meeting, Pacific Coast Dist., Los Angeles, May 7, 1954.

Proceedings, Deep Well Injection Symposium, Seventh Industrial Water and Waste Conference, Texas Water Pollution Control Assn., June 1967.

"Production Value of Scratchers and Centralizers," *World Oil* (Nov. 1948) 184.

Proposed Specifications for Large Capacity Rotary Drilled Gravel Packed Wells, 11th ed., NWWA Specifications Committee, Columbus, Ohio (1958).

Pugh, T. D.: "A Design for Cementing Deep Delaware Basin Wells," *Proc.,* Eleventh Annual Southwestern Pet. Short Course, Texas Technological College, Lubbock, April 23-24, 1964.

Pugh, T. D.: "What To Consider When Cementing Deep Wells," *World Oil* (Sept. 1967) 52.

Putman, L.: "A Progress Report on Cement Bond Logging," *J. Pet. Tech.* (Oct. 1964) 1117-1120.

"Quebracho as a Retarder for Portland Cements," Project No. 111, Halliburton Oil Well Cementing Co., Duncan, Okla. (Aug. 1942).

"Quick-Setting Gypsum Cement Aids in Landing Surface Pipe," *Oil and Gas J.* (Dec. 4, 1941) 40.

Radenti, G.: "High Fluid Loss Slurry Plugs Thief Zones," *World Oil* (May 1969).

Radenti, G., and Ghiringhelli, L.: "Cementing Materials for Geothermal Wells," *Geothermics* (1972) **I,** No. 3.

Raymond, L. R.: "Temperature Distribution in a Circulating Drilling Fluid," *J. Pet. Tech.* (March 1969) 333-341; *Trans.,* AIME, **246.**

"Recommended Practice for Testing Oil-Well Cements and Cement Additives," *API RP 10B,* API Div. of Production, Dallas (1959); 18th ed. (1972); 19th ed. (1973); 20th ed. (1974).

Regulations Pertaining to Mineral Leasing, Operations, and Pipelines in the Gulf Coast Region, Code of Federal Regulations, U. S. Dept. of the Interior, Washington (Aug. 28, 1969).

Regulations Pertaining to Mineral Leasing, Operations, and Pipelines on the Outer Continental Shelf, Code of Federal Regulations, U. S. Dept. of the Interior, Washington (April 1971).

Reid, A.: "Bibliography on Oil Well Cement," *J. Inst. Pet. Tech.* (Aug. 1935) **18,** No. 142, 734-740.

Reid, A., and Evans, J. T.: "Cements and Well Cementing; Effect of Chlorides on the Setting and Hardening of Cement," *J. Inst. Pet. Tech.* (Dec. 1932) **18,** No. 110, 992.

Reistle, C. E., Jr., and Cannon, G. E.: "Cementing Oil Wells," U. S. Patent No. 2,421,434 (June 3, 1947), filed Nov. 27, 1944. See also K. E. Wright, "Rotary Well Bore Cleaner," U. S. Patent No. 2,402,223 (June 18, 1946).

"Report of Special Subcommittee on Oil-Well Cements," *API Proc. Sec. IV (Prod. Bull. 228)* (1941) 84-85.

"Report on Cooperative Tests on Sulfate Resistance of Cements and Additives," API Mid-Continent Dist. Study Committee on Cementing Practices and Testing of Oil-Well Cements (1955).

"Report on Cooperative Tests on Thickening Times of Slow-Set Oil-Well Cements," API Committee on Standardization of Oil-Well Cements (1955).

"Retrievable Cementer Used for Squeezing and Acidizing," *Drilling* (Dec. 1, 1950) **12,** No. 2A, 144.

Reynolds, J. T.: "Use of Gel-Type Cement To Cement Casing," *Drill. and Prod. Prac.,* API (1942) 130-134.

Richardson, J. G., and Stone, H. L.: "A Quarter Century of Progress in the Application of Reservoir Engineering," *J. Pet. Tech.* (Dec. 1973) 1371-1379.

Rike, J. L.: "'Obtaining Successful Squeeze-Cementing Results," paper SPE 4608 presented at SPE-AIME 48th Annual Fall Meeting, Las Vegas, Nev., Sept. 30-Oct. 3, 1973.

Rike, J. L.: "Workover and Completion Technology — A Survey," *J. Pet. Tech.* (Nov. 1971) 1375-1385.

Rike, J. L., and McGlamery, R. G.: "Recent Innovations in Offshore Completion and Workover Systems," *J. Pet. Tech.* (Jan. 1970) 17-24.

Roberts, J. B.: "API Code 32: Testing Oil-Well Cements — Applications, Limitations, and Versatility," *Drill. and Prod. Prac.,* API (1951) 250-256.

Robinson, R. L., Herrmann, U. O., and DeFrank, P.: "How Well Conditions Influence Perforations," paper presented at CIM 12th Annual Tech. Meeting, Calgary, Alta., Canada (1961).

Robinson, W. W.: "Cement for Oil Wells: Status of Testing Methods and Summary of Properties," *Drill. and Prod. Prac.,* API (1939) 567-591.

Rockwood, N. C.: "Symposium on Chemistry of Cement," *Rock Products,* Part IV (Feb. 1956) 35; Part V (March 1956) 17.

Rogers, C. L., and Patterson D. R.: "Production Throughput Efficiency Model for Prudhoe Bay," paper SPE 4686 presented at SPE-AIME 48th Annual Fall Meeting, Las Vegas, Nev., Sept. 30-Oct. 3, 1973.

Rogers, L. C.: "For Multiple-Tubingless Wells: New Primary Cementing Technique Shows Promise," *Oil and Gas J.* (July 12, 1956) 86-90.

Rogers, W. F.: *Composition and Properties of Oil Well Drilling Fluids,* 3rd ed., Gulf Publishing Co., Houston (1963) 327.

Root, R. L., and Calvert, D. G.: "Properties of Expanding Cements for Subsurface Applications," paper SPE 3346 presented at SPE-AIME Rocky Mountain Regional Meeting, Billings, Mont., June 2-4, 1971.

Rordam, S., and Willson, C.: "Sulphate-Resistant Cement," *Petroleum Technology* (Feb. 1939) **2,** No. 1, 3pp.

Ross, W. M.: "Low Rate Displacement Solves Tough Cementing Jobs," *Pet. Eng.* (Nov. 1965) 74-79.

Ross, W. M., and Moore, E. W.: "Field Application of Low-Fluid-Loss Cement Systems," paper presented at API meeting, Denver, Colo., April 11-13, 1962.

Ruedrich, R. A., and Perkins, T. K.: "A Study of Factors Influencing the Mechanical Properties of Deep Permafrost," *J. Pet. Tech.* (Oct. 1974) 1167-1177; *Trans.,* AIME, **257.**

Ruedrich, R. A., and Perkins, T. K.: "The Mechanical Behavior of Synthetic Permafrost," *Soc. Pet. Eng. J.* (Aug. 1973) 211-220; *Trans.,* AIME, **255.**

Rule 13, Casing and *Rule 14, Plugging,* Railroad Commission of Texas, Oil and Gas Div., Austin (Oct. 1, 1970).

"Running and Cementing Liners in the Delaware Basin, Texas," *API Bull. D17,* 1st ed. (Dec. 1974).

Rust, C. F., and Wood, W. D.: "Laboratory Evaluations and Field Testing of Silica-CMHEC-Cement Mixtures," *J. Pet. Tech.* (Nov. 1960) 25-29.

Saunders, C. D., and Brown, B. D.: "New Materials for Cementing Wells," *Pet. Eng.* (Oct. 1952).

Saunders, C. D., and Nassbaumer, F. W.: "Trends in Use of Low-Weight Slurries," *Drill. and Prod. Prac.,* API (1952) 189-200.

Saunders, C. D., and Walker, W. A.: "Strength of Oilwell Cements and Additives Under High Temperature Well Conditions," paper 390-G presented at SPE-AIME 29th Annual Fall Meeting, San Antonio, Tex., Oct. 17-20, 1954.

Savins, J. G.: "Generalized Newtonian (Pseudoplastic) Flow in Stationary Pipes and Annuli," *Trans.,* AIME (1958) **213,** 325-332.

Savins, J. G., and Roper, W. F.: "A Direct-Indicating Viscometer for Drilling Fluids," *Drill. and Prod. Prac.,* API (1954) 7.

Sawdon, W. A.: "California Practice in Casing Deep Wells," *Pet. Eng.* (April 1941) 23-26.

Sawdon, W. A.: "Cementing Casing Under Adverse Conditions Is Routine Practice in Foreign Fields," *Pet. Eng.* (Oct. 1940) 67.

Sawdon, W. A.: "Cementing Practices: Casing Equipment and Cement Require Attention," *Calif. Oil World* (June 25, 1936) **28,** No. 49, 13-15.

Sawdon, W. A.: "Cementing Practices in California Oil Fields — Multiple Zone and Squeeze Jobs (II)," *Calif. Oil World* (Aug. 1940) **33,** 3.

Sawdon, W. A.: "Electrically Operated Gun Proves Effective for Perforating Cemented Casing in Hole," *Pet. Eng.* (Feb. 1933) 41-42.

Sawdon, W. A.: "Methods of Cementing Wells To Reduce Gas-Oil Ratio," *Pet. Eng.* (March 1939) 33-36.

Sawdon, W. A.: "New World's Drilling and Production Depth Records Established in California," *Pet. Eng.* (April 1938) 58-60.

Sawdon, W. A.: "Pressure-Time Chart Taken During Cementing Provides Valuable Well Record," *Pet. Eng.* (Jan. 1935) 27.

Sawdon, W. A.: "The Use of Sand in Oil-Well Cement," *Pet. Eng.* (March 1941) 23-26.

Sawdon, W. A.: "Use of Oil-Well Cement Containing Prepared Bentonite," *Pet. Eng.* (March 1939) 33-36.

Saye, F.: "A New Completion Tool To Prevent Casing and Cement Sheath Damage," API Div. of Production, Dallas (March 1963).

Saye, F.: "Current Cementing Practices in Pecos, Reeves Counties, Delaware Basin," *The Drilling Contractor* (April-May 1962).

Schmalz, J. P.: "Use of Pozzolan Cement in Prentice Field," *Pet. Eng.* (Oct. 1953) B88.

Schoenfeld, C. W.: "American Oil Well Cementing Practice," *Petroleum Bull. 18,* ASME (Sept. 1952) 13 pp.

Schremp, F. W., Chittum, J. F., and Arczynski, T. S.: "Use of Oxygen Scavengers To Control External Corrosion of Oil-String Casing," *J. Pet. Tech.* (July 1961) 703-711; *Trans.,* AIME, **222.**

Schuh, F. J.: "Failures in the Bottom Joints of Surface and Intermediate Casing Strings," *J. Pet. Tech.* (Jan. 1968) 93-101; *Trans.,* AIME, **243.**

"Scientific Oil Well Cementing," Company Bulletin, Halliburton Oil Well Cementing Co., Duncan, Okla. (July 1925).

Sclater, K. C.: "Gulf Coast Conditions Render Difficult the Cementing of Large Diameter Casing," *Pet. Eng.* (Feb. 1933) 31.

Scott, J. B., and Brace, R. L.: "Cement Bonding Is Improved With Coated Casing," *Oil and Gas J.* (Aug. 1, 1966) 136-142.

Scott, J. B., and Brace, R. L.: "How Coated Casing Reduces Primary Cementing Failures," *World Oil* (Aug. 1, 1966) 66-69.

Scott, J.: "Completion Technology Includes Plugging, Too," *Pet. Eng.* (Feb. 1971) 46.

Scott, J.: "Deep Hole Upsurge Continues Despite Rise in Drilling Costs," *Pet. Eng.* (March 1967) 51.

Scott, J.: "U. S. Deep Drilling Continues at Record Pace," *Pet. Eng.* (March 1974) 26.

Scott, P. P., Jr., Lummus, J. L., and Howard, G. C.: "How To Use Bentonite-Diesel Oil Cement Mixtures To Restore Lost Circulation," *Oil and Gas J.* (Aug. 2, 1954) 105.

Scott, P. P., Jr., Lummus, J. L., and Howard, G. C.: "Methods for Sealing Vugular and Cavernous Formations," *The Drilling Contractor* (Dec. 1953) 70-74.

Shaver, F. J.: "Cementing," *J. Inst. Pet. Tech.* (Feb. 1929) **15,** No. 72, 131-132.

Sheldon, D. H.: "Factors in a Successful Cement Job," *Petroleum World* (April 1936) **33,** No. 4, 50.

Shell, F. J.: "The Effect of Salt on DE Cement," *Cdn. Oil and Gas Ind.* (March 1957) **10,** No. 3, 64-67.

Shell, F. J., Hurley, J. R., Bergman, W. E., and Fisher, H. B.: "Low Density Oil Well Cements," *World Oil* (Sept. 1956) 131.

Shell, F. J., and Wynne, R. A.: "Application of Low-Water-Loss Cement Slurries," paper 875-12-I presented at API Rocky Mountain Dist. Div. of Production Spring Meeting, Denver, Colo., April 1958.

Shidel, H. R.: "Cement Plugging for Exclusion of Bottom Water in the Augusta Field, Kansas," *Trans.,* AIME (1919) **61,** 598-610.

Shryock, S. H., and Cunningham, W. C.: "Low Temperature (Permafrost) Cement Composition," *Drill. and Prod. Prac.,* API (1969) 48-55.

Shryock, S. H., and Slagle, K. A.: "Problems Related to Squeeze Cementing," *J. Pet. Tech.* (Aug. 1968) 801-807.

Shumate, H. J.: "Lost Circulation — Causes and Remedies," *Pet. Eng.* (Nov. 1951) B81.

Silcox, D. E., and Rule, R. B.: "Special Factors Must Be Considered in Selection, Specification, and Testing of Cement for Oil Wells," *Oil Weekly* (July 29, 1935) **78,** No. 7, 21; "Cement for Oil Wells," *Petroleum Times* (Aug. 24, 1935) **34,** 195-197.

Simons, H. F.: "Calcium Chloride in Cement Speeds Setting Time," *Oil and Gas J.* (Feb. 18, 1943) 48.

Skinner, W. C.: "A Quarter Century of Production Practices," *J. Pet. Tech.* (Dec. 1973) 1425-1431.

Slagle, K. A.: "Rheological Design of Cementing Operations," *J. Pet. Tech.* (March 1962) 323-328; *Trans.,* AIME, **225.**

Slagle, K. A., and Carter, L. G.: "Gilsonite — A Unique Additive for Oil-Well Cements," *Drill. and Prod. Prac.,* API (1959) 318-328.

Slagle, K. A., and Smith, D. K.: "Salt Cement for Shale and Bentonitic Sands," *J. Pet. Tech.* (Feb. 1963) 187-194; *Trans.,* AIME, **228.**

Slagle, K. A., and Stogner, J. M.: "Techniques of Disposing of Industrial Wastes in Deep Wells," paper presented at National Pollution Control Exposition and Conference, Houston, April 1968.

Smith, D. K.: "A New Material for Deep Well Cementing," *J. Pet. Tech.* (March 1956) 59-63.

Smith, D. K.: "Cementing Procedures and Materials," *Pet. Eng.* (April 1955) B101-B108; also in *Fundamentals of Rotary Drilling,* Petroleum Publishing Co., Tulsa, Okla. (1956) 71.

Smith, D. K.: "Physical Properties of Gel-Cements," *Pet. Eng.* (April 1951) B7-B12.

Smith, D. K.: "Report on Tests Made To Determine the Effect of Salt Water on Bentonite Cement Slurries," Halliburton Oil Well Cementing Co., Duncan, Okla. (May 1948).

Smith, D. K.: "Utilization of Fly Ash in the Cementing of Wells," *Mineral Processing* (Aug. 1967).

Smith, D. K.: "Well Cementing Method," U. S. Patent No. 3,227,213 (Jan. 4, 1966).

Smith, D. K., and Leon, L.: "Solving Hot Hole Cementing Problems," *Drilling* (July 1957) **18,** No. 9, 64.

Smith, J. H.: "Production and Utilization of Geothermal Steam," *N. Z. Engineering* (Oct. 1958) **13,** No. 10, 354-375.

Smith, R. C., and Calvert, D. G.: "The Use of Sea Water in Well Cementing," *J. Pet. Tech.* (June 1975) 759-764.

Sneddon, R.: "New Multi-Stage Cementing Technique," *Pet. Eng.* (June 1946) 152.

"Specifications for Casing, Tubing, and Drill Pipe," *API Spec. 5A,* API, Dallas, 31st ed. (April 1971); 32nd ed. (March 1973); Supplement 2 (March 1975).

"Specifications for Oil-Well Cements and Cement Additives," *API Standards 10A,* 20th ed., API, New York (1975).

Spicer, H. C.: "Rock Temperatures and Depths to Normal Boiling Point of Water in the U.S.," *Bull.*, AAPG (1936) **20,** No. 3, 270-279.

"Squeeze Cementing and Relative Merits of Well Completion by Gun Perforation and Screens," *API Proc. Sec. IV (Prod. Bull. 222)* (1938) 89-92.

"Squeeze Cementing Wells Permanently Completed," *World Oil* (March 1954) 190-198.

"Standard Procedure for Evaluation of Well Perforators," *API RP 43,* 2nd ed., API Div. of Production, Dallas (1971).

"Standards on Cement," Spec. No. C115-58, American Society for Testing Materials (1958).

"Standards on Cement, Specification for Portland Cement," Spec. No. C-150-56, American Society for Testing Materials (1958).

Stearns, G. M.: "Chemical Treatment of Mud for Cement Contamination," *Oil and Gas J.* (May 4, 1944) 79.

Stearns, G. M.: "Large Casing May Require Anchorage When Cementing," *Oil and Gas J.* (July 8, 1944) 151.

Stearns, G. M.: "Plugging Back With Cement," *Oil and Gas J.* (Sept. 2, 1944) 63.

Stearns, G. M.: "Stage-Cementing Advantages," *Oil and Gas J.* (April 16, 1942) 53.

Stearns, G. M.: "Theory of Squeeze Cementing," *Oil and Gas J.* (Aug. 12, 1944) 73.

Stearns, G. M.: "Two-Plug Cementing Process," *Oil and Gas J.* (Aug. 5, 1944) 73.

Steinour, H. H.: "The Reactions and Thermochemistry of Cement Hydration at Ordinary Temperatures," *Proc.,* Third Intl. Symposium on the Chemistry of Cement, London (1952) 261-265.

Sterne, W. P.: "Pozzolans for Oil-Well Cements," *Oil and Gas J.* (July 7, 1952) 62.

Stogner, J. M.: "Underground Disposal of Solid Wastes," paper presented at 21st Oklahoma Industrial Waste Conference, Stillwater, March 1970.

Stone, W. H., and Christian, W. W.: "The Inability of Unset Cement to Control Formation Pressure," paper SPE 4783 presented at SPE-AIME Symposium on Formation Damage Control, New Orleans, La., Feb. 7-8, 1974.

Stonley, R.: "Discussion of Thermal Considerations in Permafrost," *Proc.,* Geol. Seminar on the North Slope of Alaska, Pacific Section AAPG, Los Angeles (1970).

Stout, C. M., and Wahl, W. W.: "A New Organic Fluid-Loss-Control Additive for Oilwell Cements," *J. Pet. Tech.* (Sept. 1960) 20-24.

"Strata-Crete for Lighter Cement Slurries," Great Lakes Carbon Corp., Houston (1951) 10 pp.

"Stresses on a Centralizer," Weatherford Oil Tool Co., Houston (1974).

Striebel, W.: "Development in the Scope of Deep Well Cements," Dyckerhoff Zementwerke, Wiesbaden, Germany (1962).

Stude, D. L.: "High Alumina Cement Solves Permafrost Cementing Problems," *Pet. Eng.* (Sept. 1969) 64.

Swayze, M. A.: "Effects of High Temperatures and Pressures on Strengths of Oil Well Cements," *Drill. and Prod. Prac.,* API (1954) 72; *Oil and Gas J.* (Aug. 2, 1954) 103-105.

Swigert, T. E., and Schwarzenbek, F. X.: "Petroleum Engineering in the Hewitt Oil Field, Oklahoma," USBM, State of Oklahoma, and Ardmore Chamber of Commerce (Jan. 1921).

"Symposium on Grouting," *Coal Technology,* AIME (Aug. 1948) 23 pp.

"Symposium on Production Equipment Progress, Rotating Improves Cement Bond," *Pet. Eng.* (Oct. 1968) 46-47.

"Symposium on Use of Pozzolanic Materials in Mortars and Concretes," *Special Tech. Pub. No. 99,* ASTM, Philadelphia, Pa. (1949).

Tausch, G. H.: "Squeeze Cementing With Permanent-Type Completions," *Oil-Well Cementing Practices in the United States,* API, New York (1959) 161-175.

Tausch, G. H., and Kenneday, J. W.: "Permanent-Type Dual Completions," *Pet. Eng.* (March 1956) B24-B31.

Tausch, G. H., and McDonald, P.: "Permanent-Type Completions and Wireless Workovers," *Pet. Eng.* (Sept. 1956) B39.

Taylor, F. B.: "Acidizing Before Cementing Helps Form an Effective Bottom-Hole Water Block," *Oil Weekly* (Aug. 4, 1941) **102,** 17.

Technical Data on Oilwell Cements, Lone Star Cement Corp., Dallas (1937).

Teplitz, A. J., and Hassebroek, W. E.: "An Investigation of Oil Well Cementing," *Drill. and Prod. Prac.,* API (1946) 76-101; *Pet. Eng. Annual* (1946) 444.

Testing Oil-Well Cements and Cement Additives, 17th ed., *API RP-10B,* API (April 1971).

Texter, H. G.: "Why Oil Well Tubing and Casing Fail — By Wear, Erosion, Buckling, Torsion and Corrosion," *Oil and Gas J.* (Aug. 29, 1955).

Texter, H. G.: "Why Oil Well Tubing and Casing Fail in Tension and Why They Collapse," *Oil and Gas J.* (July 4, 1955).

Texter, H. G.: "Why Oil Well Tubing and Casing Fail — Why They Burst, Leak, Crush, and Why Last-Engaged Thread Fails," *Oil and Gas J.* (Aug. 1, 1955).

"The Amounts of Cement (94-Lb. Sacks) Necessary To Form a Cement Plug," *Pet. Eng.* (July 1937) 117.

"The Deep Ones," *Drilling-DCW* (July 1973) 26.

"The Effects of Drilling-Mud Additives on Oil-Well Cements," *API Bull. D-4,* API, Dallas (1951).

The Producing Industry in Your State, IPAA Yearbook (1972).

"The Use of Portland-Pozzolan Cement by the Bureau of Reclamation," *J. American Concrete Institute* (Oct. 1949).

Thomas, P. D.: "Steels for Oilwell Casing and Tubing — Past, Present, and Future," *J. Pet. Tech.* (May 1963) 495-499.

Thompson, G. D.: "Effects of Formation Compressive Strength on Perforator Performance," *Drill. and Prod. Prac.*, API (1962) 191-197.

Thorvaldson, W. M.: "Low Temperature Cementing," paper presented at CIM 13th Annual Tech. Meeting, Calgary, Canada (1962).

Timoshenko, S. P., and Gere, J.: *"Theory of Elastic Stability,* 2nd ed., McGraw-Hill Book Co., Inc., New York (1961).

Tinker, N. A., Jr.: "Planning, Running, and Cementing Casing Strings," *Pet. Eng.* (July 1949) B11-B16.

Tixier, M. P., Loveless, G. W., and Anderson, R. A.: "Estimation of Formation Strength From the Mechanical-Properties Log," *J. Pet. Tech.* (March 1975) 283-293.

Torrey, P. D.: "Correcting Cementing Failures," *International Oil* (Oct. 1940) **2,** 24-35.

Torrey, P. D.: "Progress in Squeeze Cementing Application and Technique," *Oil Weekly* (July 29, 1940) **98,** 68.

Torrey, P. D.: Recent Developments in the Technique of Oil Well Cementing," *Drilling* (May 1941) **2,** 30.

Torrey, P. D.: "Squeeze Cementing and Relative Merits of Well Completion by Gun-Perforation and by Screens," *API Proc. Sec. IV (Prod. Bull. 222)* (1938) **19,** 89-92.

Tough, F. B.: "Method of Shutting Off Water in Oil and Gas Wells," *Bull. 163,* USBM; *Petroleum Technology* (1918) **46,** 122.

Tough, F. B.: "Some Fundamentals To Be Observed in the Cementing of Oil Wells," *Natl. Petroleum News* (Jan. 3, 1923) **15,** 75-78.

Tragesser, A. F., Crawford, P. B., and Crawford, H. R.: "A Method for Calculating Circulating Temperatures," *J. Pet. Tech.* (Nov. 1967) 1507-1511; *Trans.,* AIME, **240.**

Tragesser, A. F., and Parker, P. N.: "Using Improved Technology To Obtain Better Cement Jobs on Deep, Hot Liners," *J. Pet. Tech.* (Nov. 1972) 1307-1313.

Underwood, D., Broussard, P., and Walker, W.: "Long Life Cementing Slurries," paper presented at API Southwestern Dist. Div. of Production Spring Meeting, Dallas, March 10-12, 1965.

"U. S. Deep Wells Decline in 1974," *Pet. Eng.* (March 1975) 27.

"Use of Bulk Cement in Wells," *Pet. Eng.* (July 1940) 74.

"Use of Gel Cement in the Cementing of Casing," *Calif. Oil World* (March 30, 1944) **347,** No. 6, 27.

Van Tuyl, I.: "Plugging Back by Squeeze-Cementing in Open Limestone Formation," *Pet. Eng.* (Oct. 1941) 29-30.

Vellinger, E.: "Retarded-Setting Slag, Clinker Cements and Their Behavior in Deep Drilling," *Petroleum Times* (May 18, 1951) 398; paper presented at Third World Pet. Cong., The Hague, May 28-June 6, 1951.

Vietti, W. B., and Oberlin, W. A.: "The Action of Accelerators in Cementing Off Bottom-Hole Water," *Oil Field Eng.* (Feb. 1928) **3,** 25-28.

Von der Ahe, K. L.: "Operating Problems in Oil Exploration in the Arctic," *Pet. Eng.* (Feb. 1953) B12.

Waggoner, H. F.: "Additives Yield Heavy High-Strength Cements with Low Water Ratios," *Oil and Gas J.* (April 13, 1964) 109-111.

Walker, A. W.: "Study of Oil Well Squeeze Cementing Operations," thesis, U. of Missouri, Rolla (1949).

Walker, T.: "A Full-Wave Display of Acoustic Signal in Cased Holes," *J. Pet. Tech.* (Aug. 1968) 811-824.

Walker, W. A.: "Cementing Compositions for Thermal Recovery Wells," *J. Pet. Tech.* (Feb. 1962) 139-142.

Warner, D. L.: "Deep-Well Disposal of Industrial Wastes," *Chem. Eng.* (Jan. 4, 1965) 73-78.

Water Well Services, company brochures, Halliburton Services, Duncan, Okla. (1961).

Webber, J. C.: "A Report on Gel Cementing," *Drill. and Prod. Prac.,* API (1952) 21-35.

Weber, G.: "Cementing Methods Designed To Prevent Cement Contamination in Casing," *Oil and Gas J.* (May 24, 1965) 94.

Weiler, J. E.: "Apparatus for Testing Cement," U. S. Patent No. 2,122,765 (July 5, 1938), filed May 15, 1937.

Weisend, C. F.: "Method of Composition for Cementing Wells," U.S. Patent 3,132,693 (May 12, 1964) filed Dec. 26, 1961.

Weisend, C. F.: "Method and Composition for Cementing Wells," U.S. Patent 3,359,225 (Dec. 19, 1967) filed Aug. 26, 1963.

"Well Completions — A Look Forward," (Panel Discussion) *J. Pet. Tech.* (Jan. 1971) 128-134.

West, E. R.: "Casing, Cementing Are Keys to Successful Drilling," *Oil and Gas J.* (May 24, 1965).

West, E. R.: "Engineered Casing, Primary Cementing Programs, Delaware Basin," *Pet. Eng.* (Feb. 1967) 71.

West, E. R.: "In Deep Delaware Basin: Casing, Cementing are Keys to Successful Drilling," *Oil and Gas J.* (May 24, 1965) 94.

West, E. R., and Lindsey, H. E., Jr.: "How To Run and Cement Liners in Ultra-Deep Wells," *World Oil* (June 1966) 101-106.

"What We're Learning From Steam," *Pet. Eng.* (Oct. 1961).

Wheeler, R., Jr., and Moriarty, D. G.: "World's Longest/Strongest Casing Set," *Pet. Eng.* (May 1969) 105.

White, F. L.: "Casing Can Be Cemented in Permafrost Area," *World Oil* (Dec. 1953) 119.

White, F. L.: "Setting Cements in Below Freezing Conditions," *Pet. Eng.,* Part 1 (Aug. 1952) B7; Part 2 (Sept. 1952) B59.

White, R. J.: "Lost Circulation Materials and Their Evaluation," *Drill. and Prod. Prac.,* API (1956) 352-359.

Wieland, D. R., Calvert, D. G., and Spangle, L. B.: "Design of Special Cement Systems for Areas with Low Fracture Gradients," paper presented at API Southwestern Dist. Div. of Production Spring Meeting, Lubbock, Tex..

March 12-14, 1969.

Wilde, H. D., Jr.: "The Cementing Problems on the Gulf Coast," *Trans.,* AIME (1930) **86,** 371-381.

Wilkinson, W. J., and Alexander, G. T.: "How To Get Good Primary Cement Jobs," *World Oil,* Part I (July 1964) 114; Part II (Aug. 1964) 65.

Willhite, G. P., and Dietrich, W. K.: "Design Criteria for Completion of Steam Injection Wells," *J. Pet. Tech.* (Jan. 1967) 15-21.

Williams, N.: "Concrete Pipe Used for Surface Casing in Two Mid-Continent Tests," *Oil and Gas J.* (Aug. 12, 1948) 66.

Williams, N.: "New Drilling Record Set for Deep Oklahoma Well," *Oil and Gas J.* (May 19, 1952) 134.

Willson, C: "Gel Cements and Their Application in Oil-Well Cementing Operations," *Drill. and Prod. Prac.,* API (1941) 293-299.

Winn, R. H., Anderson, T. O., and Carter, L. G.: "A Preliminary Study of Factors Influencing Cement Bond Logs," *J. Pet. Tech.* (April 1962) 369-372.

Wittekindt, W.: "Deep Drilling Cements: The Behavior of Deep Drilling Cement in the Borehole," *Erdol u. Hohle* (1954) **7,** 203-207; *Chem. Abstr.* (1954) **48,** 11765-e.

"World's Longest Casing String — To 22,919 Ft.," *Pet. Eng.* (July 15, 1958) B60.

Youngblood, H. L.: "New Cement Amplifies Improved Performance," *Pet. Eng.* (March 1967) 86-95.

Young, C. M.: "Casing and Cementing," *Natural Gas Course,* U. of Kansas, Lawrence (1934).

Young, V. R.: "Well Workover With Remedial Rig," Pet. Eng. Refresher Course 4 — Individual Well Analysis, Los Angeles Basin Section of SPE-AIME (1967).

"Yowell Service Announces New Shoe Squeeze Cementer," *Calif. Oil World* (Jan. 26, 1940) **33,** No. 2, 46-47.

Zaba, J., and Doherty, W. T.: "Oil Well Cementing," *Practical Petroleum Engineer's Handbook,* 2nd ed. (1939) 257-284.

Zinkham, R. E., and Goodwin, R. J.: "Burst Resistance of Pipe Cemented Into the Earth," *J. Pet. Tech.* (Sept. 1962) 1033-1040.

Author Index

(List of authors, companies, and organizations referred to in the Monograph text, references, and the bibliography.)

A

Aagaard, P. M., 154
Abadie, H. G., 95, 154
Adkins, J. E., 154
Akins, D. W., Jr., 154
Albright, J. D., 137, 154
Alexander, G. T., 173
Allen, J. H., 136, 154
Allen, T. O., 123, 154
Alquist, F. N., 154
American Assn. of State Highway Officials, 6
American Concrete Institute, 6
American Society for Testing Materials, 6, 14, 155, 156, 172
American Water Works Assn., 130, 131, 136
Anderson, D. N., 137, 154
Anderson, E. T., 137, 154
Anderson, F. M., 43, 102, 154
Anderson, K. E., 154
Anderson, R. A., 173
Anderson, T. O., 123, 154, 162, 163, 174
Anderson, W. L., 123, 154
Annis, M. R., 154
API: Bulletin D-4, 172
  Committee on Standardization of Oil-Well Cements, 169
  Division of Production, 14, 43, 73, 96, 102, 123, 163, 169, 172
  Mid-Continent District Study Committee on Casing Programs, 48, 158
  Mid-Continent District Study Committee on Cementing Practices and Testing of Oil-Well Cements, 169
  Publication 2564 (Metric System), 146-153
  RP 10B, 14, 43, 73, 96, 102, 111, 172
  RP 43, 123
  Specification 10D, 56, 158
  Specification 5A, 171
  Standard 5A, 48
  Standard 10A, 7, 14, 31, 43, 171
  Standardization Committee 10, 168
Arczynski, T. S., 32, 171
Arnold, R., 154
Aspdin, J., 6, 14, 154
Atterbury, J. H., 123, 154
Auberlinder, G. A., 156

B

B. J. Services, 56
B & W Incorporated, 56
Bachman, J. T., 4
Bachman, L. T., 155
Bade, J. F., 155
Bailey, J. E., 155
Baker, M. L., 155
Baker Oil Tools, Inc., 4, 55, 155
Baker, R. C., 4, 5, 155
Baker, R. T., 155
Baker, W. L., 155
Baker, W. M., 137, 155
Ball, W. F., 155
Ball, W. W., 136, 155
Banister, J. A., 102, 155
Barkis, B., 4, 155
Barnes, J., 155
Barnes, K. B., 155
Barnsdall Oil Co., 4
Barrios, J. R., 84, 166
Bateman, S. J., 82, 84, 165
Bawcom, J. W., 136, 155
Bayliss, B. P., 137, 155
Bayliss, J., 155
Beach, H. J., 4, 5, 31, 32, 88, 95, 96, 102, 155
Beale, A. F., 155
Bearden, W. G., 31, 43, 122, 128, 155, 156
Beaupre, C. J., 84, 156
Beck, R. W., 156
Becker, H., 122, 156
Beckstrom, R. C., 156
Beecroft, W. H., 169
Beeson, C. M., 156
Beirute, R., 156
Bell, H. W., 156
Bell, W. T., 156
Belter, W. G., 137, 156
Bender, A., 156
Benge, J., 156
Berdine, D., 5, 73, 122, 164
Bergman, W. E., 31, 156, 171
Berry, P. M., 165
Besse, C. P., 154
Best, W. E., 156
Bignell, L. G. E., 156
Billington, S. A., 156
Bingham, I. F., 156
Binkley, G. W., 31, 96, 156, 161
Birch, D., 4
Blanks, R. F., 156
Bleakley, W. B., 32, 156
Blenkarn, K. A., 48, 166
Blount, E. M., 137, 156
Bogue, R. H., 156
Boice, D., 32, 95, 156
Bombardieri, C. C., 137, 165
Boon, A. R., 156
Boone, D. E., 156
Boorman, R. D., 160, 166
Born, S., 157
Bosley, T. G., 15, 137, 166
Boughton, L. D., 32, 157
Bouman, C. A., 157
Bowden, C. B., 157
Bower, F. M., 166
Bowers, C. N., 48, 157
Brace, R. L., 171
Bradshaw, J. R., 157
Brice, J. W., Jr., 32, 73, 111, 157
Briggs, G. E., 157
Brighton, J. A., 157
Brisac, J., 73, 166
Brooks, F. A., Jr., 31, 96, 161
Bureau of Mines, 4
Broussard, P., 32, 73, 173
Broussard, P. J., 157
Brown, B. D., 136, 157, 170
Brown, R. J. E., 137, 157
Brown, R. W., 157
Bruist, E. H., 157
Bucy, B. D., 157
Bryan, B., 157
Bugbee, J. M., 4, 5, 157
Burba, J. L., 162
Burch, D. D., 5, 157
Burdyn, R. F., 168
Burke, R. G., 136, 157
Burke, R. H., 157
Burkhardt, J. A., 48, 157
Buster, J. L., 73, 111, 157
Byrd, E. E., 157

C

Cain, J. E., 157
Cain, P. A., 136, 157
Caldwell, B. M., 123, 157, 169
Caldwell, H. L., 136, 157
California Dept. of Conservation, 128, 157
California Research Corp., 28
Calvert, D. G., 31, 32, 170, 171, 173
Cameron, R. D., 167
Campbell, J. L. P., 157
Cannon, G. E., 5, 55, 128, 157, 158, 169
Cardwell, W. T., Jr., 158
Carnahan, D. A., 158
Carney, L. L., 158
Carter, L. G., 5, 32, 73, 84, 96, 103, 111, 122, 123, 137, 157, 158, 161, 163, 169, 171, 174
Carter, M. A., 15, 137, 166
Case, J. B., 158
Caswell, C. A., 137, 158
Chatterji, J., 73
Cheatham, J. B., Jr., 48, 73, 158
Chem Bureau, 158
Chen, R. T. H., 167
Childers, M. A., 73, 111, 158
Chittum, J. F., 32, 171
Christian, W. W., 73, 172
Christman, S. A., 158
Clark, C. R., 5, 73, 84, 111, 158
Clark, E. H., Jr., 48, 158
Clark, J. B., 4, 5, 32, 73, 95, 111, 164
Clark, R. C., Jr., 31, 158, 159
Clason, C. E., 32, 159
Clegg, J. D., 159
Coffer, H. F., 31, 159
Cole, F. W., 48, 166, 167
Coleman, J. R., 159
Collings, B. J., 162
Collins, R. E., 96, 156
Cook, C., 73, 84, 158
Copland, G. V., 159
Coppinger, J. E., 159
Corley, C. B., Jr., 73, 159, 164
Corrigan, G. L., 159
Couch, E. J., 165
Courtney, F. G., 159
Covell, K. A., 159
Covlin, R., Jr., 159
Cox, W. R., 159
Craft, B. C., 4, 5, 31, 43, 55, 159
Crawford, H. R., 157, 173
Crawford, P. B., 159, 173
Crews, S. H., 136, 159
Cromer, S., 167
Cromling, J., 137, 159
Crone, B. L., 159
Cunningham, W. C., 15, 32, 137, 159, 166, 171
Curtis, L. B., 159

D

Dale, O. O., 32, 159
Daneshy, A. A., 159
D'Arcy, N. A., Jr., 159
David, H., 159
Davie, F. E., 159
Davies, B. F., 160, 166
Davis, E. L., 5, 160
Davis, L. F., 160
Davis, R. E., 160
Davis, S. H., 30, 43, 84, 128, 160
Dawson, C. W., 160
Dawson, D. D., Jr., 160, 162
Dean, C. J., 160
DeBlanc, F. X., 160
DeFrank, P., 123, 170
DeGreer, D., 160
DeHetre, J. P., 160
Dellinger, T. B., 136, 157, 160
Denning, R. A., 160
Desch, C. H., 43, 165
DeWitt, G. B., 160
Dietrich, W. K., 160, 174
Diller, J., 32, 95, 156
Dimit, C. E., 155
Doherty, W. T., 5, 160, 174
Donaldson, A. L., 137, 168
Donaldson, E. C., 137, 160

Donaldson, F., 160
Dorsch, K. E., 160
Double, E., 5, 169
Dowell, 4
Dublin, J. R., III, 164
Dubrow, M. H., 84, 161
Dumbauld, G. K., 31, 96, 102, 156, 161, 167
Dunlap, I. R., 32, 161, 168
Dunn, T. H., 156

E

Ebel, R. E., 161
Edison, J. E., 161
Edwards, R. E., 161
Eilers, L. H., 155
Einarsen, C. A., 32, 161
Elkins, L. E., 161
EnDean, H. J., 155
Enloe, J. R., 73, 161
Esso Production Research Co., 4
Evan, G. W., 122, 158, 161
Evans, J. T., 169

F

Fagin, K. M., 161
Farris, R. F., 3-5, 30, 43, 128, 161, 168
Fast, C. R., 95, 164
Faulk, J. H., 30, 43, 128, 160
Fehrenbach, J. R., 137, 159
Fertl, W. H., 123, 161
Fisher, H. B., 31, 171
Flumerfelt, R. W., 161
Folinsbee, J. R., 136, 157
Folmer, L. W., 161
Foster, W. R., 166
Frank, W. J., Jr., 73, 161
Frawley, F. E., Jr., 155
Frosch, A., 164

G

Garfias, R., 154
Garvin, T. R., 111, 161
George, C. R., 158
Gere, J., 136, 173
Ghiringhelli, L., 169
Gibbs, M. A., 32. 73, 84, 161
Gidley, J. L., 158
Glenn, E. N., 84, 161
Glenn, W. E., 161
Goddard, R. D., 162
Godfrey, W. K., 123, 162
Goins, W. C., Jr., 5, 32, 56, 73, 88, 95, 96, 102, 155, 160, 162
Goldsmith, R. G., 162
Gollwitzer, L. B., 122, 168
Gonnerman, H. F., 162
Goode, J. M., 43, 159, 162, 169
Gooding, P., 162
Goodman, M. A., 162
Goodwin, R. J., 173
Goolsby, J. L., 32, 95, 162
Gordon, G. M., 162
Graham, C. R., 162
Graham, H. L., 73, 111, 162
Graves, E. D., Jr., 43, 159
Gray, G. R., 162
Great Lakes Carbon Corp., 32, 172
Greer, J. B., 162
Greer, R. C., 162
Gretener, P. E., 162
Griffin, W. H., 162
Griggs, G. E., Jr., 157
Grimm, M. W., 156
Grosmangin, M., 122, 162
Grossman, W., 162
Guest, R. J., 43, 162
Gulf Oil Corp., 20

H

Hall, D., 162
Hall, J. A., 162
Halliburton Co., 14, 122, 137, 154, 162, 168, 169
Halliburton, E. P., 2, 4, 5, 162
Halliburton Services, 29, 56, 84, 173
Halstead, P. E., 162
Handin, J. O., 162
Hansen, W. C., 14, 15, 163
Harcourt, G., 163
Harmon, J. A., 163
Harriman, D. W., 123, 161, 169
Harris, F. N., 96, 158, 163
Hartweg, D. G., 163, 167
Hassebroek, W. E., 4, 5, 55, 73, 103 163, 172
Hathorn, D. H., 84, 165
Hawkins, M. F., 56, 159
Hedstrom, B. O. A., 111, 163
Heimlick, W. H., 166
Heise, H., 163
Helmick, W. E., 163
Henderson, R. A., 163
Henderson, R. K., 163
Hendrickson, J. F., 48, 163
Herrington, C. G., 163
Herrmann, U. O., 123, 170
Hewitt, C. H., 32, 163
Hill, F. F., 1, 4, 163
Hills, J. O., 48, 163
Hilton, A. G., 55, 163
Hilton, P. E., 163
Hnatiuk, J., 163
Hoch, R. S., 73, 163
Hodges, C., 163
Hodges, J. W., 96, 163
Hoisington, G. T., 163
Holden, W. R., 43, 159
Holliday, G. H., 163
Holmes, B. C., 32, 73, 111, 157
Holmes, C. S., 163
Holmquist, D., 163
Holmquist, J. L., 48, 163
Hooper, M. S., 73, 169
Hopkins, H. F., 163
Horton, H. L., 102, 163
Hough, J. F., 164
Howard, G. C., 4, 5, 32, 73, 95, 102, 111, 156, 164, 171
Howell, E. P., 164
Hower, W. F., 14, 164
Huber, F. W., 5, 164
Huber, T. A., 73, 164
Hulsey, B. W., 111, 165
Humble Oil and Refining Co., 3, 4, 30, 164
Hunt, J. O., 163
Hurley, J. R., 31, 156, 171

I

Ihrig, H. K., 164
Imco Drilling Mud, Inc., 111
Ingram, R., 164
Irvine, H. R., 4

J

Jahn, D. F., 164
Jenkins, R. C., 158
Jennings, H. Y., Jr., 164
Jessen, F. W., 157, 164
Johnson, T. J., 5, 159
Jones, E. N., 164
Jones, J. B., 157
Jones, P. H., 5, 73, 122, 164
Jones, R. W., 164
Jones, W. M., 164

K

Kail, R. S., 159, 164
Kalousek, G. L., 32, 164
Kastrop, J. E., 84, 164
Keefover, S. A., 159
Keller, H. H., 165
Kellerman, W. F., 165
Kelly, E. F., 165
Kenneday, J. W., 172
Kennedy, H. B., 156, 159
Kennedy, H. T., 165
Kennedy, J. L.. 165
Kettenburg, R. L., 165
King, R. H., 165
King, T. G., 165
Kinley, M. M., 4
Kirby, C. E., 155
Kirchner, W., 165
Kirk, W. L., 84, 165
Kirkpatrick, H. L., 5, 159
Klein, A., 4, 5, 15, 165
Klinkenberg, A., 165
Kljucec, N. M., 137, 165
Knapp, I. N., 165
Knowles, C. R., 137, 169
Koch, R. D., 165
Koenig, J. B., 137, 165
Kokesh, F. P., 122, 162
Korepanov, V. N., 165
Krueger, R. F., 123, 165

L

Ladd, B. J., 32, 73, 111, 168
Lafarge Cement Co., 14
Lafuma, H., 15, 165
Land, J., 164
Lane, R. D., 31, 122, 128, 155, 156
Lane, W., 4
Larminie, F. G., 137, 165
Laudermilk, J. L., 165
Lawrence, D. K., 84, 165
Lea, F. M., 43, 165
Leon, L., 84, 165, 171
Leonardson, E. G., 165
Lerch, W., 162
Leutwyler, K., 165
Lindsey, H. E., Jr., 82, 84, 165, 173
Lord, D. L., 111, 165
Lone Star Cement Co., 4, 172
Longley, A. J., 56, 163, 166
Louthan, J. H., 169
Loveless, G. W., 173
Lubinski, A., 48, 165, 166
Ludwig, N. C., 32, 43, 137, 166
Ludwig, N. D., 14
Lummus, J. L., 32, 43, 166, 171

M

MacClain, C., 166
MacDonald, G. C., 166
Machinskii, E. K., 165
Mahoney, B. J., 84, 166
Maier, L. F., 15, 43, 122, 128, 137, 159, 166
Majani, P., 122, 162
Mallinger, M. A., 166
Mangold, G. B., 166
Manning, M., 5, 160
Manry, C. W., 32, 73, 111, 166
Manson, P. W., 167
Marquaire, R. R., 73, 166
Massey, J. F., 166
Master, R. W., 168
Mater, B. E., 166
McCray, A. W., 166
McDonald, P., 73, 172
McDowell, J. M., 123, 166
McEver, J. W., 48, 73, 158
McGhee, E., 166
McGlamery, R. G., 73, 170
McKinley, R. M., 166
McLaughlin, C., 164
McLean, J. C., 160
McLean, R. H., 32, 73, 111, 166
McLeod, H. O., 166
McNatt, J. C., 157
McNeil, J. S., Jr., 32, 166, 167
McRee, B. C., 128, 166
Melmick, W. E., 56, 166
Melrose, J. C., 166
Melton, L. L., 111, 165, 166
Menzel, D., 166
Merriam, R., 166

Messenger, J. U., 32, 103, 166
Messer, P. H., 166
Methven, N. E., 123, 162
Metzner, A. B., 106, 111, 167
Michigan Dept. of Natural Resources, 155
Miller, D. G., 167
Miller, H. H., 154, 167
Miller, W. C., 169
Millikan, C. V., 5, 95, 167
Mills, B., 5, 167
Missouri Water Well Driller Assn., 136
Moeller, R. E., 167, 168
Monaghan, P. H., 154
Montgomery, P. C., 14, 31, 95, 164, 167
Moore, E. W., 170
Moore, J. E., 32, 167
Moore, P. L., 48, 111, 161, 167
Moore, W., 167
Moran, J. P., 167
Morgan, B. E., 31, 102, 161, 167
Morgan, J. H., 122, 168
Moriarty, D. G., 83, 173
Morris, E. F., 15, 102, 137, 163, 167
Morris, R. L., 122, 168
Moses, P. L., 167
Motley, H. R., 167
Murphy, B., 167
Murphy, W. C., 167
Murray, A. S., 158
Muskat, M., 123, 166

N

Nadai, A., 48, 163
Nance, G. E., 159
Nash, F., Jr., 162
National Ready Mixed Concrete Assn., 159
National Water Well Assn., 130, 131, 136, 169
Neighbors, G. R., 167
Nelson, T. W., 167
Newman, K., 15, 167
New Zealand Geological Survey, 137, 161
Nickles, S. K., 168
Nuss, W. F., 156
Nussbaumer, F. W., 161, 170

O

Oatman, F. W., 4, 5, 168
Oberg, C. H., 168
Oberlin, W. A., 173
O'Brien, T. B., 5, 32, 88, 95, 96, 155, 162, 168
Oklahoma Corporation Commission, 4, 128, 160
Oliphant, S. C., 168
O'Neal, J. E., 168
Osborne, J. H., 159
Ostroot, G. W., 73, 137, 168
Outmans, H. D., 168
Owen, H. D., 157
Owsley, W. D., 56, 73, 168

P

Pacific Portland Cement Co., 4
Pardue, G. H., 122, 168
Parish, E. R., 166
Parker, P. N., 32, 73, 111, 168, 173
Parson, C. P., 102, 159, 168
Patchen, F. D., 32, 161, 168
Patterson, D. R., 170
Pavlich, J. P., 32, 157, 168
Pence, S. A., 43, 137, 166
Peret, J. W., 48, 169
Perkins, A. A., 1, 4, 5, 169
Perkins Cementing Co., 4
Perkins, T. K., 137, 164, 169, 170
Perry, D., 96, 161
Peterson, G., 122, 156
Petroleum Extension Service, 73
Petroleum Industry Training Service, 168
Pettiette, R., 169
Pew, T. W., 4
Phillips Petroleum Co., 4
Pickett, G. R., 123, 169
Piercy, N. A. V., 73, 169
Pilkington, P. E., 123, 161
Pittman, F. C., 123, 169
Pitts, C. A., 169
Pollock, R. W., 166, 169
Porteous, W. R., 166
Porter, E. W., 31, 169
Porter, W., 157
Portland Cement Assn., 14, 160
Potter, A. R., 169
Poulter, T. C., 123, 169
Prats, M., 169
Presley, C. K., 160
Prueger, N. J., 137, 156
Pruitt, G. T., 157
Pugh, T. D., 84, 103, 169
Putman, L., 169

Q

Quitana Petroleum Co., 4

R

Radenti, G., 169
Raghavan, R., 166
Railroad Commission of Texas, 4, 128, 170
Ramey, H. J., Jr., 166
Ramos, J., 164
Raymond, L. R., 169
Reed, J. C., 106, 111, 167
Reid, A., 169
Reistle, C. E., Jr., 5, 169
Reynolds, J. J., 31, 159, 170
Richardson, J. G., 170
Rike, J. L., 73, 95, 159, 170
Roberts, H., 167
Roberts, J. B., 170
Robinson, R. L., 123, 170
Robinson, W. W., 5, 170
Rochon, J. A., 137, 169
Rockwood, W. W., 170
Rogers, C. L., 170
Rogers, L. C., 170
Rogers, W. F., 111, 170
Root, R. L., 170
Roper, W. F., 111, 170
Rordam, S., 170
Ross, W. M., 168, 170
Ross, W. N., 32, 73, 111
Ruedrich, R. A., 170
Rule, R. B., 5, 171
Rumble, R. C., 166
Rust, C. F., 170

S

Santa Cruz Cement Co., 4
Saunders, C. D., 43, 84, 165, 166, 170
Savins, J. G., 111, 166, 170
Sawdon, W. A., 170
Saye, F., 170
Schlumberger Well Services, 4, 123, 158
Schmalz, J. P., 171
Schmieder, F. R., 165
Schoenfeld, C. W., 171
Schremp, F. W., 32, 171
Schuh, F. J., 48, 171
Schwarzenbek, F. X., 5, 172
Sclater, K. C., 171
Scott, J., 84, 166, 171
Scott, J. B., 123, 161, 171
Scott, P. O., Jr., 32
Scott, P. P., Jr., 102, 164, 171
Seth, M. S., 164
Shaver, F. J., 171
Sheldon, D. H., 171
Shell, F. J., 31, 32, 96, 156, 171
Shidel, H. R., 171
Shryock, S. H., 95, 137, 157, 162, 168, 171
Shumate, H. J., 171
Silcox, D. E., 5, 171
Silcox, E. F., 4
Simons, H. F., 171
Skinner, W. C., 171
Slagle, K. A., 32, 73, 95, 111, 137, 155, 158, 161, 171
Smeaton, J., 6
Smith, D. K., 14, 31, 32, 73, 95, 96, 103, 137, 158, 159, 167, 171
Smith, J. H., 171
Smith, R. C., 31, 171
Sneddon, R., 171
Snelson, L., 73, 84, 158
Spangle, L. B., 32, 173
Spicer, H. C., 172
Spurlock, J. W., 156
St. John, J. C., 123, 169
Standard Oil Co. of California, 4
Stanolind Oil & Gas Co., 3, 4
Stearns, G. M., 172
Steel, J. H., 158
Steinour, H. H., 172
Stephenson, A. H., 31, 159
Sterne, W. P., 172
Stogner, J. M., 137, 171, 172
Stone, H. L., 170
Stone, W. H., 172
Stonley, R., 137, 172
Stout, C. M., 32, 96, 172
Striebel, W., 172
Stude, D. L., 137, 167, 172
Sullins, R. S., 158
Superintendent of Documents, 161
Swayze, M. A., 43, 172
Swift, S. C., 163
Swigert, T. E., 5, 172

T

Tausch, G. H., 73, 95, 164, 172
Taylor, F. B., 172
Telford, A. S., 137, 165
Teplitz, A. J., 4, 5, 55, 73, 103, 162, 172
Texas Iron Works, Inc., 84, 160
Texas Water Pollution Control Assn., 137, 160, 169
Texter, H. G., 48, 172
Thomas, P. D., 172
Thompson, G. D., 123, 173
Thorvaldson, W. M., 137, 173
Timoshenko, S. P., 136, 173
Timur, A., 164
Tinker, N. A., Jr., 173
Tixier, M. P., 173
Toland, T., 84, 165
Torrey, P. D., 173
Tough, F. B., 5, 173
Tragesser, A., 156, 173
Troxell, G. E., 4, 5, 15, 165

U

Underwood, D., 32, 73, 173
Union Oil Company, 1, 5, 168
Universal Atlas Cement Co., 4, 13
U.S. Atomic Energy Commission, 28
U.S. Dept. of Interior, 128, 169
U.S. Government Printing Office, 161
U.S. Gypsum Co., 4

V

Van Tuyl, I., 173
Vellinger, E., 173
Vietti, W. B., 173
Von der Ahe, K. L., 137, 173

W

Waggoner, H. F., 31, 103, 158, 173
Wahl, W. W., 32, 73, 96, 102, 111, 157, 163, 168, 172

Walker, A. W., 173
Walker, T., 123, 154, 173
Walker, W., 32, 73, 163, 173
Walker, W. A., 43, 137, 168, 170, 173
Warner, D. L., 137, 173
Weatherford Oil Tool Co., 73, 172
Webber, C. E., 164
Webber, J. C., 173
Weber, G., 173
Wechsler, O., 166
Weichert, J. P., 162
Weiler, J. E., 4, 5, 173
Weisend, C. F., 25, 32, 173
Wells, W., 4
West, E. R., 84, 173
Westbrook, S. S. S., 167
Wheeler, R., Jr., 83, 173
Whitaker, W. W., 32, 73, 111, 166
White, F. L., 15, 137, 173
White, G. L., 159
White, R. J., 32, 173
Wieland, D. R., 32, 173
Wilde, H. D., Jr., 173
Wilkinson, W. J., 173
Willhite, G. P., 160, 174
Williams, N., 174
Willson, C., 170, 174
Winn, R. H., 123, 154, 174
Winney, H. F., 73, 169
Wittekindt, W., 174
Wood, D. B., 162, 170
Woodard, H. D., 163
Worzel, H. C., 123, 154
Wright, K. E., 4, 155
Wynne, R. A., 32, 96, 171

Y

Yates, D., 155
Young, C. M., 174
Young, F. S., Jr., 162
Young, V. R., 95, 174
Youngblood, H. L., 174

Z

Zaba, J., 174
Zimmerman, C. W., 43, 162
Zinkham, R. E., 174

Subject Index

A

Abandonment, use of cement plugs, 97, 101
Abu Dhabi, drilling and cementing regulations, 126
Accelerators, 9, 16-19, 30, 99
Acetic acid solution, 92
Acoustic:
  Energy travel, 117
  Log, single-curve, 117
  Logs, cement bond evaluations, 123
  Signals, 116
Additives:
  Cementing, 1-4, 14, 16-32
  Drilling fluid, 38, 39
Africa, early cements, 6
Aircraft bulk cement unit, 59
Alaska:
  Cook Inlet, sea water analyses, 19
  Naval Reserve area, 133
  Northern, aspects of petroleum development, 137
  Permafrost zones, 16, 134
Algeria, North Hassi-Messaoud field, 73
American Petroleum Institute (API):
  Classes of cement, A through H, 7-9
  Classes of cement, A, B, C, G, H, 12
  Classes of cement, A, G, H, K, S, M, 13
  Code 32, 3-5
  Committee on Cement Standardization, 3
  Committee on Standardization of Oil-Well Cements, 154
  Filter loss, 41, 91
  Metric Practice Guide, 2564, 146
  Mid-Continent Committee on Oil Well Cements, 3, 43
  Physical requirements for cements, 10
  Recommended Practice for Testing Oil-Well Cements and Cement Additives, 3, 73, 108, 111
  Settling tests, Class G cement, 36
  Specifications, 3, 6, 7, 9, 12, 14, 34, 45, 48
  Standards 5A, 45, 48
  Standards 10A and 10B, 34, 108, 111
  Standards 10D, 54
American Society for Testing Materials (ASTM), specifications, 7
Amount of Substance, customary units and conversion factors, 150
Andarko Basin, cementing, 79
Annular velocity, cementing operations, 104
Annulus cementing, 66
Apparent viscosity, 104, 106, 107
Applications of API classes of cements, 8
Arctic Islands, permafrost environments, 134
Arctic, permafrost cement, 14, 137
Australia:
  Cement classification, 12
  Drilling and cementing regulations, 126
Austria, drilling and cementing regulation, 126
Automatic fillup equipment, 56

B

Backpressure valve:
  For casing equipment, 49, 50, 66, 130
  Poppet type, 92
  Drillable packer, 94
Bactericides, mud additives, 39
Baffle tool, 100
Balanced method placement technique of cement plug, 99, 142
Barite:
  Material for weighting API Class D or E cement, 23, 30
  Plugs, 101-103
  Slurries, weight/volume relationships, 101
Barium sulfate, mud additives, 39
Batch mixer, for cementing, 61, 62
Belting discs, 1
Bentonite:
  As cement additive, 16, 19, 24, 25, 30, 31, 39, 72, 91, 133, 138
  Cement, 130
  Diesel oil, 32
  Effects on composition and properties of Class H cements, 20
  In delayed-set cementing, 66
  Physical and chemcal requirements, 20
  Prehydrated, 20
Big hole drilling, 136
Bingham-plastic model, 104-107, 109-111
Black oxide, for coloring cement, 28
Block squeezing, 5, 86
Blowout:
  Preventing by quickly forming a seal, 1
  Skelly's No. 1 Dillard, 2, 4
Bond logs, 116-118, 122
Bond strength, after mud contamination, 113
Bonding:
  Cement to formation, 114, 122
  Cement to pipe, 112, 122
  Considerations, 112-114
  Hydraulic, 113
  Properties of cement, 3
  Properties of various pipe finishes, 113
Bottom-hole pressure, 35, 46, 78, 86, 87, 90
Bottom-hole temperature, 35, 37, 74, 81-83, 90
Bradenhead squeeze method, 87, 90
Breakdown pressure, 44, 86, 88, 120
Bridging:
  Agents, 16, 30, 70, 89
  Effect on primary cementing, 71
  In linear/open-hole annulus, 83
Bulk blending equipment, 16
Bulk cement:
  Blending plant, 59
  Introduced, 4
  Rail transports, 59
  Station, 3, 58
  Storage, land-based, 59
  Units, aircraft and marine, 59
Bulk handling:
  Equipment, 1, 58
  Of cement, 2, 3, 58, 59
Bulk-transport truck, 3, 60
Bullets, for perforating, 119, 120, 122
Buoyant force, 67
Burst analysis of casing string, 45, 134
Butadiene styrene emulsions, 14

C

Calcium aluminate cements, 13, 14
Calcium chloride solutions:
  As cement accelerators, 3, 4, 16, 17, 30, 138
  Effect on high early strength cement, 31
  Effect on thickening time and compressive strength of API Class A cement, 17
Calcium sulfate, in cement, 9
Calcium lignosulfonate, 21, 23-25
Calcium sodium lignosulfonate, 23
Calcium sulfoaluminate admixtures, 5, 15
California:
  Bulk cement station, 3
  Cement scratcher, 4
  First cement job, 6
  Geyser area, 133
  Gun perforating, 4
  Imperial Valley, 133
  Laws for Conservation of Petroleum and Gas, 128
  Lompac field, 1
  Production test of casing, 72
  Rules controlling the cementing of wells, 125, 126
  Salton Sea area, 132
  Two-plug method, 4
  Water intrusion prevention, 5
  Water shutoff tests, 4
Caliper:
  Logs, 57, 98
  Survey instruments, 4
  Surveys, 66, 79
Caloric value, customary units and conversion factors, 151
Calorimeter, 40
Canada:
  Cement classification, 12
  Drilling and cementing regulations, 127
  Lynn Lake, 136
  Mackenzie Delta, 134

Permafrost zones, 16, 133, 137
Weight of sack of cement, 58
Capacity, customary units and conversion factors, 146
Capsule-type jet guns, 120
Carboxymethyl hydroxyethyl cellulose (CMHEC), 17, 18, 23-25, 30
Carpenter 20 alloy, 132
Casing:
Bending by salt flow, 72
Capacity, 140
Cementing, 24, 34, 35, 56, 58, 61, 73, 91, 95, 111, 138-141
Damage, 121
Deep-well, 48
Failure, 46-48, 71, 93
Intermediate, 44, 47, 48, 61, 69, 72, 77, 124, 127, 138
Landing procedures, 46-48
Leaks, 85
Loss down hole, 47
Movement during cementing, 64
Placed in compression, 46
Production, 44, 50, 61, 69, 72, 97, 124, 127, 135, 139
Programs for large-hole cementing, 129
Programs for some record wells, 75
Programs for steam-well cementing, 132
Programs for water-well cementing, 130
Protecting from shock loads, 1
Running, 46, 48
Seal, 48
Shoe, 5, 66
String design, 44-48, 50
Surface, 44, 47, 48, 50, 66, 72, 79, 97, 102, 124, 128, 132, 134, 138
Casing/cement bonding, 112
Cellophase, to control lost circulation of cements, 24, 30
Cement:
Accelerators, 3, 9, 16-19, 30, 31, 99
API, classes of, 2
Baskets, 55, 56, 100
Behind pipe, methods of locating, 114-118
Bentonite, *see* Bentonite, cement
Bonding properties, 3, 112-114
Bridging, 69
Chemical requirements, 8
Chemistry, 6, 14, 15, 43
Classifications, 6-14
Clinker, 6
Committee to study, 3
Controlled-fluid-loss, 88, 139
Curing time, 29
Dehydration, 70, 87, 91
Densified, 19, 25, 98, 102, 132
Densities, 40
Displacement, 4, 52, 76, 78
Early types, 6
Ettringite crystal growth in, 13
Expanded, 4, 5
Filter cake, 94
Filtrate analysis, 26
Filtration, 16, 94
Flow in an eccentric, 67
Flow properties, 16
Gel, 20, 27, 89, 138
Grout, 129
Gypsum, 4, 6, 9, 12, 14, 17, 18, 24, 30, 32, 135
High-early strength, 7, 8, 11, 12, 135
Incor-high-fineness, 4
Liners, 1, 35
Low-water-loss, 5, 95, 96
Measuring thickening time, 4
Mixing, 61
Node building effect, 89
Oil emulsion, 31
Placing down hole, 65, 67
Pozzolon, 6, 11, 20, 27, 81, 130
Pressure profile, 79
Properties, 4, 7, 9, 11, 16, 31, 43, 133, 136
Quantity for squeeze operations, 91, 92
Ranges for API classes, 58
Retainers, 4, 5
Retarded, 4, 11, 115
Retarders, 9, 16, 17, 23, 24, 27, 30, 31, 66, 67, 82, 99, 135
Salt, 4, 26, 21
Scratchers, 3, 4
Special additives, 27-30
Specialty types, 9-14, 30
Specifications, 3, 6, 43
Standards outside the U.S., 9
Sulfur resisting, 7, 8, 12
Slurry design, factors that influence, 33-43, 83, 90-92
Strength to support pipe, 37, 38
Selection of, to suit well requirements, 57, 58
Testing, 3, 57
Types, 91
Used at 2,000 feet, 3
Cement bond log:
Factors influencing
Schematic, 117
Study of cement and casing variable, 122
Cement plugs:
Barite plugs, 101, 102
Cement volume and slurry design, 99
Mud system, 98
Open-hole, 97-103
Placement precautions, 98
Placement techniques, 99-101
Testing, 101
Uses, 97, 98
Cementing:
Annulus, 66
Arctic, 137
Casing, 24, 34, 35, 56, 58, 61, 73, 91, 95, 111, 138-141
Charges, 138, 139
Compositions, 16, 135, 137
Deep liner, 42, 76-81
Deep-well, 14, 74-84
Definition, 1
Delayed-set, 66, 67, 80, 81
Equipment, 3, 56, 59-62, 83
Factors influencing bond, 122
Failures, factors that contribute, 57
Gas wells, 46
Historical background, 1
In permafrost environments, 133-137
Inner-string, 66
Job execution and planning, 1
Large-hole, 129, 130
Liner, 1, 35, 76-84
Materials for large-hole cementing, 130
Materials for steam-well cementing, 132
Materials for waste disposal wells, 132
Materials for water-well cementing, 131
Methods, 52
Methods: liners in high-pressure zone, 80
Methods: two-plug, 1, 4, 100, 101
Objectives of monograph, 1
Of casing, factors to be considered, 32
Offshore platform system, 60
Outside or annulus, 66
Permafrost, 137
Plug-back, 35, 102
Plugged perforations, 96
Plugs, 1-3, 36, 46, 48, 50-53, 97-103
Primary (*see* Primary Cementing)
Problems, 3, 5
Regulations applied to oil and gas wells, 128
Regulatory bodies controlling, 124
Research, 4, 73, 84
Reverse circulation, 65, 66, 100, 101
Rheological design, 111
Scope of monograph, 1
Slo Flo, 4
Special applications, 129-137
Squeeze, 1, 3-5, 24, 28, 30, 32, 34-36, 41, 42, 47, 61, 79, 85-96, 140, 141
Stage, 51, 52, 56, 66, 78, 79
Through pipe and casing, 65
Two-stage, 52, 65
Water-well, 130, 131, 136
Well-depth relations, 137
Centralizers:
Casing, API specifications and types, 54
In running pipe and placing cement, 49, 69, 76, 83, 98, 101
In typical primary cementing job, 2, 3, 138, 139
Influence on gas failure pressure, 113
Locating with caliper logs, 57
Location in casing string, 50, 68
Need for, 53
Patented device to hold on pipe, 4
Proper placement of cementing plug, 102
Spacing of, deviated hole, 54
Stresses on, 73
To remove filter cake, 46
Channeling, contributes to cement failure, 57, 73
Chemical acceleration, 3
Chemical requirement for API cements, 8
Ciment Fondu, 13
Clean Strip (Kleen Jet), 120
Clinker, in cement, 9
Coal, in cements, 4, 20, 24
Collapse analysis, of casing string, 45
Collapse failure, of casing, 132, 134
Collars, 46
Colombia, drilling and cementing regulation, 126
Completions:
Cased-hole, 131
Development of shaped charges, 123
Disposal wells, 132
For deep permafrost, 137
Offshore, 73
Open-hole, 131, 133
Permanent-type, 95
Techniques, 130-132, 134, 135
Tubingless, 73
Compressive forces, 48
Compressive strength:
Bentonite cement, 21
Cements used in high-temperature steam wells, 133
Class A and H cements, effects on diatomaceous earth, 22
Class A, C, D-E, G, H, API cement, 7-9
Class G and H cement, 99
Class H cement, influence of time and temperature, 38
Class H cement slurries, effect of bentonite on the composition and properties, 20
Comparison of effects of sea and fresh water, 19
Effect of calcium chloride, API Class A cement, 17
Effect of mud contamination, 99
Effect of polymer dispersants, API Class G cement, 26
Effect of potassium chloride, API Class A cement, 27
Effect of salt on API Class G cement, 24
Effect of sodium chloride, API Class A cement, 18
Effect of sodium silicate, API Class A cement containing CMHEC, 18
Formation, effect on penetrating efficiency of bullet and jet perforators, 119

Formation, effect on perforator performance, 123
High-gel salt cements prepared with API Class A cement, 21, 58
Hydrated API Class H cement, 28
Of cement, 9-11, 16, 58, 120, 121
Testing, 37
Vs hydraulic bond strength, 118
Vs mechanical bond strength, 118
Vs shear bonding strength, 113
Concentration, customary units and conversion factors, 150
Conductor casing, 44, 61, 66, 72, 74, 133
Conductor pipe, 124, 135
Consistency index, 105, 107-111
Contact time:
Definition, 63
When cementing in laminar flow, 67
Contaminants, in mixing waters, 38
Controlled pressure completion, 120
Conversion factors for metric system, 146-153
Correlation:
Reynolds-number/friction-factor, 110
Rheological complex fluids, 111
Corrosion:
By down-hole formation fluids, 48
Holes in pipe, 57
Protecting casing from, 1
Resistance down-hole, 44
Corrosive formation waters, 42
Critical pump rate, effect of dispersants, 26
Critical velocity:
Departure from laminar flow, 107, 109, 110, 111
Effect of dispersants, 26
Curing time of cements, 29, 37

D

Deep wells:
AWWA standard for, 136
Cementing considerations, 74-76, 103
Cementing liners through abnormally-pressured formations, 79, 80
Cementing liners through fractured formations, 78, 79
Cementing liners with low fluid levels, 80, 81
Check lists for running and cementing liners, 82, 83
Completion techniques, 84
Equipment used in hanging liners, 77, 78
Liner hanging cuts casing cost, 84
Factors to be considered in designing cement slurries, 81, 82
U.S. decline in 1974, 83
Use of liners, 76, 77
Dehydration:
Effect on primary cementing, 71
Effect on squeeze cementing, 88
Delaware Basin:
Cementing, 73, 79
Running and cementing, 84
Wells, casing program, 75
Densified cement, 19, 25, 29
Density:
Control during primary cementing, 61, 62
Customary units and conversion factors, 149
Densometer, 2
Dept. of the Interior, 124
Design, rheological, of cementing operations, 32
Diacel systems, 30
Diatomaceous earth:
As light-weight cement additive, 20, 30, 31, 39
Effect of salt on apparent velocity of slurries, 26
Effects on API Class A and H cements, 22
Introduced to industry, 4
Dicalcium silicate, 7
Diesel oil:
Cements, 9, 12
To reduce swabbing time, 64
Differential fillup equipment, 66
Differential sticking, of casing, 64, 75
Diffusivity of cements, 136
Directional drilling, use of cement plug, 98
Dispersants:
Cements with, 17, 19, 29, 30, 81, 82, 139
Chemical, 63
Effect on critical flow rates of slurries in turbulence, 26, 108
In mixing water, 99
To control filter loss, 25, 66
To improve placement techniques, 32, 135
To increase slurry density, 30, 39, 132
To reduce viscosity, 21
To thin cement, 76
Displacement stage-cementing method, 52
Displacement studies, 78
Doglegs, 46
Dolomite reservoir, squeeze cementing technique, 95
Down-hole plugging, 97, 142
Drake well, 1
Drillable squeeze packer, 92
Drill collars, 37, 47
Drilling fluid additives, 38
Drilling mud, 3
Drillstem tests, 81, 98
Dump bailer, 1, 100
Dyes, special additive for cement, 27, 28

E

Economics of perforating, 122
Egypt, gypsum cement, 6
Elastic stability, 136
Emulsifiers, mud additives, 39
Energy, customary units and conversion factors, 152, 153
Enforcement and penalties by regulatory bodies, 127
England, early cements, 6
Entropy, customary units and conversion factors, 151
Equalization point formula, 142
Equations:
30-minute fluid-loss-value, 41
Apparent viscosity, 107
Bingham-plastic model, 106
Bottom-hole temperature, 34
Casing/open-hole annular area, 109
Critical velocity, 109
Displacement velocity, 109
Flow behavior index, 109, 143
Fluid consistency index, 109, 143
Frictional pressure drop, 109, 143
Height of cement column, 142
Hydrostatic pressure, 95, 109
Hydrostatic pressure of squeeze column, 95
Maximum allowable surface squeeze pressure, 95
Maximum annular backup pressure, 95
Power-law model, 107
Pumping rate required for turbulence, 143, 144
Reynolds number, 109, 144
Sacks of cement for given plugging operation, 142
Shear rate, 105, 106
Shear stress, 105, 107
Support pressure to resist collapse above packer, 95
Velocity at some specific Reynolds number for generalized calculations, 110
Velocity of cement slurry, 64
Volume of fluid for cement, specific contact time, 63
Ettringite crystal growth, 13
Europe:
Early cements, 6
Weight of sack of cement, 58
Exothermic reaction, 73
Expanded perlite, 20, 22, 24
Expanding cements, 4, 9, 12, 13, 15, 29
External casing packer, 55, 56
External pressure stress, safety factor, 44
External yield pressure, 45

F

Facility throughput, customary units and conversion factors, 146
Fanning friction factor, 110, 143
Far East, early cements, 6
Fibers, special additive for cement 27, 29
Fillup:
Cement slurry, 141
Efficiency, 114
Equipment, 83
Factor, 130
Filtration control, 41, 42, 66, 70, 83, 91, 94
Filtration-control agents, 16, 24, 30, 67
Filtration properties, of bentonite, 20
Fineness analysis, of cement, 9, 10
Float collar:
Delayed-set method of cementing liner, 81
Differential fillup, 51, 56
For stabbing in tubing, 50
In conventional method of cementing liner, 77
In primary cementing job, 138, 139
In running and cementing liners, 83
Super seal, 2, 50
With automatic fillup value, 49
With external casing packer, 55
With latch-down plug, 50, 53
Float shoe, 49-51, 53, 55, 56, 83, 138, 139
Floating equipment:
Charges, 138, 139
For casing, 3, 47, 49, 51, 66
For running and cementing liners in deep wells, 83
Flocele, 138, 139
Flotation factor, in running large-diameter pipe, 64
Flow behavior index, 105, 107-111
Flow calculations:
Comparison using Bingham-plastic and power-law models, 111
Equations used, 109, 110
Flow properties of wellbore fluids, 104-107
Instruments used to predict fluid flow properties, 107, 108
Plug flow vs turbulent flow, 108, 109
Primary cementing jobs, 143, 144
Flow rates:
Customary units and conversion factors, 149
For turbulent and plug flow without cement dispersants, 108
Fluid-density balance, 39, 40
Fluid-loss-control agents, 4, 32, 39, 71, 81, 89, 91, 94, 96, 139
Fluid migration, 85, 86
Fluidity of cement slurries, 3
Fluorescein, for coloring cement, 28
Fly ash, 11, 12, 16, 20
Forces on mud channel during cementing, 108
Formation:
Abnormal pressure, 79
Breakdown, 94
Evaluation, acoustic character logs applications, 123

Fracture gradients, 78, 79, 87, 90
Fractures, plugging off water, 102
Packer collars, 50, 56
Packer shoes, 50, 56
Testing, 98
Fracture gradient: *See* Formation: fracture gradient
Fractured formations:
Carbonates, 95
Liner cementing through, 78
Plugging off water, 102
Primary and remedial cementing, 84
France:
Cement classification, 12
Drilling and cementing regulations, 126
Free-fall stage cementing method, 52
Fresh water:
Aquifer, 130, 131
As displacing fluid, 64
As workover fluid, 92
Effect on thickening time and compressive strength, API Classes A and H cements, 19
Salt-water interface, 126
Friction reducers, 4, 16, 25, 26, 30, 108

G

Gamma ray log, 116
Gas channeling, 70
Gas intrusion, 86
Gas leakage:
From top of liner, 80
In annulus, 57, 70
In primary cementing, 73
Gas migration, 70
Gas-oil ratio, controlling, 85
Gel cement, 20, 27, 89, 138
Gelation, 71
Geothermal gradient, 135
Geothermal steam:
For power in New Zealand, 137
Production, 9
Recovery of, 132
Wells, 133, 137
Germany, drilling and cementing regulations, 126
Gilsonite, in cements, 4, 20, 22, 24, 30, 32, 138, 139
Gravel pack, 100, 131, 136
Greece, calcined limestone cement, 6
Guar, 63
Guiding equipment:
For casing, 49-51, 66
Shoe, 2, 49, 50, 56, 58, 129, 130, 138
Gulf Coast:
Casing programs, 75
Cement failures, 4
Cementing problems, 5
Gun perforating, 4
Mineral leasing, 128
Running casing, 46, 68
Setting casing through salt formations, 71
Squeeze cementing, 5
Texas and Louisiana, temperature studies, 33
Wells, 3, 34
Gulf of Mexico:
Rules controlling the cementing of wells, 126
Sea water analyses, 19
Statewide WOC requirements, 127
Gulf of Suez, sea water analyses, 19
Gun perforating, 4, 123
Gypsum:
Cement, 4, 6, 9, 12, 14, 17, 18, 24, 30, 32, 135
Special additives, 27, 29, 30

H

Hasteloy alloy, 132
Heat and Heat Capacity, customary units and conversion factors, 151
Heat of hydration, 11, 16, 37, 39-41, 113, 114, 116, 130, 134-136
Heat transfer coefficient, 41, 133
Hematite, material for weighting API Class D or E cement, 23, 30, 39
Hesitation squeeze, 88
Hewitt field, 2, 4, 5
High-alumina cements, 14, 15
High-pressure gas well, 44
High-gel salt cement, 20-22
Horsepower:
Displacing, 3
Hydraulic, 104, 105, 143, 144
Mixing, 3
Of cementing trucks, 3
Of pumping equipment, 3
Hydrating cement, 5
Hydraulic bond:
Effect of perforating, 118
Failure, 113, 114
In a wellbore, 112
Strength, 113, 118
Hydraulic cutters, 122
Hydraulic jetting, 119, 120
Hydraulic perforators, 120
Hydraulic shear strength, 114
Hydraulically set liner hanger, 78
Hydrazine, special additive for cement, 27-29
Hydrochloric acid solution, 92
Hydromite, special cement additive, 30
Hydrostatic head, 69, 70
Hydrostatic pressures:
At time of perforating, 121
Encountered in wells, 34
Squeeze job example, 141

I

Ice:
In drilling mud, 3
Lenses in permafrost, 134
Iceland, electrical power from steam wells, 132
Illinois, first bulk cement station, 3
Impact pressure, 118
Incaloy 825 alloy, 132
Incor high-fineness cements, 4
Inorganic materials, in cement mixing waters, 38
Insert flapper valve, 50, 51
In-situ combustion wells, 14
Interfacial gelation, in annulus, 62
Intermediate casing, 44, 47, 48, 61, 69, 72, 77, 124, 127, 138
Intermediate liner, 74, 77
Internal pressure stress, safety factor, 44
Internal yield pressure, 45
International regulatory bodies, 124
Iodine 131, 115
Ireland, drilling and cementing regulations, 126
Italy:
Drilling and cementing regulations, 126
Electrical power from steam wells, 132
Pozzolanic-lime cement, 6

J

Japan:
Cement classification, 12
Drilling and cementing regulations, 126
Electrical power from steam wells, 132
Jet guns, 119, 120
Jet mixer, 4, 39, 61, 62
Jet perforation, 118, 119, 122, 123
Jet Research Center, 120
Job:
Planning, 1, 90
Evaluation, 1
Execution, 1

K

Kansas, 42, 87

L

Large-hole cementing:
Casing programs, 129
Materials, 130
Techniques, 129
Latex cements, 9, 14, 25, 30
Libya, drilling and cementing regulations, 126
Lignin:
Dispersant, 21
Retarders, 23, 24
Lignosulfonic acid, 23
Limenite, heavy-weight cement additive, 23, 30
Limit plug, 100
Linear expansion tests, 13, 29
Linear cementing, 1, 35
Liner hanger, 76-78, 81-83
Link Jet, 120
Lompac field, 1
Los Angeles Basin, 4
Louisiana:
Gulf Coast, casing programs, 75
Gulf Coast, temperature studies, 33
South, 81
Lost circulation:
Control agents, 16, 22, 24, 25, 30, 35
Control with cement plug, 98, 99, 101
During drilling, 88
During running of casing, 48, 68
Fighting with gypsum cement, 12, 30
First published material, 4
Major problem in exploration and development, 5
Materials, 31, 39, 63
Salt as preventive, 26
Sealing off, 1, 29
While cementing, 32 97
Zone, 47, 100
Low-heat-of-hydration cement, 14
Low-pressure squeezing, 88
Lumnite, 13

M

Mackenzie Delta, Canada, 134
Marine bulk cementing unit, 59, 60
Marsh funnel viscosity, 98
Mass, customary units and conversion factors, 150
Measuring lines, 3
Mechanical:
Bond, 118
Bond strength, 118
Cutters, 120, 122
Equipment, 1
Mechanically set liner hanger, 78
Mechanics, customary units and conversion factors, 148
Mediterranean area, earliest hydraulic cements, 6
Methylene blue, for coloring cement, 28
Metric practice guide, API 2564, 146-153
Mexico, electrical power from steam wells, 132
Microannulus, 113
Middle East, early cements, 6
Mineral leasing, regulations in Gulf Coast region, 128
Minimum elongation, of casing and tubing, 45
Misconceptions, about squeeze cementing, 89, 90
Mississippi:
Deep drilling practices, 84
Gulf Coast, casing program, 75
Mixers:
Batch, 61, 62
Coverage, 1
Jet, 61, 62
Recirculating, 61, 62

Mixing waters:
Check list for running and cementing liner in deep wells, 83
Contaminants, 38
Effect on performance of API Class H cement, 38
For primary cementing, 59, 60, 62, 138, 139
Inorganic materials in, 38
Required for slurring viscosity, 34, 35
Sea water, 38
Squeeze jobs, 140
Modified cement, 20, 21
Monograph, objectives and scope, 1
Montana, 47, 71
Morrow shale, 27
Mozambique, drilling and cementing regulations, 126
Mud additives, effects on cement, 39
Mud contamination, 32, 47, 50, 51, 63, 97, 99, 101, 102, 113, 114
Mud decontaminants, 27, 28, 39
Mud displacement for cement slurries, 111
Mud Kil, special cement additive, 30
Mud preflushes, 63, 69
Mud-treating chemicals, effects on oil well cementing, 102
Multiple-stage cementing tool, 52, 56, 65

N

Naval Reserve area, Alaska, 133
Neat cement, 4, 17, 19, 29, 63, 70, 88, 89, 91
Netherlands, drilling and cementing regulations, 126
New Mexico, temperature gradients, 33
New Zealand:
Geothermal steam for power, 137
Wairakei project, 132
Newtonian fluids, 104, 106, 109, 110, 144
Nitrogen, as light-weight cement additive, 20, 22, 23
Non-Newtonian fluids, 104, 106-110
North Dakota, 47, 71
North Hassi-Messaoud field, 73
North Sea, 42, 47
North America, areas of permafrost, 14
Nylon, to control circulation of cements, 24, 29, 30

O

Offshore platform cementing system, 60
Oil well:
Cementing, definition, 1
Cements, depth ranges for API classes, 58
Repair by scabbing methods, 95
Oil-emulsion cements, 31
Oklahoma:
Bottom-hole treating pressures and fracture gradients, 87
Hewitt field, 1, 2, 4 5
Lone Star Baden unit well, 75
Historic test bests 30,000 feet, 84
Rules controlling the cementing of wells, 126
Oklahoma Corporation Commission, 126
Open-hole plug-back:
Operations, 102
Studies, 98
Opposed-piston pendulum pump, 2
Organic polymers, as filtration-control additive, 24, 30
Ottawa sand, material for weighting API Class D or E cement, 23
Outer continental shelf, 126-128
Oxygen scavengers, 32

P

Pacific region, rules controlling the cementing of wells, 126
Paraformaldehyde, as mud decontaminant for cement, 27, 28, 30
Perforating:
Devices and methods, 118
Effects on cement sheath, 118-120
Factors influencing, 121, 122
In gas-producing zones, 120
Time, 121
Perforation density, 121, 122
Performance properties of casing and tubing, 45
Perlite in cement to control lost circulation, 20, 22, 24, 30, 133
Permafrost:
Cements, 9, 14, 15, 16
Environments, 129
Permeability:
Of cement, 16, 41, 82
Of filter cake, 91
Permeators, 120, 122
Permits for drilling, 127
Persian Gulf, sea water analyses, 19
Phenophthalien, for coloring cement, 28
Physical properties:
Of steel, 45
Of tubular goods, 45
Pipe centralization, 67
Placement time, cement slurry, 75, 78
Plaster of Paris, early cement, 6
Plastic cements, 9, 11, 12
Plastic flow, 106
Plastic pipe, 113
Plastic viscosity, of mud, 46, 67, 75, 76, 98, 102, 105-107, 109, 111
Plug-back:
Cementing, 35, 102
Check list, 102
Operations, 101
Plug flow, 106, 108, 109
Plugging operations, 1, 3-5, 12, 97, 128, 142
Plugs:
Bottom, 2, 48, 50, 52, 53, 56, 63, 66, 98, 100, 131
Bridge, 55, 56, 88, 100
Cast iron, 1
Container, 2, 46, 50, 52, 53, 56
Displacement, 52
Female, 77
Free-falling, 52
Latch-down, 50, 51, 53, 56
Male, 77
Top, 2, 48, 50, 52, 53, 56, 63-66, 69, 100, 104, 138, 139, 143
Wiper, 46, 52, 53, 56, 63, 64, 68, 83, 98, 100
Plunger triplex pump, 2
Pneumatic batch mixer, 62
Polymer fluid-loss additive, effect on API Class G cement, 42
Polymers, as dispersants or thinners, 25-27, 30, 42, 91
Polyurethane, 134
Polyvinyl acetate emulsions, 14
Polyvinyl chloride emulsions, 14
Pore pressure, 79, 80
Portland cement:
As plugging material, 12
As scavenging preflush, 63
Chemical compounds found in, 7
Effect of liquid latex on slurries, 14
Effect of some additives on physical properties, 32, 94, 137
For cementing water wells, 131
For controlling blowouts, 2
First use, 1
Hydrates, 11
Manufacture of, 7
Oxide analysis of, 7
Standards, 14
Tricalcium aluminate in, 7, 8, 13
Typical composition and properties of API classes, 7
Portland Cement Association, 13
Potassium chloride, effects on strength of cement, 27
Power, customary units and conversion factors, 152, 153
Power-driven duplex pump, 2
Power-law model, 104, 106-111
Pozzolans:
Additive for controlling cement properties, 39-72
Artificial, 11
Cements, 6, 11, 13, 20, 26, 27, 30, 81, 130
Definition, 9
Effect on physical properties of cement, 31
When introduced, 3
With gel, 115
Preflushing, during primary cementing, 62, 63
Prehydrated benonite, 20, 21
Pressure:
Backup, 93
Bottom-hole, 35, 46, 78, 86, 87, 90
Breakdown, 44, 86, 88, 120
Collapse 129, 130
Customary units and conversion factors, 151
Differential sticking, of drill pipe, 56, 64
Down-hole treating, 93
Fracturing, 78, 79
Frictional, developed during pumping, 68
Impact, 118
Influence on down-hole performance of cement slurries, 33, 34
Maximum for cementing, 60, 141
Pore, 79, 80
Surface, 35, 86
Surges, 48
Primary cementing:
Conditioning hole, 46
Considerations, after cementing, 72, 73
Considerations, during cementing, 59-65
Considerations, in planning a cementing job, 57-59
Correcting defective job, 85
Directional holes, 69
Displacement, 66-68, 111
Field practices, 3
Flow calculations, 143, 144
Gas leakage after cementing, 69-71
High-strength granular sealing material increases efficiency, 32
Jobs, examples of, 138, 139
Mechanical devices to improve jobs, 3
Multiple strings, 68, 69, 111
Placement techniques, 65, 66
Reverse circulation, 73
Through soluble formation, 71, 72
Typical job, 2
Production casing, 44, 50, 61, 69, 72, 97, 124, 127, 135, 139
Production liner, 74, 77
Protection squeeze, 5
Prudhoe Bay, Alaska, 133
Pump types:
Centrifugal, 60, 62
Duplex double-action piston, 60
Opposed-piston pendulum, 2
Plunger triplex, 2
Positive displacement, 60
Power-driven duplex, 2
Single-acting triplex plunger, 60
Vertical double-acting duplex, 2
Pumpability:
Effect of pressure, 82
Limits of, 36
Of cement slurries, 3, 33, 75, 131

Of permafrost cement, 14
Temperature and pressure, influence cement set, 81, 82
Pumping equipment:
Conventional, 100
For bulk handling of cement, 58-61
For primary cementing, 59-61
For water supply, 59
Horsepower, 3, 60
Marine, 59
Pumping rate, to achieve turbulent flow in annulus, 108
Pumping wells, buckling of tubing, 48

R

Radioactive densometer, 40
Radioactive-tracer surveys, 114-116, 122
Radioactive tracers, special additive for cement, 27, 28, 30, 73
Radioactive wastes, 137
Railroad Commission of Texas, 126, 128
Rayfrac, radioactive tracer in cement, 28
Recirculating mixer, for cementing 61, 62
Red iron oxide, for coloring cement, 28
References, 5, 14, 30, 43, 48, 56, 73, 83, 95, 102, 111, 122, 128, 136
Refractory cements, 9, 13, 14
Refrigerant fluids, 134
Regulatory bodies controlling the cementing of wells, 124
Resilence of cement slurries, 16
Resin cements, 9, 11, 12
Retarded cement, 4, 11, 115
Retarders:
As cement additive, 16, 30
CHMEC, 17, 24
Effect on physical properties of cements, 31
In API Class G cement, 9, 135
In mixing water, 99
Lignin, 23
Saturated salt water, 24
To delay setting of cement, 66, 67, 82
Retrievable bridge plug, 93
Retrievable squeeze packer, 92, 93
Retrievable squeeze retainer, 4
Retrievable tool, 119
Rheological complex fluids, 111
Rheological parameters of cement slurry and drilling fluids, 104
Rheological properties of cement slurries, 5
Rheometer, for evaluating drilling fluids at elevated temperatures, 111
Ribbon batch mixer, 62
Rocky Mountain, bottom-hole treating pressures and fracture gradients, 87
Roman Empire, early cements, 6
Rotary speed, maximum for drilling out cement, 47, 48

S

Sable Island, sea water analyses, 19
Sack cement, 2, 3
Safety factor:
For external pressure stress, 44, 47
For internal pressure stress, 44, 48
For tensile stress, 44, 47
For squeeze job, 141
Salt cement, 4, 26, 27, 30, 32, 73
Salt concentration vs freezing temperature, 136
Salt loading, behavior of casing subjected to, 48, 71, 73
Salt water:
As displacing fluid, 140, 141
As workover fluid, 92
Disposal of, 131, 137
Fresh water interface, 126
Sample calculations:
For high-pressure squeeze without filtration control, 93
For low-pressure squeeze with filtration control, 93
Saturated salt water as cement retarder, 23, 24, 29
Scab liner, 77
Scandium, 46, 115
Scanning election microscope, 70, 71
Scavenger slurries, 69
Scratchers:
First commercial use, in California, 4
In running pipe and placing cement, 49, 56, 76, 83, 98, 101
Influence on gas bond failure pressure, 113
Locating with caliper logs, 57
Proper placement of cementing plug, 102
Reciprocating types, 55, 56
Rotating types, 54-56, 68
To improve primary cementing jobs, 3
To remove filter cake, 46
Sea water:
Analyses, 19
As a cement accelerator, 17, 19, 30
As a displacing fluid, 64
Effect on thickening time and compressive strength, API Classes A and H, 19
In cement mixing waters, 38
Use in well cementing, 31
Sealants, mud additive, 39
Setting time of cement, 16
Sealing casing seat, 97
Sealing efficiency, maximum, 94
Shear bond:
In a wellbore, 112
Strength, 113
Vs closed-in pressure, 114
Shear rate, 105-107
Shear stress, 105-107
Shock loads, protecting the casing from, 1
SI units for metric system, 146-153
Sidetracking, 97, 98
Sidewinder, 120
Silica flour, special additive for cement, 27, 28, 30, 41, 82, 133
Silica-lime cements, 11, 30
Single stage method of cementing liners in high-pressure zones, 80
Skid-mounted cementing unit, 60
Slo Flo cementing, 4
Slurry density, 39
Slurry design, squeeze operation, 90-92
Strata-Crete, for lighter cement slurries, 32
Sodium chloride:
As cement accelerator, 17, 30, 31
As cement thinner, 25, 26
Effect on thickening time and compressive strength of API Class A cement, 18
Effect on setting properties of oil-well cements, 32
Sodium chromate, with paraformaldehyde as mud decontaminate in cement, 27, 28
Sodium montmorillonite, 19
Sodium silicate:
As light weight cement additive, 20
Effect on thickening time and compressive strength, API Class A cement with CHMEC, 18
For controlling lost circulation, 24, 30
Used to accelerate cement slurries, 17
Solid wastes, underground disposal, 137
Soundness of API cements, 10
South Dakota, bentonite as cement additive, 19
Space, customary units and conversion factors, 147
Spacers:
In annulus, 62
Mud/cement, 72, 74
Specifications for oil-well cement, 3, 5, 31
Specific volume, customary units and conversion factors, 150
Spring-tension hanger, 78
Squeeze cementing:
Basis for well-stimulation test schedules, 35
Bottom-hole circulating conditions, 61
Controlling filtrate in the cement slurry, 41
Cooling during, 34
Erroneous theories, 89
Example of jobs, 140, 141
Filtration-control additives used, 30
Fluid-loss values of cement slurries, 42
Fractured zones, 89
Fracturing pressure information, 79
Helpful formulas, 95
High-strength granular sealing material increases efficiency, 32
Job planning, 90
Packers, 92
Pressure calculations, 93
Pressure requirements, 88
Slurry design, 90
Techniques, 87
Terminology, 86
Testing jobs, 94
Thickening time requirements, 36
Where required, 85
With calcium lignosulfonate as retarder, 24
WOC time, 93
Worksheet, 94
Squeeze packer method, 88, 92, 93
Squeeze pressure, 86, 93
Squeezing, 3-5
Stage:
Cementing, 51, 52, 56, 66, 78, 79
Collars, 3
Standardization of cement testing, 3
Starcor retarded cement, 4
Steam duplex pump, 2
Steam recovery projects, 132
Steam-well cementing, 132, 133
Storage, of cement, 58, 59
Straddle packers, 92, 93
Strength stability, 82
Strengths of set cements and dehydrated cement cores, 94
Stub liner, 77
Subsurface:
Casing equipment, 49-56
Injection for waste disposal, 131
Tools used to place cement, 1
Sulfate resistance, of hardened cement, 42
Sulfate resistant cement, 7, 8, 12
Super Dyna Jet, 120
Surface active agent, in diesel oil cement, 12
Surface:
Areas, influence on the volume of set cement, 36
Casing or pipe, 44, 47, 48, 50, 66, 72, 79, 97, 102, 124, 128, 132, 134, 138
Casing equipment, 49-56
Pipe, 127
Surge tanks, for cement handling, 59
Swing Jet, 120
Symbols, metric practice guide, 146

T

Tagging, in offshore operations, 97

Temperature:
Bottom-hole circulating, 35, 37, 74, 81-83, 90
Customary units and conversion factors, 151
Down-hole, 115
Gradient, 81, 82, 133
Of cement slurries, 33, 34
Survey instrument, 4
Surveys, 114, 115, 122
Tensile force, 44, 48
Tensile strength, of tubular goods, 45
Tensile stress, safety factor, 44
Tension analysis, of casing string, 45
Tension failure of casing, 132
Testing:
Early techniques, 4
For standardization of cement, 3, 5
Oil-well cements and cement additives, 96
Procedures for WOC time, 3
Squeeze jobs, 94
Water shutoff, 3
Tetracalcium aluminoferrite, 7, 8
Texas:
Bulk cement station, 3
Casing programs for Gulf Coast wells, 75
Gulf Coast temperature studies, 33
North, 4, 87
Pressure test of casing, 72
Rules controlling the cementing of wells, 126
Southwest, 87
West, 33, 42, 74-76, 81, 87
Thaw profile, 135
Thermal conductivity:
Of earth, 135
Of gypsum-portland cement, 136
Of polyurethane, 134
Of refractory cement
Thermal model studies, 134
Thermistor cables, well temperature monitoring through permafrost, 137
Thief zones, sealing off, 1, 85
Thickening-time:
Analysis of cements, 9-11
API Class A and H cements, effects on diatomaceous earth, 22
API Class G or H cements, retarding effect of calcium lignosulfonate, 23
API Class H cements, effects of bentonite on composition and properties, 20
Calcium lignosulfonate retarder in API Classes G and H cements, 24
Casing cementing vs squeeze cementing, 91
Cement slurry design, 83
Composition of effects of sea and fresh water, 19
Controlled to allow placement of cement, 80
Effect of calcium chloride, API Class A cement, 17
Effect of densification, API Class G cement, 19
Effect of polymer dispersants, API Class G cement, 26
Effect of salt on API Class G cement, 24
Effect of sodium chloride, API Class A cement, 18
Effect of sodium silicate, API Class A cement containing CHMEC, 18
Effect of temperature, of various API cements at atmospheric pressure, 33
Effect of varying pressure, API Class H cement with retarder, 34
Excessive, over retardation, 78
High-gel salt cement prepared with API class A cement, 21
Of cement at high temperature, 82
Primary cementing jobs, examples of, 138, 139
Squeeze cementing jobs, examples of, 140
Tester, 3, 4, 36
Thinners, 25, 39
Tie-back liner, 77
Time, customary units and conversion factors, 147
Tobermorite (calcium silicate), 82
Torque impulses, 47
Torsional strength, 47
Transport Properties, customary units and conversion factors, 152
Triaxial loading tests, 71
Tricalcium aluminate, 7, 8, 13
Tricalcium silicate, 7, 8
Trinidad, West Indies, sea water analyses, 19
Tubingless completions, 68
Tubular goods:
Buckling, 48
Physical properties, 45
Turkey, drilling and cementing regulations, 126
Two-plug method placement technique of cement plug, 100, 101
Two-stage cementing, 52, 56, 65, 80
Two-stage placement pressure profile, 79

U

Unaflo retarded cement, 4
United Kingdom:
Cement classification, 12
Drilling and cementing regulations, 126
United States:
Bottom-hole treating pressures and fracture gradients, 87
Casing cementing practices, 56
Deep-well completion, 76
Deep wells drilled, 1971 through 1974, 74, 83
Early cements, 6
Federal regulations for outer continental shelf, 126, 127
Improved casing cementing practices, 73
Oil-well cementing practices, 84
Regulatory bodies, 124, 125
Rules controlling the cementing of wells, 125
U.S. Geological Survey, 124, 127

V

Vacuum, customary units and conversion factors, 151
Velocity:
Annular, 68
Fluid, in multiple-string cementing, 68
Of cement slurry, 64
Venezuela, drilling and cementing regulations, 126
Vertical bond failure, 113
Vertical double-acting duplex pump, 2
Viscometer:
Direct-indicating for drilling fluids, 111
Pipeline (capillary tube), 107
Rotational (Fann V6 Meter), 104-109, 111
Viscosity, of cement slurries, 34-36, 83, 101

W

Wagner fineness, 7
Wagner turbidimeter, 10
Wairakei project, New Zealand, 132
Waiting on cement time (WOC):
After placement of open-hole plugs, 101
Cement slurry design, 83
Engineered cementing operations to eliminate, 122
For water wells, 131
Minimum, method of determining, 43, 128
Oklahoma Corporation Commission rule, 4
Primary cementing, 72, 138, 139
Reducing with accelerators, 30
Selecting, 37
Short time, 3
Squeeze cementing, 93, 94
State Regulations, 124, 127
Walnut shells to control lost circulation of cements, 24, 30
Waste-disposal wells, 131, 132, 137
Water-base muds, 63, 98, 134
Water content:
Analysis, of cements, 9, 10
Of cement slurries, 34-36
Water intrusion, 86
Water ratio:
Excessive amount results in weak set cement, 37
Influence on volume of set cement, 36
Low, additive yield heavy high-strength cements, 31, 103
Water shutoff, 3-5, 12
Water sources and supply, 59
Water spacer, 3
Water-sensitive reservoir rocks, 32
Water-well cementing:
Casing programs, 130
Cementing materials, 131
Cementing techniques, 130
Completion techniques, 130
Methods, 131
Wellbore conditioning, 46, 76
Wellbore hydraulics, computer calculation, 105
Well completions: *See* Completions
Well perforators, evaluation of, 123
Well productivity, effect of formation penetration beyond perforating casing, 123
Well stimulation:
Fracturing pressure in formation, 78
Test schedules, 35
Well workovers, 95
West Germany, cement classification, 12
Whipstock plug, 19
Whipstocking, 97-99, 101
Wildcat well, 44
Williston Basin area, 71
Wiper plug, 46, 52, 53, 56, 63, 64, 68, 98, 100
Wireline workovers, 73
Work, customary units and conversion factors, 152, 153
Workover fluids, 92
Wyoming, bentonite as cement additive, 19

Y

Yellow oxide, for coloring cement, 28
Yield of cement slurry, 58
Yield point:
Bingham, 106, 107, 111
Of barite slurry, 101
Of bentonite, 20
Of mud, 75, 76, 98, 102, 105, 106, 109
True, 106
Yield strength:
Of casing and tubing, 45
Of mud, 46, 67
Yowell tool, 5

Z

Zero tricalcium aluminate, 132
Zonal isolation:
Achieving, 1
Failure can be costly, 57
Hydraulic bonding, 112
Open-hole cement plugs, 98